T0206048

Global Navigation Satellite Systems

Global Navigation Satellite Systems

New Technologies and Applications

Second Edition

Basudeb Bhatta

CRC Press
Taylor & Francis Group
Boca Raton London New York

CRC Press is an imprint of the
Taylor & Francis Group, an **informa** business

Second edition published 2021
by CRC Press
6000 Broken Sound Parkway NW, Suite 300, Boca Raton, FL 33487-2742

and by CRC Press
2 Park Square, Milton Park, Abingdon, Oxon, OX14 4RN

First edition published by CRC Press 2011

CRC Press is an imprint of Taylor & Francis Group, LLC

ISBN: 978-0-367-47408-9 (hbk)
ISBN: 978-0-367-70972-3 (pbk)
ISBN: 978-1-003-14875-3 (ebk)

Typeset in Times
by Deanta Global Publishing Services, Chennai, India

Visit the Support Material: https://www.routledge.com/9780367474089

Contents

Preface

Global Navigation Satellite System (GNSS) is the standard generic term for satellite navigation systems that provide autonomous geospatial positioning information with global coverage. A GNSS allows small electronic receivers to determine their locations (latitude, longitude, and altitude) and precise time information using radio signals transmitted from navigation satellites along a line of sight. There are currently several layers of satellite navigation systems. The United States' GPS, Russian GLONASS, European Union's Galileo, and Chinese BeiDou are for global coverage. Several systems are also available for regional coverage by several countries—such as Indian IRNSS and Japanese QZSS. In addition, augmentation systems on these core systems are also being offered by several government and private agencies. Development of multiple GNSS constellations and their augmentations have resulted in increased accuracy and reliability; as a result, new branches of science and engineering are adopting this technology rapidly.

The need to determine precise locations for use in a variety of applications is inevitable; areas such as surveying, navigation (automobile, aviation, maritime), tracking, mapping, earth observation, mobile-phone technology, rescue applications and many more. For example, the International Civil Aviation Organization and the International Maritime Organization have accepted GNSS as essential in their navigation. GNSS is revolutionizing and revitalizing the way nations operate in space, from guidance systems for the International Space Station's return vehicle, to the management of tracking and control for satellite constellations. Military applications of GNSS are extremely widespread from the mobilization of troops to supply of arms and amenities; it is also used in rescue operations and missile guidance.

Vehicle manufacturers now provide navigation units that combine vehicle location and road data to avoid traffic jams and reduce travel time, fuel consumption, and therefore pollution. Road and rail transport operators are able to monitor the goods' movements more efficiently and combat theft and fraud more effectively by means of GNSS. Taxi companies now use these systems to offer a faster and more reliable service to customers. Delivery service providers are increasingly dependent on GNSS.

Incorporating the GNSS signal into emergency-service applications creates a valuable tool for those lifesaving entities (fire brigade, police, paramedics, sea and mountain rescue), allowing them to respond more rapidly. There is also potential for the signal to be used to guide the blind; monitor Alzheimer's sufferers with memory loss; and guide explorers, hikers, and sailing enthusiasts. GNSS tracking system is also being used for many wildlife applications. It can be used for the control of epidemic/pandemic as it was used during the 2020 Covid-19 pandemic.

Surveying systems incorporating GNSS signals are being used as tools for many applications, such as urban development and coastal management. GNSS can be incorporated into geographical information systems for the efficient management of agricultural land and for aiding environmental protection. Another key application is the integration of third-generation mobile phones with Internet-linked applications.

Nowadays, everyone of us carry GNSS receivers in our smartphones and smart watches for many different purposes.

The role played by the GNSS systems in our everyday lives is set to grow considerably with new demands for more accurate information along with integration into more applications. Some experts regard satellite navigation as an invention that is as significant in its own way as that of the watch—no one today can ignore the time of day, and in the future, no one will be able to do without knowing their precise location.

ABOUT THE BOOK

This book begins with the fundamentals of GNSS. As readers go through the chapters of this book, they will learn about the functional segments of GNSS, working principles, signals, accuracy related issues, different navigation and positioning methods, various GNSSs and their augmentations, satellite geodesy, and applications of GNSS. This book provides a clear idea of how a GNSS works, what are the error-related issues involved, and how to deal with these errors. Not only will readers gain an insight into the techniques, trends, and applications of GNSS, they will also develop knowledge on selecting an appropriate GNSS instrument and suitable method for a specific application from the practical point of view. Guidance on GNSS surveying, navigation, and mapping has also been incorporated in great detail.

The book is written in a manner that makes it both a starting point for someone approaching the subject of satellite navigation and positioning systems for the first time as well as a reference for those already familiar with this technology. It was done by organizing the subject matters into four areas: basic overview, technology description, mathematical explanations, and finally practical implementations. Thus, for the uninitiated, the text provides a complete understanding of GNSS. For the initiated readers, it becomes a reference in which one can easily locate formulas, concepts, guidance, or other relevant information and, in particular, a very complete and thorough list of references and citations.

The primary purpose of this book is to be a learning resource for college and university students, as well as for individuals now in the industry who require indoctrination in the basics of GNSS and its applications. Practicing surveyors will appreciate the provided detailed guidance on various surveying operations. It is hoped that this book will attract and inspire individuals who might consider a specialized career in this field or in the broader fields allied with navigation, positioning, tracking, satellite-communication, space technology, and earth sciences.

CONTENT AND COVERAGE

The book is comprised of 12 chapters. Each chapter in the book commences with an introduction, which briefly outlines the topics covered in the chapter, and ends with exercises that help the students to assess their comprehension of the subject matter studied in the chapter. The chapters also contain many illustrations and notes that complement the text.

Chapter 1 covers the basic concepts of GNSS and provides a brief outline of positioning and navigation, history of navigation, and introduction to satellite-based navigation systems. **Chapter 2** describes the functional segments of GNSS and provides system descriptions for GPS, GLONASS, Galileo, and BeiDou. **Chapter 3** introduces the geometric concepts of positioning and the basic working principle of GNSS. **Chapter 4** deals with GNSS signals, how they are transmitted, what coded information is carried by these signals, and how these signals are used to determine the distance from the satellite to the receiver. **Chapter 5** explains the errors and accuracy issues involved in GNSS. Different positioning and navigation methods, such as static/kinematic, standalone/differential, real-time/post-processed, etc., are discussed in **Chapter 6**.

Chapter 7 is furnished to introduce several regional satellite navigation systems (such as QZSS, IRNSS), satellite-based augmentation systems (e.g., EGNOS, WAAS, MSAS, and GAGAN), ground based augmentation systems (e.g., LAAS, DGPS) and others (e.g., Inertial Navigation System and Pseudolite). **Chapter 8** furnishes the details of GNSS receivers—architecture, signal acquisition and positioning, classification of receivers, and other receiver related relevant information. **Chapter 9** deals with geodesy—different coordinate systems, datums, and projections, while **Chapter 10** focuses on the numerous applications of GNSS. **Chapter 11** and **12** aim to address practical issues involved in surveying and mapping respectively. These two chapters describe how surveying and mapping is done with GNSS and the factors one should consider for surveying and mapping applications.

A very rich glossary and list of references at the end are included for the benefit of the students and researchers.

MATLAB® is a registered trademark of The MathWorks, Inc. For product information, please contact:

The MathWorks, Inc.
3 Apple Hill Drive
Natick, MA 01760-2098 USA
Tel: 508 647 7000
Fax: 508-647-7001
E-mail: info@mathworks.com
Web: www.mathworks.com

Acknowledgements

I am grateful to all the authors of the numerous books and research publications mentioned in the list of references in this book for their classic and innovative contributions to the field of GNSS. I have consulted this varied literature, picked up the relevant materials, synthesized them, and put them in an organized manner with simple language in this book, mainly for the benefit of the students. I express my gratitude to the teachers, researchers, and organizations who have contributed tremendously to the quality and quantity of information in this book.

I am very much thankful to Dr. Aditi Sarkar who has edited the chapters of this book. I am also grateful to Prof Chiranjib Bhattacharjee, Director, CAD Centre, Jadavpur University for extending necessary facilities to write this book. I express my gratitude to my colleagues; without their help and cooperation writing of this book was never possible.

I would like to express my gratitude to my parents who have been a perennial source of inspiration and hope for me. I also want to thank my wife, Chandrani, for her understanding and full support while I worked on this book. My daughter, Bagmi, deserves a pat for bearing with me during this rigorous exercise.

Basudeb Bhatta

Author

Basudeb Bhatta earned a PhD in engineering from Jadavpur University, Kolkata. He is currently the course coordinator of the Computer Aided Design Centre, Jadavpur University. He has more than 25 years of industrial, teaching, and research experience in the domains of remote sensing, GNSS, GIS, and CAD. He has published many research papers, monographs, and textbooks on remote sensing, GIS, GNSS, and CAD. Dr. Bhatta has been instrumental in initiating a large number of courses on geoinformatics, GNSS, CAD, and related fields. He is a life member of several national and international societies.

Author

Acronyms

2D	Two-dimensional
3D	Three-dimensional
AAI	Airports Authority of India
AFB	Air Force Base
AGNSS	Assisted Global Navigation Satellite System
AltBOC	Alternate Binary Offset Carrier
AM	Amplitude Modulation
APL	Applied Physics Laboratory (Laurel, Maryland, USA)
AS	Anti-Spoofing
BDS	BeiDou Navigation Satellite System
BDT	BeiDou system Time
BIPM	Bureau International des Poids et Measures
BLUE	Best Linear Unbiased Estimate
BOC	Binary Offset Carrier
BPSK	Binary Phase Shift Keying
C code	Civilian code
C/A code	Coarse Acquisition code
CDMA	Code Division Multiple Access
CGCS2000	China Geodetic Coordinate System 2000
CL code	Civil-Long code
CM code	Civil-Moderate code
CNSS	Compass Navigation Satellite System
CPU	Central Processing Unit
CS	Commercial Service
DGNSS	Differential Global Navigation Satellite System
DGPS	Differential Global Positioning System
DOD	Department of Defence (United States)
DOP	Dilution of Precision
drms	distance root mean square
DVD	Digital Versatile Disk
EC	European Commission
ECEF	Earth-Centred Earth-Fixed (coordinate system)
ECI	Earth- Centred Inertial (coordinate system)
EGNOS	European Geostationary Navigation Overlay Service
EHF	Extremely High Frequency
EMR	ElectroMagnetic Radiation
ESA	European Space Agency
FAA	Federal Aviation Administration (United States)
FDMA	Frequency Division Multiple Access
FM	Frequency Modulation
GAGAN	GPS-Aided Geo Augmented Navigation

GBAS	Ground Based Augmentation System
GCS	Geographical Coordinate System
GDOP	Geometric Dilution of Precision
GIOVE	Galileo In-Orbit Validation Element
GLONASS	GLObal'naya NAvigatsionnaya Sputnikovaya Sistema (English: GLObal NAvigation Satellite System, also called GLobal Orbiting NAvigation Satellite System)
GMDSS	Global Maritime Distress Safety System
GNSS	Global Navigation Satellite System
GPS	Global Positioning System
GRAS	Ground-based Regional Augmentation System
GRS80	Geodetic Reference System 1980
GST	Galileo System Time
GSTB-V1	Galileo System Test-Bed Version 1
GTRF	Galileo Terrestrial Reference Frame
HAS	High Accuracy Service
HDOP	Horizontal Dilution Of Precision
HF	High Frequency
HPS	High Precision Service
ICAO	International Civil Aviation Organization
IGS	International GNSS Service (formerly International GPS Service)
IMU	Inertial Measurement Unit
INLUS	Indian Land Uplink Station
INMCC	Indian Master Control Centre
INRES	Indian Reference Stations
INS	Inertial Navigation System
IODC	Issue Of Data Clock
IODE	Issue Of Data Ephemeris
IOV	In-Orbit Validation
IRNSS	Indian Regional Navigation Satellite System
ISRO	Indian Space Research Organization
ITRF	International Terrestrial Reference Frame
ITU	International Telecommunications Union
JCAB	Japan Civil Aviation Bureau
LAAS	Local Area Augmentation System
LBS	Location-Based Service
LF	Low Frequency
LFF	ultrakurzwellen-LandeFunkFeuer (German)
LOP	Line of Position
LORAN	LOng-range RAdio Navigation
LOS	Line Of Sight
M code	Military code
MBOC	Multiplex Binary Offset Carrier
MEO	Medium Earth Orbit
MF	Medium Frequency

MSAS	Multi-functional Satellite Augmentation System
MTSAT	Multi-functional Transport SATellite
NASA	National Aeronautics and Space Administration
NavIC	Navigation with Indian Constellation
NAVSAT	NAVigation SATellite
NAVSTAR	NAVigation Satellite Timing And Ranging
NDGPS	Nationwide DGPS
NGA	National Geospatial Agency
NLOS	Non Line Of Sight
NNSS	Navy Navigation Satellite System
NTSC	National Time Service Centre (China)
OCS	Operational Control Segment
OD&TS	Orbit Determination and Time Synchronisation
OEM	Original Equipment Manufacturer
OS	Open Service
OTF	On-The-Fly
P code	Protected/Precise code
PDA	Personal Digital Assistant
PDL	Position Data Link
PDOP	Position Dilution Of Precision
PE-90	Parameters of the Earth 1990
PLL	Phase Locking Loop
PPP	Precise Point Positioning
PPS	Precise Positioning Service
PRN	Pseudo-Random Noise
PRS	Public Regulated Service
PZ-90	Parametry Zemli 1990 (English: Parameters of the Earth 1990, PE-90)
QPSK	Quadrature Phase-Shift Keying
QZSS	Quasi-Zenith Satellite System
R&D	Research and Development
RAIM	Receiver Autonomous Integrity Monitoring
RDF	Radio Direction Finder
RDOP	Relative Dilution Of Precision
RF	Radio Frequency
RINEX	Receiver INdependent EXchange
RNSS	Regional Satellite Navigation Systems
RNSS	Regional Navigational Satellite System
RS	Restricted Service
RTCM	Radio Technical Commission for Maritime (Services)
RTK	Real-Time Kinematic
SA	Selective Availability
SAR	Search-And-Rescue
SBAS	Satellite Based Augmentation System
SDCM	System for Differential Corrections and Monitoring
SHF	Super High Frequency

SI	International System of Units
SoL	Safety of Life (service)
SPS	Standard Positioning Service
SPS	Standard Positioning Service (for GPS); Standard Precision Service (for GLONASS)
SU	Soviet Union
TAI	Temps Atomique International (English: International Atomic Time)
TDMA	Time Division Multiple Access
TDOP	Time Dilution Of Precision
TEC	Total Electron Content
TRF	Terrestrial Reference Frame
TT&C	Telemetry, Tracking, and Control
UERE	User Equivalent Range Error
UHF	Ultra High Frequency
URA	User Range Accuracy
US	United States (of America)
USA	United States of America
USB	Universal Serial Bus
USNO	United States Naval Observatory
UTC	Coordinated Universal Time
UTM	Universal Transverse Mercator
VDOP	Vertical Dilution Of Precision
VHF	Very High Frequency
VLF	Very Low Frequency
VOR	Very high frequency Omnidirectional Range
WAAS	Wide Area Augmentation System
WGS84	World Geodetic System 1984

1 Overview of GNSS

1.1 INTRODUCTION

Global Navigation Satellite System (GNSS) is a satellite-based navigation and positioning system that provides autonomous spatial positioning. A GNSS allows a small electronic receiver to determine its location using signals transmitted from navigation satellites. For anyone with a GNSS receiver, the system can provide location (and time) information in all weather conditions, day or night, anywhere in the world. Currently, four GNSS constellations are in operation with global coverage. Other than these, some regional or local systems are also available.

Nowadays, all of us carry one or more GNSS receivers in our smartphones or smartwatches. Equipped with these receivers, users can accurately locate where they are and can easily navigate to where they want to go, whether walking, driving, flying, or sailing. GNSS has become a mainstay of transportation systems worldwide, providing navigation for aviation, ground, and maritime operations. Disaster relief and emergency services also depend upon GNSS for location and timing capabilities in their life-saving missions. Activities such as banking, mobile phone operations, and even the control of power grids, are facilitated every day by the accurate timing provided by GNSS. Engineers, surveyors, geologists, geographers, and various professionals can perform their work more efficiently, safely, economically, and accurately using GNSS technology.

1.2 DEFINITION OF GNSS

The applications of GNSS for positioning, navigation, and timing services are enormous; as a result, different definitions of GNSS can be found in the existing literature. We have yet to come up with a commonly accepted and actionable definition of GNSS (Swider 2005). Swider (2005) defined GNSS as:

> GNSS collectively refers to the worldwide civil positioning, navigation, and timing determination capabilities available from one or more satellite constellations.

The International Civil Aviation Organization (ICAO 2005) defines GNSS as:

> GNSS is a worldwide position and time determination system that includes one or more satellite constellations, aircraft receivers, system integrity monitoring augmented as necessary to support the required navigation performance for the intended operation.

Another simple definition is:

> GNSS is a satellite-based system that is used to pinpoint the geographic location of a user's receiver anywhere in the world.

The above definition is short, simple, and memorable; however, it is technologically weak. A better definition of GNSS is:

> GNSS is a system consisting of a network of navigation satellites monitored and controlled by ground stations on the earth, which continuously transmit radio signals that are captured by the receivers to process, thus making it possible to precisely geolocate the receiver by measuring distances from the satellites and to provide precise time information anywhere in the world at any time.

Geolocation refers to identifying the real-world geographic location of a GNSS receiver.

1.3   NAVIGATION AND POSITIONING

Navigation is the process of planning, reading, and controlling the movement of a craft, vehicle, person, or object from one place to another (Bowditch 1995). In GNSS literature, the act of determining the course or heading of movement is also called *navigation*. The word navigate is derived from the Latin roots *navis*, meaning 'a ship', and *agere*, meaning 'to move', or 'to drive', the art and science of conducting a craft as it moves about its ways (Richey 2007). The word 'navigation' perhaps was originally used by the mariners. All navigational techniques involve locating the navigator's position compared to known locations or patterns. Most modern navigation relies primarily on positions determined electronically by receivers collecting information from satellites.

Positioning is a process used to determine the location of one position relative to other defined positions. Obtaining locations in real-time, (i.e., with no delay between appearance in a location and availability of positional information about the very same location) is known as *real-time positioning*. The word positioning, in the GNSS community, is commonly used to refer to 'finding the position' rather than 'getting into the position'. Therefore, positioning may include not just the location, but also the *bearing* (direction) of the positioned object. A positioning system determines the location of an object in space, thus, also referred to as *spatial positioning*. Spatial means 'having to do with space', or 'related to space'. Since GNSS deals with positioning in a geographic sense or in terms of geographic coordinates, the word 'positioning' is also referred to as *geospatial positioning*.

1.4   POINTS OF REFERENCE

A *reference point* (also known as *control point* in surveying) is a location used to determine (or express) the location or position of another one, by giving the relative position. To determine the location of one object we need to refer to reference points or locations of other objects. For example, to define the location of a glass, one might say 'on the table'; hence the table is a reference point which has a location. In this example, however, the position of the glass is not very precise. One may ask 'where on the table?' Therefore, for precise determination of position, we need to adopt a geometrical approach like, '30 cm from left edge of the table, 20 cm from front edge of the table, and on top of the table'; and thus, it defines the exact position.

From the preceding discussion, it is clear that we need to use references to determine the position of objects. But how many references do we need? Let us start with a single reference point. Assume that a tower is installed at a known point, A, on the earth (Figure 1.1a). We are somewhere at a distance of 5 km from A. This does not tell us where we are, but it narrows our position to a point on the circle with the radius of 5 km from the tower, as shown in Figure 1.1a. It can be said that we are anywhere on the circumference of this circle.

Next, let us assume that a second tower is installed at another known point, B, on the earth. We are at a distance of 7 km from B. This tells us that we are somewhere on a circle with the radius of 7 km from the tower B. We now have two pieces of information: our distance to point A is 5 km and our distance to point B is 7 km. So, we are on circle A and circle B at the same time. Therefore, we must be at the intersection of the two circles, one of the two points P or Q, as shown in Figure 1.1b.

Measuring our distance to a third tower C, in the same manner, would identify exactly where we are. Figure 1.1c shows that we must be at the point P, where three circles intersect. This process of determining one's location with the distance measured from three reference points located on the ground is known as *two-dimensional trilateration* (refer to Chapter 3, Section 3.2). However, this is a case for determining two-dimensional (2D) position, where we need at least three reference points. In this example, we have assumed that we are on the earth's surface. Therefore, the surface of the earth would act as an additional reference. But, if we are at some height from the earth's surface, we need at least four reference points for determining three-dimensional (3D) locations.

Suppose we measure our distance from a point and find it to be 10 km. Knowing that we are 10 km from a particular point narrows down all the possible locations we could be in the whole universe to the surface of a sphere that is centred on this point and has a radius of 10 km (Figure 1.2a). Next, let us say we measure our distance to a second point and find out that it is 11 km away. This conveys that we are not only on the first sphere but we are also on a sphere that has an 11 km radius centred at the second point. In other words, we are somewhere where these two spheres intersect. Such an intersection is indeed a circle (Figure 1.2b). If we then make a measurement from a third reference point and find that we are 12 km from that one, our position is narrowed down even further to the two points (P and Q as shown in Figure 1.2c) where the 12 km sphere cuts through the circle that is the intersection of the first two spheres. So, by ranging from three reference points we can narrow down our position to just two points in the space.

Even though there are two possible positions, they differ greatly in their location. However, by adding a fourth reference point, we can determine our precise three-dimensional position. Let us say our distance from a fourth point is 15 km. Now, we

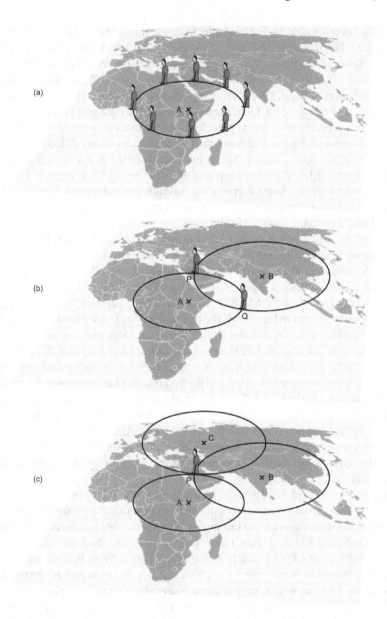

FIGURE 1.1   (a) Single-point referencing, (b) double-point referencing, and (c) triple-point referencing.

have a fourth sphere intersecting the first three spheres at one common point, and this is the precise location. This technique is known as *3D trilateration*.

In summary, we need points of reference to determine or define the location or position of any object. At least three reference points are required to define 2D position and four are needed to define positions in 3D. However, it is worth mentioning

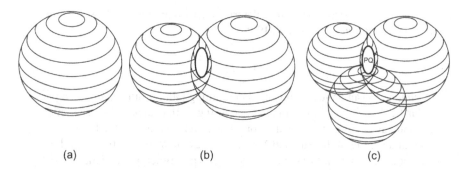

(a)   (b)   (c)

FIGURE 1.2 Principle of 3D positioning: (a) distance measurement from one reference point establishes position on a sphere; (b) distance measurement from two reference points establishes circle of intersection of two spheres; (c) distance measurement from three reference points narrows position to only two positions (P and Q).

that GNSS can determine 3D positions with reference to only three points by means of mathematical and geometrical tricks (discussed in Chapter 3, Section 3.5).

1.5 HISTORY OF NAVIGATION SYSTEMS

Where are we? How do we get to our destination? These questions are as old as the history of mankind. The long history of navigation and positioning and the myriad of techniques and instruments humans used in these efforts are beyond the scope of this book. However, the following sections briefly describe the evolution of navigation and positioning with important breakthroughs.

1.5.1 THE CELESTIAL AGE

Identifying and remembering objects and landmarks as points of reference was how early humans used to find their way through jungles and deserts. Leaving stones, marking trees, referencing mountains were the early navigational aids. Stones, trees, and mountains were the early examples of 'points of reference'. Today, we still use the same concept for positioning and navigation on the land—often, we use several landmarks for our daily navigational purpose.

Identifying points of reference was easy on land. But it became a matter of life and survival when man started to explore the oceans, where the only visible objects were the Sun, the Moon, and the stars. Naturally, they became the 'points of reference' and the era of *celestial navigation* began. Celestial navigation and positioning was the first serious solution to the problem of finding one's position in unknown territories, where the Sun, the Moon, and stars were used as points of reference. Celestial navigation is the process where angles between objects in the sky (celestial objects) and the horizon were used to locate one's position on the earth. At any given instant of time, any celestial object (e.g., the Sun, the Moon, or stars) can be located directly over a particular geographic position on the earth. This geographic position (on the earth's surface) is known as the celestial object's *subpoint*, and its location (e.g., its latitude and longitude) can be determined by referring to tables in a *nautical almanac* or *air almanac* (Bowditch 1995).

NOTE

An *almanac* is an annual publication containing tabular information in a particular field or fields often arranged according to the calendar. Astronomical data and various statistics are also found in almanacs, such as the times of the rising and setting of the Sun and Moon, eclipses, hours of full tide, stated festivals of churches, terms of courts, lists of all types, timelines, and many more.

A *nautical almanac* is a publication describing the positions and movements of celestial bodies for the purpose of enabling navigators to use celestial navigation to determine the position of their ship while at sea with reference to the sun, moon, planets, and 57 stars chosen for their ease of identification and wide spacing.

The measured angle between the celestial object and the horizon is directly related to the distance between the subpoint and the observer, and this measurement is used to define a circle on the surface of the earth, called a celestial *line of position* (LOP), the size and location of which can be determined using mathematical or graphical methods (Bowditch 1995). The LOP is significant because the celestial object would be observed at the same angle above the horizon from any point along its circumference at that instant. The observer is located anywhere on this LOP. Two LOPs, calculated from two celestial objects, can limit the observer's position to only two points, each located where these two LOPs intersect (Figure 1.1b). In most cases, it is easy to determine which of the two intersection points is the observer's correct location. Sometimes the two intersection points are thousands of kilometres apart, making it easy to reject one of the two points. Similarly, a mariner may find that one of the points is on land, and that point can be rejected. Generally, observation of two celestial objects can provide precise location. The observation of a third celestial object provides a third LOP and only one intersection point.

Early celestial navigation was based on angular measurement between the horizon and a common celestial object. The relative position of these celestial objects and their geometrical arrangement look different from various locations on the earth. Therefore, by observing the configuration of this arrangement, one could intuitively estimate their position on the earth and the direction that one should take for the destination. While this early celestial navigation technique was useful for travelling short distances from land, it was problematic for long voyages, and often navigators were lost at sea.

Later, the geometrical configuration of stars, from the observer's point of view, was more accurately determined by measuring the relative angles between them. This led to the development of a number of more accurate instruments, including the kamal, astrolabe, octant, and sextant (Figure 1.3). The measured angles were then used to determine the position of the observer with the aid of published pre-calculated charts that eased the tedious computation task. The magnetic compass was also used to find the direction for navigation purposes.

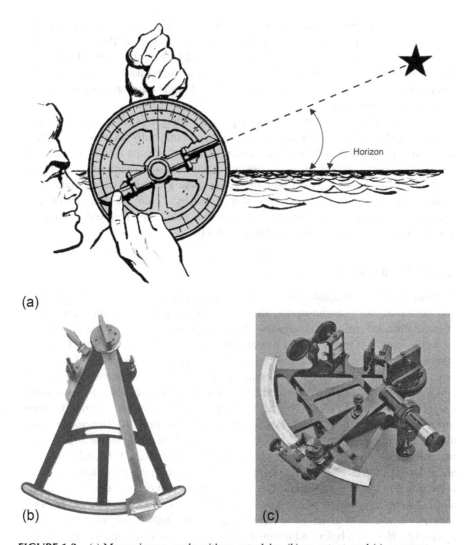

FIGURE 1.3 (a) Measuring an angle with an astrolabe, (b) an octant, and (c) a sextant.

NOTE

Navigators measure distances on the globe in arcdegrees, arcminutes, and arcseconds. A nautical mile is defined as 1.1508 mi (1852 m), which is also one minute of angle along a meridian on the earth. Sextants can be used to read angles accurately to within 0.2 arcminutes. Therefore, theoretically, the observer's position can be determined within 0.23 mi (370 m). Most ocean navigators, measuring from a moving platform, can achieve a practical accuracy of 1.5 miles (2.8 km), which is enough to navigate safely when one is on the sea.

The process of measuring the angles of the celestial objects with the aforesaid instruments was time-consuming and inaccurate. For instance, they could not be used during the day (because in the daytime the sun is the only visible object in the sky) or on cloudy nights. Furthermore, the measured angles had to be transferred to special charts, and, after tedious calculations, the derived position was good only to about several kilometres.

Faced with difficulties of determining a position, navigators of long ago must have dreamt of gadgets that would perform such a task automatically and more accurately. There were probably people who had pictured a device, or even worked on building one, that aligned itself with celestial objects quickly, measured angles to these points of reference, and computed their positions automatically. The idea of automatic computation of position through measurement of distances to points of reference became a reality only in the mid-20th century when radio signals were deployed, and the era of radio navigation began (Javad and Nedda 1998).

1.5.2 THE RADIO AGE

Radionavigation is the application of radio-frequency signals to determine one's position on the earth. Near the middle of 20th century, scientists discovered a way to measure distances using radio signals. The concept was to measure the time it took for radio signals to travel from a transmitting station to a special device (receiver) designed to receive them. Multiplying the signal travel time by the speed of the signal gives the distance between the transmitter and the receiver. The geometrical concept of positioning was simple trilateration. If we consider three radio-signal transmitting towers located at A, B, and C (Figure 1.1c), we can determine our location by measuring the distances from these three towers (refer to Section 1.4). Hence, the transmitting towers are acting as points of reference. A person needs to use a receiver to measure the distances from these transmitters. The transmitters A, B, and C together are called a transmitter 'chain'. A chain may have four or more transmitters in order to have better accuracy, and multiples of such chains can cover a greater area. The range of a radio transmitter was generally about 500 km (Javad and Nedda 1998). This concept was the most advanced technique used for navigation and positioning in the Radio Age. However, before achieving this advanced technique, radio navigation and positioning techniques had passed through several other phases, as outlined below.

The first system of radio navigation was the Radio Direction Finder (RDF). By tuning in a radio station and then using a directional antenna to find the direction to the broadcasting antenna, radio sources replaced the stars and planets of celestial navigation with a system that could be used in all weather and times of day (Dutton and Cutler 2004; Appleyard et al. 1998). This system was widely used in the 1930s and 1940s.

In the 1930s, German radio engineers developed a new system, called the 'Ultrakurzwellen-Landefunkfeuer' (LFF), or simply 'Leitstrahl' (guiding beam). However, it was referred to as *Lorenz* outside of Germany (Bauer 2004) because it was the name of the company that manufactured the equipment. Initially, it was used as a landing system for aircraft.

In Lorenz, two signals were broadcast at 38 MHz frequencies from two highly directional antennas in the same line with beams a few degrees wide (Figure 1.4). One of these two antennas was pointed slightly to the left of the other, with a small angle in the middle where the two beams overlapped. The left and right antennas were switched on and off in turn. The broadcast was switched so that the left antenna was turned on only briefly, sending a series of 'dots' 1/8-second long repeating once every second. After switching off the second antenna, the signal was then sent from the right antenna, broadcasting a series of 7/8-second long 'dashes'. The signals could be detected for some distance off the end of the runway, as far as 30 km.

An aircraft approaching the runway would tune its radio to the broadcast frequency and listen for the signals. If they heard a series of dots, they knew they were off the runway centreline to the left (the *dot-sector*) and had to turn to the right to line up with the runway. If they were off to the right, they would instead hear a series of dashes (the *dash-sector*) and turn left (Figure 1.4). Key to the operation of the system was an area in the middle where the two signals (left and right) overlapped, and the dots of the one signal 'filled in' the dashes of the other resulting in a steady tone known as the *equi-signal*. By adjusting their path until they heard the equi-signal, the pilot could align their aircraft with the runway for a safe landing.

The next major advancement in radio navigation system was the use of two signals that varied not in sound, but in phase. The VHF Omnidirectional Range (VOR), is a type of radio navigation system for aircraft. This system was designed to broadcast a very high frequency (VHF, ranging from 108.0 MHz to 117.95 MHz) radio composite signal. The concept of VOR is rather complicated and beyond the scope of this book to discuss. However, in VOR, a single master signal was sent out continually from the station, and a highly directional second signal was sent out that varies in phase 30 times a second compared to the master (refer to Chapter 4, Section 4.7 for details on phase difference). This signal was timed so that the phase varies as the secondary antenna spins, such that when the antenna was at 90 degrees from the north, the signal was 90 degrees out of phase of the master. By comparing the phase of the secondary signal to the master, the angle could be determined without any physical motion in the receiver. This angle was then displayed in the cockpit of

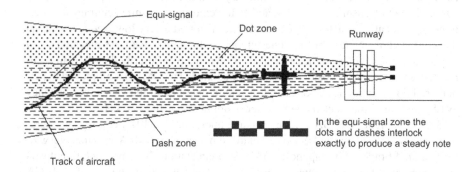

FIGURE 1.4 Lorenz beams for aircraft landing.

the aircraft, and could be used to take a fix just like the earlier RDF systems. This system was, in theory, easier to use and more accurate (Clausing 2006).

RDF, Lorenz, and VOR were the early radio navigational aids. They, however, did not use the concept of signal travel time to measure the distance from the transmitter. The first system based on the measurement of the difference of signal arrival times from two or more reference locations, the British GEE system (Gee means 'grid', i.e., the electronic grid of latitude and longitude), was first used during World War II (Hecks 1990). GEE transmitters sent out precisely timed pulses. There were three GEE stations, one master, and two slaves. The master sent a pulse of two milliseconds followed by a double pulse. The first slave station sent a single pulse of one millisecond after the master's single pulse, and the second slave sent a single pulse of one millisecond after the master's double pulse. Onboard the aircraft, the signals from the three stations were received. The onboard equipment would display the two slaves' signals as blips on an oscilloscope type display (usually as a two-dimensional graph). Since the display timing was controlled by the pulses from the master station, the display equipment gave the difference in reception time of the pulses and hence the relative distance from the master and each slave. The aircraft carried a navigation chart with several hyperbolae (lines of position) plotted on it (Bowditch 1995). Each hyperbolic line represented a line of constant time difference for the master and one slave station. All that the navigator had to do was find the intersection of the two hyperbolae representing the two slave stations to know their location. GEE was accurate to about 165 yards (150 m) at short ranges, and up to a mile (1.6 km) at longer ranges.

NOTE

GEE, OMEGA, and LORAN were all called hyperbolic systems due to the shape of the lines of position on the chart.

Building on the British GEE, the OMEGA radio navigation system was developed. It was the first truly global radio navigation system for aircraft, operated by the United States (US) in cooperation with six partner nations. The OMEGA was originally developed by the US Navy for military aviation use. It was approved in 1968 with only eight transmitters and the ability to achieve a four-mile accuracy to determine one's location. Each OMEGA station transmitted a very low frequency signal that consisted of a pattern of four tones unique to the station that were repeated every 10 sec. Because of this and the radio navigation principles described earlier, an accurate fix of the receiver's position could be calculated. The OMEGA employed hyperbolic radio navigation techniques and the chain operated in the very low frequency (VLF) portion of the electromagnetic spectrum between 10 to 14 kHz.

Six of the eight OMEGA stations became operational in 1971; day-to-day operations were managed by the US Coast Guard in partnership with Argentina, Norway, Liberia, and France. The Japanese and Australian stations became operational several years later. Due to the success of GPS satellite navigation, the use of OMEGA declined during the 1990s to a point where the operating cost of OMEGA could no longer be justified. OMEGA was permanently terminated on 30 September 1997, and all stations ceased operation. By the end of its life, it was used primarily by civilians.

Several other similar radio navigation systems were also initiated by a number of countries during this period, such as Alpha (the Russian counterpart of OMEGA), American LORAN, CHAYKA (the Russian counterpart of LORAN), and DNS (British/US). The most capable of these, LOng-range RAdio Navigation (LORAN), became operational in the 1950s (Bowditch 1995; Clausing 2006) and is a low frequency terrestrial radio navigation system still in use in some parts of the world (https://www.loran.org). The latest version of LORAN in common use is LORAN-C (COMDTPUB 1992), which operates in the low frequency portion of the electromagnetic spectrum from 90 to 110 kHz.

In LORAN, a single 'master' station would broadcast a series of short pulses, which were picked up and re-broadcast by a series of 'slave' stations, together making a 'chain'. By exercising precise control over the time between the reception and re-broadcast of the pulses by the slaves, the time it took for the radio signal to travel from station to station could be measured by listening to the signals. Since the time for the re-broadcasts to reach a remote receiver varies with its distance from the slaves, the distance to each slave could be determined. By plotting the hyperbolae representing the ranges on a map, the area where they overlapped formed a fix location. Each LORAN chain consists of at least four transmitters and typically covers an area of about 500 miles. To provide LORAN coverage for larger areas, several LORAN chains are used. As the sophistication of computer systems grew to the point where they could be placed on a single chip, LORAN suddenly became very simple to use and quickly appeared in civilian systems starting in the 1980s. However, like the beam systems, civilian use of LORAN was short-lived when newer technology quickly drove it from the market.

NOTE

With the perceived vulnerability of GNSS systems and their own propagation and reception limitations, renewed interest in LORAN applications and development has surfaced. Enhanced LORAN, also known as eLORAN or E-LORAN, comprises advancements in receiver design and transmission characteristics which have increased the accuracy and usefulness of traditional LORAN. With reported accuracy as high as 8 meters, the system becomes competitive with normal GNSS. The eLoran receivers now use 'all in view' reception, incorporating time signals and other data from up to 40 stations. These enhancements in LORAN make it adequate as a substitute when GNSS is unavailable or degraded.

Although LORAN was a major breakthrough for navigation and positioning, it had the following limitations: 1) LORAN coverage was limited to about 5% of the earth's surface where the chains were established; 2) the system was operated by local governments and was generally situated near coastal areas having high traffic volume; 3) the LORAN signal suffered from electronic effects of weather and ionospheric effects; and 4) LORAN could provide only two-dimensional position information (latitude and longitude). It could not provide height information and, hence, could not be used in aviation. In general, the accuracy of LORAN was about 20–100 m (for Loran-C), and therefore it was not suitable for surveying/mapping.

1.5.3 THE SATELLITE AGE

To overcome the limitations of land-based radio navigation systems, the satellite-based radio navigation system was conceived in which improved radio transmitters were put aboard artificial satellites orbiting the earth at high altitudes to provide wider coverage. On 4 October 1957, the Soviet Union launched the world's first artificial satellite, *Sputnik*. Two scientists, William H. Guier and George C. Weiffenbach, at Johns Hopkins University's Applied Physics Laboratory (APL) (Laurel, Maryland, USA), took a series of measurements of Sputnik's *Doppler shift* (refer to Chapter 4, Section 4.10.1) yielding the satellite's position and velocity (Bedwell 2007). This team continued to monitor two other satellites Sputnik II and American Explorer I. In March 1958, Guier and Weiffenbach were summoned to meet their boss, Frank T McClure, who stunned them with his interpretation of their data. McClure, who would serve as a director of APL's Research Centre for nearly 25 years, offered a reverse premise: If measuring a satellite's Doppler shift could determine its precise orbit, then the satellite's signals should also permit a properly equipped observer to fix their location on the planet (Bedwell 2007).

Soon after his eureka moment, McClure shared his conclusions with another APL researcher, Richard Kershner. Over a long weekend McClure and Kershner worked out the preliminary details of *Transit*, the first satellite-based navigational system (Stansell 1978). The first Transit satellite was placed in polar orbit in 1960. When the system became operational in 1962, despite the opposition of US Navy officers who doubted that it would work, it had seven satellites orbiting at an altitude of about 1100 km. These satellites broadcast signals to ground-based users who could locate themselves by measuring the signals' Doppler shifts. The satellites also transmitted information about their orbital positions, which they obtained from a set of four ground-based tracking stations. The US Navy hoped for accuracy within about 1 km; but it proved to be much better at around 25 m (Bedwell 2007).

Transit was so helpful and reliable to the US Navy's submarines and surface vessels that it was released to civilian users in 1967. Under the name NAVigation SATellite (NAVSAT) or Navy Navigation Satellite System (NNSS), it would help guide both weekend sailors and commercial shipping crews until the mid-1990s.

Although Transit demonstrated that navigation satellites were helpful and reliable, it was hardly user-friendly. It required long observation times, and the small number of satellites made for spotty access; sometimes it remained silent for hours at a time (Bedwell 2007). Users on vessels that were in motion had to make time-consuming corrections; and it yielded data in only two dimensions, latitude and longitude. For the third dimension, altitude, which would be crucial for aviation, a more advanced system was needed.

(a)   (b)

FIGURE 1.5 Timation satellite—under assembly (left) and an artist's concept (right) (Courtesy: NASA).

Within a few years, tracking the satellites from earth had become a fairly routine task. Then it was time to track vessels on earth with the help of satellites. In 1964, Roger L. Easton was leading a team in Space Systems Division of Naval Research Laboratory (Washington, DC) that developed an improved space-based navigation system called *Timation* (Aldridge 1983). This was a more direct predecessor of today's GNSS. Easton visualized a constellation of signal-emitting satellites carrying clocks synchronized to a master clock on earth. Doppler shifts would not be used; instead, by measuring how long it took the signal to arrive, users could tell how far they were from the satellite. When repeated with other satellites and combined with data about their orbits, this procedure could yield the user's position in three dimensions (Bedwell 2007).

The use of time readings by means of signals transmitted from satellites was Timation's key innovation. Obviously, it would require extremely accurate clocks. In 1964, Easton received a US patent for Timation. When the first two Timation satellites (Figure 1.5) were placed in orbit, in 1967 and 1969, they carried clocks with stable quartz-crystal oscillators. Two later satellites would be equipped with *atomic clocks*, which eventually set the GPS standard. Later, several other experiments and launches besides Timation were carried out towards establishing a satellite-based global navigation system with high accuracy.

NOTE

A crystal oscillator (or quartz-crystal oscillator) is an electronic circuit that uses the mechanical resonance of a vibrating crystal of piezoelectric material to create an electrical signal with a very precise frequency. This frequency is commonly used to keep track of time (as in quartz wristwatches), to provide a stable clock signal for digital integrated circuits and to stabilize frequencies for radio transmitters/receivers.

In December 1973, bolstered by the successful experiments with Timation, the US Department of Defence approved a $104 million initial budget to develop a satellite navigation system named NAVigation Satellite Timing and Ranging Global Positioning System (NAVSTAR GPS), commonly known as GPS (Parkinson 1994). Soon after, the approval on 14 July 1974 of the first prototype GPS satellite, Navigational Technology Satellite-1 was put into orbit by the US Department of Defence, but its clocks failed shortly after launch. The Navigational Technology Satellite-2, redesigned with *caesium* atomic clocks, was sent into orbit on 23 June 1977. By 1985, the first 11-satellite GPS Block-I constellation was in orbit. GPS Block-II constellation was started with its first launch in February 1989 and completed in March 1994 (Bedwell 2007).

A similar satellite navigation system was also initiated by the Soviet Union, named GLObal'naya NAvigatsionnaya Sputnikovaya Sistema (GLONASS; English translation GLObal NAvigation Satellite System, also called GLobal Orbiting NAvigation Satellite System). In the late 1960s and early 1970s, the Soviet Union identified the need and benefits of developing a new satellite-based radio navigation system. Their existing *Tsikada* satellite navigation system (similar to US Transit), while highly accurate for stationary or slow-moving ships, required several hours of observation by the receiving station to fix a position, making it unusable for many navigational purposes and for guidance of the new generation of ballistic missiles.

From 1968 to 1969, the research institutes of the Soviet Ministry of Defence, Academy of Sciences, and Soviet Navy cooperated to develop a single system for navigation of their air, land, sea, and space forces. This collaboration resulted in a 1970 document that established the requirements of such a system. Six years later, in December 1976, a plan for developing the GLONASS was accepted. The first satellite of this system was put into orbit on 12 October 1982. From 1982 through April 1991, the Soviet Union successfully launched a total of 43 GLONASS-related satellites in addition to five test satellites. In 1991, 12 functional GLONASS satellites, in two orbital planes, were available—enough to allow limited usage of the system. Following the disintegration of the Soviet Union in 1991, continued development of GLONASS was undertaken by the Russian Federation. It was declared partially operational on 24 September 1993 by the then-president Boris Yeltsin. Although the project was scheduled to be completed by 1991, the full in-orbit constellation was completed in December 1995, and the system was declared fully operational.

Initially, GPS and GLONASS were conceived to serve military purposes; civilians were not allowed to use these signals. On 1 September 1983, Soviet fighter jets shot down a Korean Airlines Flight 007 which had strayed into Soviet airspace. This resulted in the death of all 269 passengers on the flight. To help avert such future tragedies, US President Ronald Reagan announced that GPS signals would be made available for international civilian use as the system came on line. It was decided that US military interests would be protected through coding of military signals and by applying perturbations (alterations) to reduce the precision of readings made available to civilian users through a system called Selective Availability (SA). That scheme, however, did not hold up very long, as electronics manufacturers quickly found ways to override SA with *differential systems*. On the other hand, during the

Gulf War, US soldiers and tankers recognized the value of GPS and requested many more receivers than the US Army had in stock. Officials turned to the private sector, ordering more than 10,000 units from commercial suppliers. With so many commercial receivers being used by the troops, the US government had to disengage SA on 2 May 2000 by order of US President Bill Clinton. This encouraged several private companies to pursue manufacturing of GPS receivers, and the increasing competition forced the price to drop. As a result, a new door to the world of satellite navigation opened, and numerous civilian applications mushroomed. Although the SA of GPS was detached in the year 2000, restrictions that prohibited civilian users from acquiring the high precision signal of GLONASS satellites were only withdrawn on 18 May 2007.

GPS and GLONASS both are operated and controlled by the militaries of respective nations and do not ensure continuing providence of signal to civilians for ever. This uncertainty initiated a third GNSS, Galileo (named after the Italian astronomer Galileo Galilei), for civilian use. The European Union and European Space Agency agreed on 26 May 2003 to introduce their own alternative to GPS and GLONASS, called the Galileo positioning system. The first satellite of this system Galileo In-Orbit Validation Element-A (GIOVE-A) test satellite was launched on 28 December 2005 and the second (GIOVE-B) on 27 April 2008. Several tens of satellites in this system were thereafter launched and became fully operational in the year 2020. Galileo services are free and open to everyone with limited accuracy. The higher-precision capabilities are only available for paying commercial users.

In 1983, China initiated its own independent satellite navigation system. Called Compass Demonstration Navigation System (also known as Compass Experimental Navigation System or BeiDou I), this technological demonstration initially provided only regional coverage. BeiDou 1A, the first satellite of this system, was launched in October 2000; it was followed by BeiDou 1B in December 2000, BeiDou 2A in May 2003, and BeiDou 2B in February 2007. After successful experimentation with BeiDou I, China initiated a global system, BeiDou II or Compass Navigation Satellite System (CNSS, commonly referred to as *Compass*). The first satellite of this system, Compass-M1, was successfully launched on 14 April 2007; the second, Compass-G2 (this was the first launch of the second-generation Compass satellites), was launched on 15 April 2009 followed by several other launches. The third phase of the BeiDou system—*BeiDou-III*—was initiated in the year 2015 with the launch of its first satellite BeiDou-3 I1-S. Several tens of satellites have been launched since. The BeiDou-III is also known as BeiDou Navigation Satellite System (BDS) or simply *BeiDou*. This system was developed by the Chinese Academy of Space Technology and became fully operational in the year 2020.

As of the year 2020, there were only four core GNSS systems in existence as discussed in the preceding paragraphs (GPS, GLONASS, Galileo, and BeiDou) that are meant for providing position and timing information for a variety of applications. However, for safety and critical applications, the basic constellations cannot meet the requirements in terms of accuracy, integrity, and availability. For this purpose, the basic constellations are augmented by overlay systems such as differential global navigation satellite system (DGNSS) and inertial navigation system (INS).

For example, the USA has put up its Wide Area Augmentation System (WAAS), Europe its European Geostationary Navigation Overlay Service (EGNOS), Japan its Multi-functional Satellite Augmentation System (MSAS), and India its GPS-Aided Geo-Augmented Navigation (GAGAN), in their respective regions.

It is obvious that the ability of a nation to supply signals from satellites also infers the ability to deny their availability. The operator or owner of a specific GNSS potentially has the ability to degrade or eliminate satellite navigation services over any territory it desires. Thus, as the satellite navigation becomes an essential service, countries without their own satellite navigation systems effectively become client states of those who supply these services. For this reason, various countries have sought to develop their own regional satellite navigation systems, such as Japan's Quasi-Zenith Satellite System (QZSS) and Indian Regional Navigation Satellite System (IRNSS, operational name is NavIC). These are similar in principle to the core GNSS but operate regionally rather than globally.

1.6 SATELLITE-BASED NAVIGATION AND POSITIONING SYSTEMS

The theory behind the operation of the satellite navigation systems is similar to that of the land-based radio navigation systems. In land-based radio navigation systems, the transmitting towers were the reference points located on the earth and the distances to them were measured by the receivers to compute the two-dimensional positions (latitude and longitude or x and y) by finding the intersection of several circles (or hyperbolae). In satellite-based systems, the satellites act as the reference points, and the distances to them are measured to determine the three-dimensional positions (latitude, longitude, and altitude or x, y, and z) by finding the intersection of several spheres (Figure 1.2). In a land-based system, the location of a transmitting tower is fixed, accurately known, and stored in the database of the receivers. However, the location of a satellite is not fixed, since it orbits the earth at a high speed. Therefore, it is rather more complicated to determine one's position with reference to them. However, satellites have a mechanism of giving information about their location at any instant of time.

A GNSS may have several layers of infrastructure:

- Core satellite navigation systems (or core GNSS)—currently GPS, GLONASS, Galileo, and BeiDou.
- Global satellite-based augmentation systems.
- Regional satellite-based augmentation systems, such as WAAS (United Sates), EGNOS (European Union), MSAS (Japan), and GAGAN (India).
- Regional satellite navigation systems, such as QZSS (Japan), NavIC (India), and Beidou-I of China (now retired).
- Continental-scale ground-based augmentation systems, for example, the Australian Ground-Based Regional Augmentation System (GRAS) and the US Department of Transportation National Differential GPS (DGPS) service.
- Local Ground-Based Augmentation System (GBAS), such as the US Local Area Augmentation System (LAAS).

In this chapter, we have defined navigation and positioning, provided a brief history of human developments in the areas of navigation and surveying, and introduced the most recent developments in satellite-based navigation. Going forward, we shall conduct an in-depth discussion on core GNSSs; then we'll move on to related developments such as augmentation systems and regional navigation satellite systems.

EXERCISES

DESCRIPTIVE QUESTIONS

1. What do you understand by 'navigation and positioning'? What is a GNSS? What are the 'points of reference' in a satellite-based navigation and positioning system?
2. What do you understand by 'satellite-based navigation and positioning systems'? Explain in brief.
3. How are points of reference used to determine one's position in two dimensions and three dimensions?
4. Describe celestial navigation briefly.
5. Elaborate on the age of satellite navigation.
6. What is radionavigation? Explain LORAN.
7. Write about different types of satellite-based navigation and positioning systems.

SHORT NOTES/DEFINITIONS

Write short notes on the following topics

1. GNSS
2. Subpoint
3. Nautical almanac
4. Lorenz
5. eLORAN
6. Transit
7. Timation
8. GLONASS
9. NAVSTAR GPS
10. Galileo
11. BeiDou
12. GIOVE
13. Selective availability

2 Functional Segments of GNSS

2.1 INTRODUCTION

As stated in Chapter 1, as of now, we have only four GNSS constellations with global coverage: (1) NAVigation Satellite Timing And Ranging Global Positioning System (NAVSTAR GPS, or simply, GPS), (2) GLobal Orbiting NAvigation Satellite System (GLONASS), (3) Galileo, and (4) BeiDou. This chapter provides the basic concepts of functional segments of GNSS and how these segments interact with each other.

Each GNSS system consists of a space segment (satellites in the sky), a control segment (ground stations), and a user segment (GNSS receivers) (Figure 2.1). All of the aforementioned four GNSS systems are based on more or less similar architecture and principles. Let us now consider these three segments and discuss them in detail. We shall then have a closer look at how a GNSS works in Chapter 3.

2.2 SPACE SEGMENT

The space segment of a GNSS consists of a series of satellites that transmit radio signals continuously and which comprise the heart of the system. The satellites are placed in a medium earth orbit (approximately at 20,000 km altitude), although the altitude varies for different constellations. Operating at such a high altitude allows the signals to cover a greater area. Satellites are arranged in such a manner that a GNSS receiver on the earth can receive signals or information from at least four satellites.

The satellites travel at very high speed (more than 13000 km/h, varies for different constellations), they are powered by solar energy, and are built to last, on average, for 10–12 years. If the solar energy fails (eclipses, and such factors), they have backup batteries onboard to keep them running and small rocket boosters for periodical orbit correction (to keep them flying on the correct path).

Each satellite contains at least three high-precision atomic clocks and constantly transmits radio signals using its own unique identification code (or frequency in the case of GLONASS). Each satellite transmits low-power radio signals on several frequencies in the microwave zone of electromagnetic spectrum (refer to Chapter 4). GNSS receivers are designed to receive these signals. The signals travel in the 'line of sight', which implies that it can pass through clouds, glass, and plastic, but not go through most solid objects such as buildings and mountains.

Each signal contains pseudorandom codes (a complex pattern of digital code). The main purpose of these coded signals is to allow for calculation of signal travel time from the satellite to the user's receiver. This *travel time* is also called the *time of*

19

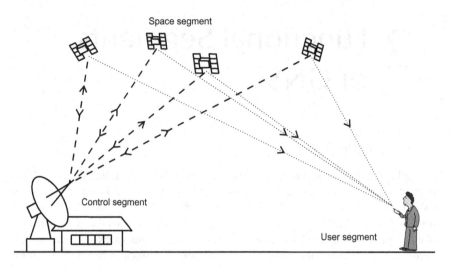

FIGURE 2.1 Functional segments of GPS.

arrival or *propagation time*. The travel time multiplied by the speed of light equals the satellite *range* (distance from the satellite to the receiver). The *navigation message* (the information that satellites transmit to a receiver; refer to Chapter 4) contains the satellite orbital and clock information, general system status messages, and an ionospheric delay model (refer to Chapter 5). The satellite signals are timed using highly accurate atomic clocks. Since the velocity of light is about 3×10^8 m/s, a tiny error in time measurement can produce a highly erroneous distance measurement.

NOTE

An *atomic clock* is an extremely accurate clock that uses an atomic resonance frequency standard as its counter (Audoin *et al.* 2001). The terms 'atomic clock' and 'atomic frequency standard' are often used interchangeably. Early atomic clocks were *masers* (microwave amplification by the stimulated emission of radiation) with attached equipment. At present, the best atomic frequency standards (or clocks) are based on more advanced physics involving cold atoms and atomic fountains. National standards agencies maintain an accuracy of 10^{-9} s/day, and a precision equal to the frequency of the radio transmitter pumping the maser. The clocks maintain a continuous and stable timescale—International Atomic Time (TAI, from the French name Temps Atomique International). The first atomic clock was built in 1949 at the US National Bureau of Standards. The first accurate atomic clock, based on the transition of the caesium-133 atom, was built by Louis Essen in 1955 at the National Physical Laboratory in UK. This led to the internationally agreed definition of the second being based on atomic time. In 1967, the International

System of Units (SI) defined the second as 9,192,631,770 cycles of the radiation, which corresponds to the transition between two energy levels. This definition makes the caesium atomic clock (often called a *caesium oscillator*) the primary standard for time and frequency measurements.

2.2.1 GPS SPACE SEGMENT

The nominal constellation of the GPS space segment is composed of 24 satellites (21 operational + 3 spares or backups) (Spilker and Parkinson 1995; Kaplan 1996). However, there are currently about 28–30 GPS satellites actually in space—at least 24 active and 4–5 spares is the current standard—offering improved precision of calculated position. GPS satellites are placed in six near-circular orbits centred on the earth. These six orbital planes (Figure 2.2a) have approximately 55° inclination (tilt relative to earth's equator or equatorial plane) and are separated by 60° right ascension (angle along the equator from a reference point to the orbit's intersection) (Samama 2008). The satellites are placed at a nominal altitude of 20,200 km above sea level. However, the distance travelled by the signal from the satellite to a receiver varies from around 20,200 km, if the satellite is in its zenith, to around 25,600 km when the satellite is at horizon (Figure 3.4, Chapter 3).

NOTE

The path (course of motion) followed by a satellite in space is referred to as its *orbit*. The plane in which a satellite moves is called its *orbital plane*. The plane containing the earth's equator is known as *equatorial plane*. The angle between the satellite's orbital plane and the earth's equatorial plane is termed the *orbital inclination*. Needless to say, the inclination of equatorial orbital plane is 0°. When the inclination is 90°, the satellite moves over the poles; that is, centre of the earth, north and south poles lie in the orbital plane. This is called *polar orbit*. Orbital planes with an inclination in between 0° and 90° are termed as *inclined orbits*.

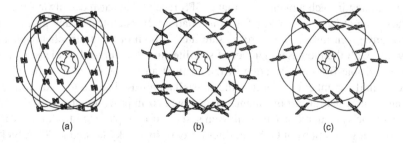

(a)   (b)   (c)

FIGURE 2.2   Constellation of satellites for (a) GPS, (b) Galileo, and (c) GLONASS.

The time taken by a satellite to make one revolution around the earth is known as *orbital period*. Changing the altitude of a satellite in a circular orbit alters the time taken for a complete orbit; the greater the altitude, the longer the orbital period. At about 35,786 km height above mean sea level, the orbital period is equal to one sidereal day. *Sidereal day* refers to the time taken by earth to rotate 360° relative to the stars, i.e., 23 h 56 min 4.091 s. If the orbital period equals one sidereal day, the orbit is called *geosynchronous orbit*; that means the orbit is synchronized with the rotational period of the earth. Geosynchronous orbits may be circular or elliptical with zero or non-zero inclination. A *geostationary orbit* is a special kind of geosynchronous orbit. If any geosynchronous orbit is circular and its inclination is 0°, then it is called *geostationary*. In this case, the relative movement between the earth and the satellite is zero, and the satellite appears stationary with respect to the earth. This allows the satellites to observe and collect information continuously over specific areas.

A satellite orbit is not fixed all the time due to the asymmetry of earth's gravitational field, the gravitational fields of the sun and the moon, the solar radiation field, and atmospheric drag (Pratt *et al.* 2003). Atmospheric drag is applicable for *low earth orbit* satellites. However, GNSS satellites are placed outside of the earth's atmosphere so they are not affected by atmospheric drag. Orbits are also not exactly circular and thus are referred to as *near-circular*. Due to these reasons, an orbital period may vary from time to time. However, the general practice is to consider the nominal (average) period. Satellites occasionally require their orbits to be corrected because of forces that occur when a satellite is in orbit that can cause them to deviate from their initial orbital path. In order to maintain the planned orbit, a control centre on the ground issues commands to the satellite to place it back in the proper orbit. A small rocket booster in the satellite is ignited periodically for orbit correction (to maintain the proper orbit) (Pratt *et al.* 2003).

The orbital period of GPS satellites is half a sidereal day, or 11 h 57 min 58 s. The ground tracks of a given satellite on the earth's surface are therefore almost identical from one day to the next. On each of the six orbital planes, four satellites (nominal) are positioned in such a manner so that a GPS receiver on earth can always receive signals from at least four of the full constellation (24 satellites) at any given time; and at any given instant there are 12 satellites in both of the earth's hemispheres. However, with the increased number of satellites in orbit, currently, at least 6 satellites are visible from any location of the earth at any given time. With the recent increase in the number of satellites, the constellation was changed to a non-uniform arrangement. Such an arrangement was adopted to improve reliability and availability of the system, relative to a uniform system, if multiple satellites fail to work.

Different generations of GPS satellites co-exist in the sky (Samama 2008; Sickle 2008), and other generations are in the queue (Table 2.1). As a result, the capability

TABLE 2.1

Various Generations of GPS Satellites

Satellite	First Launch	Last Launch	Signal
Block I	1978	1985	**L1:** C/A + P(Y) **L2:** P(Y)
Block II	1989	1990	**L1:** C/A + P(Y) **L2:** P(Y)
Block IIA	1990	1997	**L1:** C/A + P(Y) **L2:** P(Y)
Block IIR	1997	2004	**L1:** C/A + P(Y) **L2:** P(Y)
Block IIR-M	2005	2009	**L1:** C/A + P(Y) + M **L2:** P(Y) + CM + CL + M
Block IIF	2010	2016	**L1:** C/A + P(Y) + M **L2:** P(Y) + CM + CL + M **L5:** C-I + C-Q
Block IIIA	2018	2023	**L1:** C/A + P(Y) + M + C **L2:** P(Y) + CM + CL + M **L5:** C-I + C-Q
Block IIIF	2026	2034	**L1:** C/A + P(Y) + M + C **L2:** P(Y) + CM + CL + M **L5:** C-I + C-Q

and functionality varies widely. As of the year 2020, all satellites of Block I, II, and IIA have retired; other generations are in operation.

NOTE

Precision (or protected) code is known as P-code. The Y-code is used in place of the P-code whenever the anti-spoofing mode of operation is activated. We shall discuss anti-spoofing in Chapter 5.

The main functions of a GPS satellite are as follows:

- It receives and stores data from the control segment.
- It maintains a very precise time.
- It transmits coded signals to user receivers through the use of three frequencies: L1 (1575.42 MHz), L2 (1227.60 MHz), and L5 (1176.45 MHz) (Table 2.1).
- It controls both its attitude (orientation) and position in the orbit using small rocket boosters.
- It enables a wireless link among the satellites in case of GPS III (Block III) satellites.

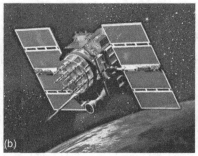

FIGURE 2.3 (a) GPS Block IIR-M satellite during assembly; (b) artist's concept of GPS Block II/IIA satellite; (c) artist's concept of GPS Block IIR-M satellite.

Initially, GPS satellites were designed to transmit coarse acquisition (C/A) code on L1 band and precision or protected (P) code on both L1 and L2 bands. The C/A code was available to the civilians, whereas P code was used by the US military. Therefore, the GPS could provide two so-called services: The Standard Positioning Service (SPS) using C/A code, and the Precise Positioning Service (PPS) for high precision positioning using P code.

The Block IIR-M satellites (Figure 2.3) added three new signals: two C code (C for 'civil') signals in L2 band and the M code in both L1 and L2. Two C codes are the civil-moderate (CM) code and the civil-long (CL) code. The M code is the new military code. The GPS system with Block IIR-M satellites is often called *Modernized GPS*. The Block IIF satellites added another frequency, in L5 band, designed mainly as a backup necessary for civil aviation purposes. The L5 signal served another purpose—interoperability with other GNSS constellations (will be detailed later).

The next generation of GPS satellite is Block III. The GPS system with Block III satellites is commonly called *GPS III*. The Block IIIA added another new signal— L1C (civil) that is compatible with the Galileo system. The Block III satellites are equipped with cross-linked command and control architecture, allowing all Block III vehicles to be updated from a single ground station instead of waiting for each satellite to appear in view of a ground antenna. Overall, these capabilities contribute to improved accuracy, integrity, and assured availability for both civil and military users.

2.2.2 GLONASS SPACE SEGMENT

The GLONASS space segment was designed for a nominal constellation of 24 satellites (21 operational + 3 spares) distributed in three orbital planes (Figure 2.2c) with an inclination of 64.8° with reference to the equatorial plane (Samama 2008). The altitude of this near-circular orbit is around 19,100 km, leading to an orbital period of 11 h 15 min 44 s. The orbital planes are 120° apart in longitude. Eight satellites are regularly spaced in each orbital plane with 45° spacing, allowing complete coverage of the earth. Over the five decades of development, the satellites themselves have gone through numerous revisions. The name of each satellite is *Uragan* (in Russian; *hurricane* in English); however, internationally, they are referred to as GLONASS satellites (Figure 2.4). Newer generations of initial GLONASS satellites are GLONASS-M (2003), GLONASS-K1 (2011), and GLONASS-K2 (2019 onwards). Future satellites are GLONASS-V (2023) and GLONASS-KM (2030).

In December 1995, the full constellation of GLONASS satellites was completed and the system was declared operational. Unfortunately, following the financial and political crisis in the Soviet Union (now Russia), the GLONASS constellation could not be maintained, and the number of operational satellites decreased dramatically to only seven in 2002. In addition, the lifetime of the satellites (3 years, compared to 10 years for the GPS satellites) exacerbated this situation. To keep the system operational, numerous launches had to be made, which led to more financial difficulties. However, in October 2011, it was fully restored. This was achieved progressively by maintaining the constellation at a minimal level and successively adding new satellites, improving the lifetime and performances of the GLONASS-M satellites, and developing new smaller GLONASS-K satellites in order to deploy a full 24-satellite constellation of both GLONASS-M and GLONASS-K for domestic and international availability. GLONASS-M satellites have a longer service life of 7 years (GLONASS had 3 years' life) and are equipped with updated antenna feeder systems. The main feature of the GLONASS-M satellites is the transmission of a new signal for civilian use in L2. The life span of GLONASS-K is about 10–12 years; and a third civilian frequency (L3) has been added. Russia has a plan to have a total of 30 (instead of 24) satellites, 10 in each orbital plane with two of them to be used as operating reserves (spares).

GLONASS satellites transmit two types of signals, Open Standard Precision Service (SPS) and obfuscated High Precision Service (HPS). The SPS and HPS signals can be thought of as Standard Positioning Service, using C/A code, and Precise Positioning Service using P(Y) code of GPS, respectively. All satellites transmit the SPS and HPS codes using L1 and L2 bands. However, unlike GPS,

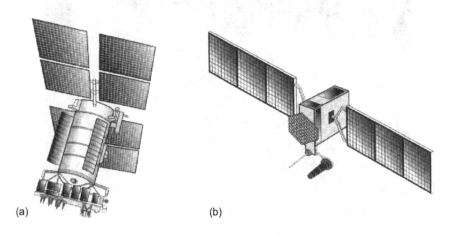

(a)   (b)

FIGURE 2.4 (a) GLONASS-M; and (b) GLONASS-K satellites.

they use different frequencies in L1 band (1598.0625–1607.0625 MHz) and in the
L2 band (1240–1260 MHz) (refer to Chapter 4, Section 4.6). Initially only HPS
signal was broadcast in the L2 band. An additional SPS signal has been added in
GLONASS-M satellites to substantially increase the accuracy of civilian applications (GLONASS ICD 2002). GLONASS-K added a third civilian signal in the L3
band (1202.025 MHz) to improve the accuracy. The approach of using different frequencies within a band, as with the L1 band, is similarly applicable for L2. However,
L3 uses a single frequency (explained later in Chapter 4).

2.2.3 GALILEO SPACE SEGMENT

The Galileo constellation is composed of 30 satellites (27 operational + 3 spares) distributed in three orbital planes (Figure 2.2b) at 23,222 km altitude. The inclination
of the orbital planes with reference to the equatorial plane is 56° (longitudinal angle
120°), allowing a better coverage than GPS for the northern countries of Europe,
for instance (Samama 2008). Each orbital plane houses ten satellites (8 operational
+ 2 spares), equally spaced at 40°. Such a configuration leads to an orbital period
of 14 h 4 min, which gives a 17/10 value in comparison to the total sidereal duration of a day. This feature means that it revolves the earth 17 times every ten sidereal days (Zaharia 2009). In the year 2004 the Galileo System Test-Bed Version 1
(GSTB-V1) project validated the on-ground algorithms for Orbit Determination and
Time Synchronisation (OD & TS). Under this project, two GIOVE satellites were
launched—GIOVE-A in 2005 and GIOVE-B in 2008 (Figure 2.5). These test bed
satellites were followed by four In-Orbit Validation (IOV) satellites (2011–2012),
which were then followed by several *Full Operational Capability* (FOC) satellites
starting in 2014.

FIGURE 2.5   GIOVE-A satellite.

NOTE

GIOVE is the Italian name of the planet Jupiter, whose natural satellites were first observed by Galileo Galilei, leading to the first precise longitude calculation. It also stands for Galileo In-Orbit Validation Element.

The Galileo signals are sent in different frequencies (Table 2.2), two of which are in common with GPS (L1 and L5) for interoperability purposes. The names used for these frequencies are sometimes confusing; different names have been used in various literature for the same frequency. Actually, the original principle of Galileo is based on services, not signals. Thus, for each service, a subtle combination of bands and signal arrangement has been developed.

There are four different services available from Galileo constellation as follows (Issler *et al.* 2003; Samama 2008): (1) *Open Service* (OS), (2) *High Accuracy Service* (HAS) (previously known as *Commercial Service*), (3) *Public Regulated Service* (PRS), and (4) *Safety of Life* (SoL) service. The OS is free to anyone and is based on E1, E5a, and E5b frequencies. Several combinations of these frequencies are also possible, such as a dual frequency service is based on E1 and E5a (or E1 and E5b together) or single frequency services (any one of E1, E5a, and E5b). Even triple frequency services using all the signals together (E1, E5a, and E5b) are also being tested. Receivers can achieve an accuracy of 8 m horizontally and 35 m vertically if they are single frequency receivers. Dual frequency receivers can provide 8 m horizontal and 15 m vertical accuracy; triple frequency is capable of achieving much better accuracy.

The HAS offers an accuracy of nearly 1 cm. The HAS allows for development of applications for professional or commercial use owing to improved performance

TABLE 2.2

Frequency Specification of Galileo

Signal	Central Frequency (MHz)
E1-I	1575.42
E1-Q	1575.42
E2-I	1561.098
E2-Q	1561.098
E5a-I	1176.45
E5a-Q	1176.45
E5b-I	1207.14
E5b-Q	1207.14
E6-I	1278.75
E6-Q	1278.75
L6	1544.71

and data with greater added value than that obtained through the OS. This service is free of charge, with content and format of data publicly and openly available on a global scale. Galileo HAS uses a combination of signals in E6 band plus the OS E1 band.

The PRS provides position and timing information restricted to government-authorised users (police, military, etc.). It is similar to OS and HAS but with some important differences that make it operational at all times and in all circumstances, including periods of crisis. The main aim is robustness against jamming. It uses E1 and E6 bands and provides horizontal accuracy of 6.5 m and vertical accuracy of 12 m, with a dual frequency receiver. The main users are European organisations like the European Police Office (Europol), the European anti-fraud office, and civilian security forces. Member states' structures such as national security services, frontier surveillance forces, or criminal repression forces are also users.

The SoL service provides integrity; this means a user will be warned when the positioning fails to meet certain margins of accuracy. Because Galileo's range is worldwide, its satellites are able to detect and report signals of search-and-rescue (SAR) beacons from *Cospas-Sarsat*, making it a part of the Global Maritime Distress Safety System. The SAR is Galileo's contribution to the Cospas-Sarsat system, an international satellite-aided search and rescue initiative. Galileo satellites are able to pick up signals from emergency beacons and relay them to national rescue centres. These emergency beacon signals can be transmitted from a ship, a plane, or even from individuals, allowing a rescue centre to determine the precise location. The SoL service uses E1 and L6 bands.

NOTE

In the field of search and rescue, *distress radio beacons* (also known as *distress beacons*, *emergency beacons*, or simply *beacons*) are for tracking the transmitters which aid in determining location of boats, aircraft, and/or persons in distress. *Cospas-Sarsat* (www.cospas-sarsat.int) is an international satellite-based search and rescue distress alert detection and information distribution system, established by Canada, France, United States, and the (former) Soviet Union in 1979. The *Global Maritime Distress Safety System* (GMDSS) is an internationally agreed-upon set of safety procedures, equipment, and communication protocols used to increase safety and make it easier to rescue distressed ships, boats, aircrafts, and even individuals.

2.2.4 BEIDOU SPACE SEGMENT

The Chinese BeiDou is designed to have five geostationary satellites, three inclined geosynchronous orbit (IGSO) satellites at 35,786 km altitude and 55°orbital inclination, and 27 medium earth orbit (MEO) satellites at 21,528 km altitude (Cao and Jing 2008; Grelier *et al.* 2007; Gao *et al.* 2007; Wilde *et al.* 2007; Gao *et al.* 2008). The

TABLE 2.3
Frequency Specification of BeiDou

Signal	Central Frequency (MHz)
B1-I	1561.098
B1-Q	1561.098
B1-C	1575.42
B1-A	1575.42
B2-I	1207.14
B2-Q	1207.14
B2a	1176.45
B2b	1207.14
B3-I	1268.52
B3-Q	1268.52
B3-A	1268.52

MEO satellites are arranged in three orbital planes with an orbital inclination of 55°, and the orbital period is about 12 h 53 min 24 s.

BeiDou offers different sets of services—global and regional. Global services include *open service* and *authorised service*. The open service is free of charge and open to all users worldwide with a positional accuracy of 10 m. The authorised service aims to ensure high reliability even in complex situations. Regional services include *wide area differential service* and *short message service*. The Chinese government did not disclose service details.

The principle of BeiDou is similar to Galileo in that it is based on services instead of signals. Therefore, BeiDou signals include a variety of modulations (Table 2.3). Table 2.3 furnishes all of the signals and their frequencies. The reader will notice that some of the signals are identical to GPS and Galileo.

2.3 CONTROL SEGMENT

The *control segment* (also referred to as *ground segment*) does what its name implies—it 'controls' the GNSS satellites by tracking them and then providing them with corrected orbital and clock (time) information. The control segment consists of a group of several ground-based *monitor stations*, a number of *upload stations* and generally one or two *master control station(s)*. Monitor and upload stations are collectively called telemetry, tracking, and control (TT&C) stations. The master control station is also known as the *system control station* or, simply, *control station*.

The main functions of the ground segment are to

- Monitor the satellites.
- Estimate the on-board clock state and define the corresponding parameters to be broadcast (with reference to the constellation's master time).

- Define the orbits of each satellite in order to predict the ephemeris (precise orbital information) together with the almanac (coarse orbital information).
- Determine the attitude (orientation) and location of the satellites in order to determine the parameters to be sent to the satellites for correcting their orbits.
- Upload the derived clock correction parameters, ephemeris, almanac, and orbit correction commands to the satellites.

Monitor stations track the satellites continuously and provide tracking information to the master control station. In the master control station, this tracking information is then incorporated into precise satellite orbit and clock correction coefficients; and the master control station forwards them to the upload stations. The upload stations transmit these data to each satellite at least once every day. The satellites then send the orbital information to the GNSS receivers over radio signals. Figure 2.6 illustrates this concept schematically. Earlier, the data from upload stations were transmitted to each satellite; therefore, it was necessary to establish many upload stations around the world. However, today satellites can communicate among them. Thus, data uploaded to a satellite within the vicinity can be sent to a satellite that is not within the vicinity of the upload station. The same applies for the monitoring stations as well. This advancement eliminated the requirement of establishing worldwide ground stations.

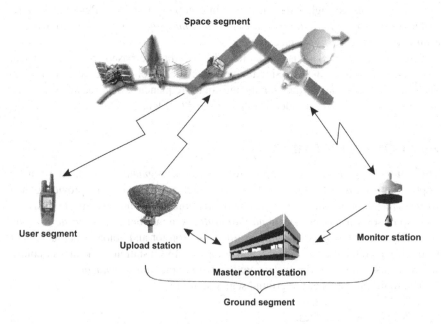

FIGURE 2.6 Operational concept of GNSS.

NOTE

Without the ability to take the huge amounts of satellite data and condense it into a manageable number of components, GNSS processors would be overwhelmed. *Kalman filtering* is used in the uploading process to reduce the data to the satellites. Kalman filtering, a *recursive solution for least-square filtering* (Kalman 1960), has been applied to the results of radio-navigation for several decades. It is a statistical method of smoothing and condensing large amounts of data (Minkler and Minkler 1993). One of its uses in GNSS is reduction of pseudoranges (approximate ranges between a monitoring station and a satellite) measured at very short time intervals (e.g., 1.5 s for GPS). Kalman filtering is used to condense a smoothed set of pseudoranges for a period of few minutes (e.g., 15 min for GPS). This filtered data is then transmitted to the master control station.

2.3.1 GPS Control Segment

For the GPS system, the ground segment is composed of a master control station located at Schriever Air Force Base (formerly Falcon AFB) in Colorado and three uploading stations located in Ascension, Diego Garcia, and Kwajalein (Figure 2.7). Sixteen monitor stations are used to carry out the measurements required for the definition of the data to be uploaded. Six Operational Control Segment (OCS) monitor stations are located in Hawaii, Colorado Springs, Ascension, Diego Garcia, Cape Canaveral, and

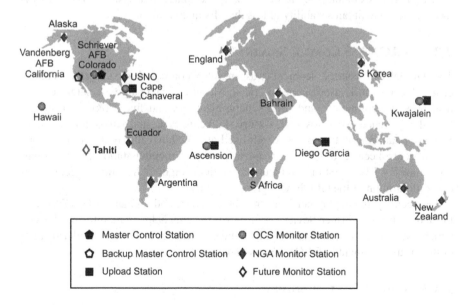

FIGURE 2.7 GPS control segment.

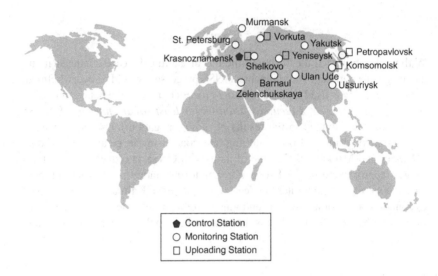

FIGURE 2.8   GLONASS control segment.

Kwajalein. Additionally, ten National Geospatial Agency (NGA) monitoring stations have been operational since September 2005. The NGA is comprised of a worldwide network of GPS monitoring stations that operates under high-performance standards. One backup master control station has also been established at Vandenberg AFB, California. Figure 2.7 shows the locations of the GPS ground segment stations. At Schriever AFB, Colorado, the 'master clock' of the United States Naval Observatory is located, which maintains stability of less than 1 s in 20 million years.

2.3.2   GLONASS CONTROL SEGMENT

The GLONASS control segment was initially composed of a system control centre located at Krasnoznamensk (near Moscow) (in charge of satellite control, orbit determination, and time synchronization); and five TT&C stations (St Petersburg, Schelkovo [Moscow], Ussuriysk, Yenisseysk, and Komsomolsk-Amur. Synchronization monitoring was centralized at Schelkovo. Today it comprises one system control centre located at Krasnoznamensk, 12 monitor stations, 8 laser ranging stations, 4 TT&C stations and 5 upload stations. Figure 2.8 shows the geographical configuration of the GLONASS ground segment.

For security and deployment reasons, the entire ground segment of GLONASS is located within the former Soviet Union territory. This helps in monitoring the system, but it reduces the range of uploading and surveillance stations. A modernization of the ground segment of GLONASS is underway.

2.3.3   GALILEO CONTROL SEGMENT

The ground control segment of the Galileo system, is composed of two control centres (master control stations) located at Oberpfaffenhofen (Germany) and Fucino

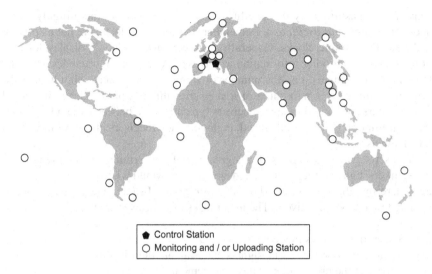

FIGURE 2.9 Galileo control segment.

(Italy). Maintaining two control centres instead of one increases the reliability by imposing redundancy. The ground control segment uses a global network of six TT&C stations to communicate with each satellite through regular, scheduled contacts, long-term test campaigns, and contingency contacts. In addition, 10 upload stations for transmitting the navigation messages allows an increased uploading rate compared to GPS; which improves accuracy by providing more accurate ephemeris data. In addition to the aforementioned stations, the ground segment of Galileo has been designed to include both regional and local components. The idea is that the satellite coverage is limited and that there will be a need, in specific cases, for additional components such as terrestrial transmitters (Samama 2008). Figure 2.9 shows the geographical distribution of Galileo ground segment stations.

2.3.4 BEIDOU CONTROL SEGMENT

BeiDou control segment includes a master control station in China, 30 monitoring and two upload stations. The master control station is responsible for satellite constellation control and processing of measurements received from the monitor stations to generate the navigation message. Monitor stations collect data for all the BeiDou satellites in view from their locations. Upload stations are responsible for uploading the orbital corrections and the navigation message to BeiDou satellites. The Chinese government did not disclose much about the ground segment of BeiDou.

2.4 USER SEGMENT

The user's GNSS receiver is the user segment of the GNSS system. In general, GNSS receivers are composed of an antenna (internal or external) tuned to the frequencies transmitted by the satellites, receiver–processors, and a highly stable clock (often a *crystal oscillator*). Remember that receiver clocks are not as precise as the satellite atomic clocks. Generally, receivers also include a display for providing location and other information to the user. A receiver is often described by its number of channels, i.e., the number of signals it can receive simultaneously. Originally limited to a maximum of four or five, this has progressively increased over the years so that, nowadays, receivers typically have between 12 and 24 channels at minimum. However, advanced, high-accuracy receivers may have more than 200 channels.

The user segment is composed of a great variety of terminals including boaters, pilots, hikers, hunters, the military, and anyone who would wish to know where they are, where they have been, or where they are going. In fact, every smartphone is equipped with GNSS receivers. The major tasks of a receiver are to

- Select the satellites in view.
- Acquire the corresponding signals and evaluate their health.
- Carry out the propagation time measurements.
- Carry out the *Doppler shift* measurements.
- Calculate the location of the terminal and estimate the *user range error.*
- Calculate the speed of the terminal.
- Provide accurate time.

Therefore, users will have at their disposal a single terminal allowing localization, time reference, altitude determination, speed indicator, and so on. GNSS receivers come in a variety of formats, from devices integrated into cars, phones, and watches, to dedicated devices such as those shown in Figure 2.10 (several other figures are furnished in Chapter 8) from various manufacturers. Nowadays, several receivers are also available that have the capability to receive signals from more than one GNSS constellation (e.g., a combined GPS/GLONASS receiver); these are called multiconstellation receivers.

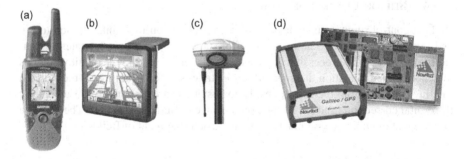

FIGURE 2.10   (a) Handheld GPS receiver from Garmin; (b) Automobile GPS navigation receiver; (c) GPS/GLONASS survey receiver from Trimble; (d) GPS/Galileo receiver from NovAtel.

2.5   SUMMARY AND COMPARISON OF THE FOUR SYSTEMS

In the early stages of development of the GPS program addressed the need for two different types of signals:

- The first, to provide robustness and potentially higher accuracy (for military purposes).
- The second, to meet civilian needs.

The development of two frequencies and different codes was the solution and gave rise to two services, the Standard Positioning Service (SPS) and the Precise Positioning Service (PPS). All GPS users have access to the SPS, which is subject to voluntary degradations depending on geopolitical or strategic issues decided by the United States government. For example, selective availability (SA) was active since GPS was first operational until May 2000, in order to decrease accuracy for civilian receivers. Upon the withdrawal of SA, the horizontal accuracy improved from typically 100 m to around 10–15 m. Some believe that this withdrawal of SA was intended to discourage the development of Galileo and to show that there was no need for a new global constellation (Samama 2008). One of the major arguments for developing Galileo was precisely that the GPS system was solely controlled by the United States military.

GLONASS was designed in a similar fashion. The radio transmission principles remain, with two different types of signals, one for civilian and another for military purposes. The corresponding GLONASS services are the Standard Precision Service (SPS) and the High Precision Service (HPS). Galileo was built around the notion of services. Four such services are available that are intended to cover the needs of mass market, professional, scientific, and governmental users. BeiDou is also based on services—an open service which is free of cost, with authorised service for restricted users.

Table 2.4 summarizes the main characteristics of these four systems. Some of the parameters outlined in this table may not be understood at this point of the discussion; they will be addressed in later chapters (Table 2.4).

NOTE

Additional information and latest updates about a specific GNSS can be obtained from the following websites:

www.gps.gov (for GPS)
www.navcen.uscg.gov (for GPS)
www.glonass-ianc.rsa.ru (for GPS)
www.glonass-iac.ru/en (for GPS, GLONASS, Galileo, BeiDou, and others)
www.galileognss.eu (for Galileo)
www.esa.int/esaNA/galileo.html (for Galileo)
www.en.beidou.gov.cn (for BeiDou)
www.insidegnss.com (for GPS, GLONASS, Galileo, BeiDou, and others)

TABLE 2.4

Comparison of GPS, GLONASS, Galileo, and BeiDou

Parameters	GPS	GLONASS	Galileo	BeiDou
Orbital plane	6	3	3	3
Number of satellites (as originally planned)	21 operational + 3 spares	21 operational + 3 spares	27 operational + 3 spares	30 MEO + 5 geostationary
Inclination	55°	64.8°	56°	55°
Altitude above sea level (km)	20,200	19,100	23,222	21,528
Orbiting speed of satellite (m/s)	3870	3950	3675	3779
Orbital period	11 h 57 min 58 s	11 h 15 min 44 s	14 h 4 min	12 h 37 min 34.45 s
Services	2	2	5	10
Master control station	1 active + 1 backup in US	1 in Russia	2 in Europe	1 in China
Ephemeris	Keplerian elements of the orbit and first derivative	Geocentric Cartesian coordinates and their derivatives	Keplerian elements of the orbit and first derivative	Keplerian elements of the orbit and first derivative
Update rate of ephemeris	2 h	30 min	3 h	1 h
Update rate of almanac	< 6 days	< 6 days	< 6 days	< 7 days
Transmission time of almanac	12.5 min	2.5 min	10 min	12 min
Geodesic reference frame	WGS 84	PZ 90	GTRF	CGCS2000
Time reference	UTC (USNO)	UTC (SU)	TAI (BIPM)	UTC (NTSC)
Multiple access scheme	CDMA	FDMA, CDMA	CDMA	CDMA
Frequency bands	3	3	5 + 1	6

(Continued)

TABLE 2.4 (CONTINUED)
Comparison of GPS, GLONASS, Galileo, and BeiDou

Parameters	GPS	GLONASS	Galileo	BeiDou
Carrier frequencies (MHz)	L1: 1575.42 L2: 1227.60 L5: 1176.45	L1: 1598.0625–1607.0625 L2: 1242.9375–1251.6875 L3: 1198–1208	E1-I: 1575.42 E1-Q: 1575.42 E2-I: 1561.098 E2-Q: 1561.098 E5a-I: 1176.45 E5a-Q: 1176.45 E5b-I: 1207.14 E5b-Q: 1207.14 E6-I: 1278.75 E6-Q: 1278.75 L6: 1544.71	B1-I: 1561.098 B1-Q: 1561.098 B1-C: 1575.42 B1-A: 1575.42 B2-I: 1207.14 B2-Q: 1207.14 B2a: 1176.45 B2b: 1207.14 B3-I: 1268.52 B3-Q: 1268.52 B3-A: 1268.52
Code	Different for each satellite	Same for all satellites, different for each satellite	Different for each satellite	Different for each satellite
Code or chip frequency (MHz or megabits per sec—Mbps or mega chips per second—Mcps)	L1 C/A: 1.023 L1&L2 P: 10.23 L1&L M: 5.115 L1 C: 1.023 L2 CM: 0.5115 L2 CL: 0.5115 L5 C-I: 10.23 L5 C-Q: 10.23	L1&L2 C/A: 0.511 L1&L2 P: 5.115 L3 C/A: 4.096 L3 P: 4.096	E1: 2.5575 E2: 1.023 E5: 10.23 E6: 5.115	B1: 2.046 B2 : 10.23 B3: 10.23 B1-BOC: 1.023 B2-BOC: 5.115 B3-BOC: 2.5575
Selective availability	Off	No	No	No
Anti-spoofing	Yes	No	No	No

EXERCISES

DESCRIPTIVE QUESTIONS

1. What do you understand by 'functional segments' of GNSS? Describe the roles of the control segment.
2. Describe the space segments of Galileo and GPS.
3. Compare the space segments of GPS and GLONASS.
4. Describe the space segment of Galileo and BeiDou.
5. Describe the services provided by Galileo and BeiDou.
6. Describe the BeiDou Navigation System.
7. In general, how does a ground segment work? Describe the ground segments of GPS and GLONASS.
8. What do you understand by 'user segment'? What are the roles of user segment? What is multiconstellation receiver?

SHORT NOTES/DEFINITIONS

Write short notes on the following topics:

1. Orbital plane
2. Orbital period
3. Block IIR GPS satellite
4. Standard Positioning Service
5. Precise Positioning Service
6. Modernized GPS
7. GPS III
8. C/A code and P code
9. Uragan
10. GLONASS-K
11. BeiDou
12. Standard Precision Service
13. High Precision Service
14. GIOVE
15. Open Service of Galileo
16. High Accuracy Service of Galileo
17. Public Regulated Service

3 Working Principle of GNSS

3.1 INTRODUCTION

As discussed in Chapter 1, a Global Navigational Satellite System (GNSS) uses satellites as reference points for positioning purposes. Positioning an object, which may be fixed or travelling on or near the earth's surface, is done through a geometric technique that uses satellites (as 'points of reference') and incredibly accurate time. While this chapter provides the basic idea of how a GNSS works, it is important to realise that the working principle of GNSS is not as simple as it is described in this chapter. The subsequent chapters will discuss several of the pitfalls and caveats involved in the calculation of one's precise position. However, before we concentrate on core technical matters, it is necessary to understand the basic working principle of GNSS.

3.2 TRIANGULATION AND TRILATERATION

Before proceeding with discussion of the GNSS, it is essential to know and understand *triangulation* and *trilateration*—geometrical concepts for determining one's position. The word 'triangulation' has several definitions. In trigonometry and geometry, for instance, triangulation is the process of finding coordinates and distance to a point by calculating the length of one side of a triangle with given measurements of angles and sides of the triangle formed by that point and two other known reference points, using the *law of sines*. In Figure 3.1, let us assume, we know the locations of A and B, but C is unknown. The distance between A and B can be determined as we know the locations of those two points. Also assume that the angles α and β are known to us, but θ is not known. Angle θ can be calculated as: $\theta = 180° - \alpha - \beta$ (the sum of three angles in any triangle = 180°).

By the law of sines

$$\frac{\sin \alpha}{BC} = \frac{\sin \beta}{AC} = \frac{\sin \theta}{AB} \tag{3.1}$$

Now we can calculate AC and BC

$$AC = \frac{AB \sin \beta}{\sin \theta} (AB, \beta, \text{ and } \theta \text{ all are known}) \tag{3.2}$$

$$BC = \frac{AB \sin \alpha}{\sin \theta} (AB, \alpha, \text{ and } \theta \text{ all are known}) \tag{3.3}$$

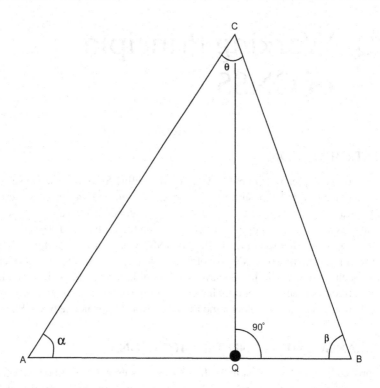

FIGURE 3.1   Determining a position based on triangulation.

The perpendicular distance from C to AB can also be calculated as either

$$QC = AC\sin\alpha \tag{3.4}$$

or,

$$QC = BC\sin\beta \tag{3.5}$$

QB and QC can also be calculated using trigonometric functions, and thus by using the coordinates of B point, the coordinates of C can be determined since the orthogonal distances (QB and QC) from B to C are known to us.

Triangulation is used for many purposes, including surveying, navigation, metrology, astrometry, binocular vision, model rocketry, establishing the direction of weapons, etc. Many of these applications involve the solution of large meshes of triangles, with hundreds or even thousands of observations.

Trilateration is a method of determining the relative positions of objects using the known locations of two or more reference (control) points and the measured distances between the subject and each reference point. To accurately and uniquely

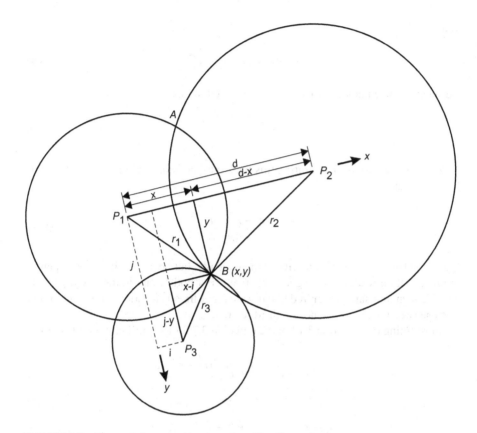

FIGURE 3.2   Determining a position based on 2D trilateration.

determine the relative location of a point on a 2D plane, using trilateration alone, generally at least 3 reference points are needed; for 3D trilateration we need 4 reference points (refer to Chapter 1, Section 1.4).

A mathematical derivation for the solution of a 2D trilateration problem can be explained with a simple example. In Figure 3.2, standing at B (having coordinates x and y), we want to know our location relative to the reference points P_1, P_2, and P_3 on a 2D plane. Measuring r_1 narrows our position down to a circle. Next, measuring r_2 narrows it down to two points, A and B. A third measurement, r_3, gives our fix at B. A fourth measurement could also be made to reduce and estimate error.

If we consider P_1 point as the origin of the coordinate system, x-axis is along P_1P_2, and y-axis is the normal to P_1 along P_3, then, starting with three equations,

$$r_1^2 = x^2 + y^2 \tag{3.6}$$

$$r_2^2 = (d-x)^2 + y^2 \tag{3.7}$$

and,

$$r_3^2 = (x-i)^2 + (j-y)^2 \tag{3.8}$$

we subtract the equation 3.7 from 3.6, and solve for x as

$$x = \frac{r_1^2 - r_2^2 + d^2}{2d} \tag{3.9}$$

Now, x is known to us. Substituting this back into the equation 3.6 produces the following equation:

$$y^2 = \frac{(r_1^2 - r_2^2 + d^2)^2}{4d^2} \tag{3.10}$$

We can solve y from equation 3.10 mathematically, but because y has been squared in this equation, we can never get a negative y. For both A and B, the y will provide positive sign and same magnitude, and we cannot fix our location on a single point of these two. Therefore, we need a third reference.

Substituting the equation 3.10 in the equation 3.7 for the third circle finds the y:

$$
\begin{aligned}
y &= \frac{r_1^2 - r_3^2 + (x-i)^2 + j^2 - \dfrac{(r_1^2 - r_2^2 + d^2)^2}{4d^2}}{2j} \\[2mm]
&= \frac{r_1^2 - r_3^2 + (x-i)^2 + j^2 - x^2}{2j}
\end{aligned} \tag{3.11}
$$

A 3D trilateration requires another reference point to solve the z coordinate. The preceding is a simple example of 2D trilateration where one of the reference points (P_1) is the origin of the coordinate system and a joining line of two reference points is an axis. In real applications, we need to determine one's position with reference to a well-defined and fixed coordinate system and origin (e.g., a geocentric coordinate system). It is, however, possible to translate any set of three points to comply with these constraints, find the solution point, and then reverse the translation to find the solution point in the original coordinate system.

Trilateration in three-dimensional space is quite complex. For ease of understanding, the term that is generally used in the industry to describe how the GNSS positioning system works, is 'triangulation'. From the preceding discussion it is clear to us that triangulation and trilateration are different. Triangulation uses angular measurements to solve one's position, whereas trilateration uses distance measurements. GNSS and ground-based trilateration both rely exclusively on the measurement of distances to fix positions. One of the differences between the ground-based trilateration and GNSS, however, is that the distances, called ranges, are measured to reference points on the surface of the earth. In GNSS, however, they are measured to satellites that are orbiting thousands of kilometres above the earth. Furthermore, performing measurements from moving references is much more challenging than from the fixed references.

3.3 ALMANAC AND EPHEMERIS

When the GNSS receiver starts its job, it should have information about where the satellites are (location) and how far away they are (distance). Let us first discuss how the GNSS receiver knows where the satellites are located in space. The GNSS receiver picks up two kinds of coded information from the satellites, *almanac* and *ephemeris*. Almanac data are course orbital model for all satellites. Every satellite of a specific constellation broadcasts almanac data for all satellites of that constellation. The almanac data are not very precise and are considered valid for up to six or seven days. The data are continuously transmitted by the satellites and stored in the memory of the GNSS receiver. Thus, the receiver knows the coarse (rough) orbits of the satellites and where each satellite is supposed to be. The almanac data are periodically updated with new information as the satellites move around.

Ephemeris (pronounced i-'fe-me-res) data, in comparison, provide very precise orbital information of each satellite and are required by the receiver to determine its precise location. Because satellites can drift slightly out of orbit due to perturbations, ground monitor stations are used to keep track of satellite orbit, attitude, location, speed, and clock drift. The ground monitor stations send this information to the master control station, where it is used to predict the precise orbit of the satellite for next few hours and to calculate the clock correction coefficients. These predicted orbital data and calculated clock corrections are then sent to the upload stations for uploading to the satellites. The satellites then transmit this information to the user's receiver. This precisely predicted orbital data are called ephemeris data. It is important to remember that although it has been emphasised that ephemeris data are 'very precise', they are never perfect—just the output of an attempt for precise 'prediction'. Ephemeris data are updated to the satellites every 2 h for GPS, 30 min for GLONASS, 3 h for Galileo and 1 h for BeiDou (Table 2.4, Chapter 2). The ephemeris data are transmitted from the satellites as coded information to the user's receiver. Each satellite broadcasts only its own ephemeris data. Subsets (pieces) of ephemeris data are broadcast by each satellite continuously to the user's receiver, which remains valid for just a few minutes.

NOTE

The orbit of a satellite is not fixed; it changes due to gravitational forces and other factors. The main perturbations are due to the asphericity (asymmetry) of the earth, lunisolar (moon and sun) gravitational attraction, and the influence of the solar radiation field. Deviations caused by these forces are corrected by periodic orbit-adjust operations. Similarly, due to internal and external torques acting on the satellite, its orientation slowly drifts. The orientation of a satellite in space is called its *attitude*. Both orbit and attitude parameters are controlled by the attitude and orbit control system to comply with specified tolerance limits. Each satellite is equipped with small internal rocket boosters that generate the thrust needed to maintain these critical parameters.

After the GNSS receiver is started, it searches for the satellites and establishes links with them. Once the initial link is confirmed with one satellite, the receiver unit downloads the almanac, i.e., data about all the other satellites' (approximate) locations. With the almanac data programmed into its computers, the GNSS receiver is able to locate the remaining satellites much faster and begin storing the ephemeris data from them. Ephemeris data is required by the receiver to know the precise location of each visible satellite and for precise determination of receiver's location.

When the GNSS unit is not turned on for a length of time, the almanac and ephemeris can get outdated or go 'cold'. In this 'cold' state, it could take longer for the receiver to *lock on* to the satellites. A receiver is considered 'warm' when the data that are stored in it are valid at that instant (not outdated). If the GNSS receiver is moved for more than a few hundred kilometres in OFF state or accurate time is lost, the almanac data will also be invalid. In such cases, the receiver will have to 'sky search', or be reinitialised, so that it can store the new and valid almanac and ephemeris data. GPS transmission time of almanac data is 12.5 min. For GLONASS, Galileo, and BeiDou it is 2.5 min, 10 min, and 12 min, respectively—the amount of time it takes for each of those receivers to go from the 'cold' state to the 'warm' state.

Almanac and ephemeris data must be acquired before starting the GNSS receiver to work. Once the GNSS receiver has locked onto enough satellites to calculate a position, we are ready to begin navigating/surveying. Most receiver units display a position page or a page showing user's position on a map (map screen) that assists the user in navigation.

3.4 TIME AND RANGE

Even though the GNSS receiver knows the precise locations of the satellites in space from the ephemeris data, it still needs to know how far away the satellites are (the distance), so that it can determine its position on the earth. There is a simple formula that tells the receiver how far it is from each satellite. The distance is equal to the velocity of the transmitted signal multiplied by the time taken by the signal to reach from a given satellite to the receiver as shown in the following equation:

$$\text{velocity} \times \text{travel time} = \text{distance} \tag{3.12}$$

We can recollect how in our childhood we tried to find out how far a thunderstorm was from us. When we saw a lightning flash, we counted the number of seconds until we heard the thunder. The longer the count, the farther away the storm was. GNSS works on the same principle, called *time of arrival* or *propagation time*. We would have noticed that during a thunderstorm, we heard the sound sometime after we saw the lightning. The reason is that sound waves travel much more slowly than light waves. We can estimate our distance from the storm by measuring the delay between the time that we see the lightning and the time that we hear the thunder. Multiplying this time delay by the speed of sound gives us our distance to the storm (assuming that the light reaches us almost instantaneously compared to the sound). Sound travels about 344 m (1130 ft) per sec in the air. So, if it takes 5 seconds between the time that we see the lightning and we hear the thunder, our distance to the storm is $5 \times 344 = 1720$ m.

Using the same basic formula, we can determine the distance from a satellite to a receiver. The receiver already knows the speed of the signal: it is the speed of a radio wave (i.e., the speed of light—299,792,458 m/sec or 186,282.03 miles/sec) less any delay as the signal travels through the earth's atmosphere (the aforementioned speed of light is true for a vacuum). However, we know that when the signal propagates through the atmosphere it gets delayed. Now the GNSS receiver needs to determine the time part of the formula. The answer lies in the coded signals the satellites transmit. The transmitted code is called *pseudorandom noise* (PRN) because it looks like a noise signal (Spilker 1980). The PRN is a fundamental part of GNSS. PRN codes are specific codes that look like random combinations of 0s and 1s; but, of course, they are not at all random. Physically, it is a very complicated and lengthy digital code—so complicated, that it almost looks like random electrical noise, hence the name *pseudorandom*. There are several good reasons for its complexity (Langley 1990): first, the complex pattern helps ensure that the receiver does not accidentally sync up to some other signal. The reason that is crucial to make the GNSS receivers economical is that the codes make it possible to use 'information theory' to 'amplify' the signal. Because of this reason, GNSS receivers do not require big dish antennae to receive the signal from a satellite.

When a satellite generates the PRN, the GNSS receiver simultaneously generates the same code and tries to match it up to the satellite's code. The receiver then compares the two codes to determine how much it needs to delay (or shift) its code to match the satellite's code. This delay or *time shift* is then multiplied by the speed of the signal to determine the distance.

How this matching of codes is achieved will be discussed in detail in Chapter 4. For now, determining the delay of time is understood by a simple example (Javad and Nedda 1998). Let us assume that our friend at the end of a large field repeatedly shouts numbers from 1 to 10 at the rate of one count per second (10 seconds for a full cycle of 1 to 10 counts). Let us also assume that we do the exact same thing, synchronised with him, at the other end of the field. Synchronisation would be achieved by both of us starting at an exact second and observing our watches to count 1 number per second (assuming that both of us have very accurate watches). Since the sound takes some time to travel from one place to another, we hear the number patterns of our friend with a delay relative to our patterns. If we hear the friend's count with a delay of two counts relative to ours, then our friend must be 688 m away from us (344 m/sec $\times$ 2 sec). This is because the counts are 2 seconds apart.

However, in GNSS, precise measurement is required; we need to measure fractions of a second. To achieve this, the satellites are equipped with very accurate atomic clocks that can measure the time in nanoseconds. This high accuracy in time measurement will ultimately help to calculate the receiver's precise position (Langley 1991a). Since the receiver does not have an atomic clock (due to size and cost constraints), its clock does not keep time as precisely as the satellite's. So, each distance measurement needs to be corrected to account for the GNSS receiver's internal clock error. For this reason, the range measurement is referred to as a *pseudorange*. To determine position, using pseudorange data, a minimum of four satellites must be tracked. Section 3.6 describes this concept in greater detail.

3.5   NUMBER OF SATELLITES

Based on the earlier discussions of how many reference points are required in different trilateration calculations (three points for 2D and four points for 3D), we could assume that GNSS 3D trilateration would require 4 reference points. But this is not the case. It requires only three reference points to determine a 3D position; however, a fourth reference point is essential to measure accurate time in the receiver (refer to Section 3.6). Let us try to understand how only three references can determine a 3D position in GNSS.

We discussed in Chapter 1 that by ranging from three satellites we can narrow down our position to just two points in the universe (refer to Section 1.4 and Figure 1.2). However, the resulting two points of the intersection of the three spheres' surfaces present specificity: they are located one above and the other below the plane containing the three satellites (a plane connecting the centres of three spheres). Terrestrial positioning is unique in the sense that 'visible' satellites are all located above the horizon. This means that the mentioned plane cannot intersect the earth, so the intersection lying above this plane is an impossible solution and can be rejected. Hence, the earth itself acts as a fourth sphere. Only one of the two possible points will actually be on or near the surface of the planet and we can reject the one in space (Samama 2008). Therefore, three satellites can give us 3D positioning information. However, the GNSS receivers generally look to four or more satellites to achieve or measure precise time (atomic accuracy time) in the GNSS receiver.

3.6   TIME SYNCHRONISATION

Although reference points are fundamental to positioning in any navigation system, in GNSS, time synchronisation between satellite and receiver is critical to achieving position accuracy. The measurement of the propagation time of a signal from a satellite to the receiver requires two variables: the time at which the signal was transmitted (the time at which it leaves the satellite) and the bias between the satellite clock and the receiver clock. Managing the send-time is the role of the ground control segment of the constellation (refer to Chapter 4).

However, the so-called *clock bias* of the receiver is very difficult to overcome, as there is the need for very precise synchronisation. One has to remember that 1 nanosecond is equivalent to 30 cm (or 1 m to 3.3 ns) since the speed of light is about 3×10^8 m/sec. A single nanosecond error in time synchronisation means 30 cm error in distance measurement. The time synchronisation needs to achieve such a level of accuracy. On the satellite side, the timing is almost perfect because they have incredibly precise atomic clocks onboard; receivers are generally equipped with *quartz crystal clocks* (also called *crystal oscillators*), which are far less accurate. But both the satellite and the receiver need to be able to precisely synchronise their PRN codes to make the system work. The designers of GNSS came up with a brilliant solution that enables receivers to achieve the accuracy of a satellite's atomic clock, ensuring that every GNSS receiver is essentially an 'atomic-accuracy' clock or a 'virtual atomic clock'.

In a GNSS system, the receiver clock bias is identical for all measurements, for all the satellite signals a receiver receives, as long as these measurements are made at the same instant. In such a case, which is typical for modern receivers, which normally have at least 12–24 parallel channels for simultaneous measurements, the clock bias, in fact, includes a common bias to all the measurements. Thus, there is a new 'unknown' of the positioning problem—*clock bias* (*t*), in addition to the three coordinates of the receiver's position

(*x*, *y*, and *z*). Thus, the solution vector of the GNSS positioning is made up of four variables: *x*, *y*, *z*, and *t*. Before going ahead with how to solve this problem, we need to remember that clock bias is not the only issue between satellites and receivers. So, while the clock bias is identical for all satellite measurements, other errors are specific to each signal. Chapter 5 deals with these errors in detail.

If our receiver's clock was perfect, then ideally all four spheres would intersect at a single point (receiver's position). But with imperfect clocks, a fourth measurement, provided as a cross-check, will not intersect with the intersecting point of first three. To eliminate this difference, the receiver's computer determines the discrepancy in measurements. Then the receiver looks for a correction factor that it can deduct from all its timing measurements that would cause them all to intersect at a single point. Because of the clock bias error added to the measurement, the distance obtained from the time measurement is called a 'pseudorange' (instead of range).

Returning to positioning, there is the need to find a way to extract the clock bias variable.

Instead of three measurements relating to the purely geometrical aspects for terrestrial positioning, there is the need for one more measurement. Since the receiver must solve its position (x_r, y_r, z_r) and the clock bias (*t*), four satellites are required to solve the receiver's position in terrestrial navigation. The correction of clock bias leads the receiver's clock back into synchronisation, achieving atomic accuracy time in our receiver. Once this correction is made, it applies to all the other measurements, thereby producing correct distance measurements. The observation equation for a *receiver-clock-biased range* (i.e., pseudorange) is (Langley 1991a):

$$\rho = R + ct \qquad (3.13)$$

where c is the velocity of signal in a vacuum (or simply the 'velocity of light'), t is the receiver clock error, ρ is the measured pseudorange, and R is the true 'geometric' range. True geometric range, in three dimensions can be calculated by a simple 3D trilateration equation:

$$R^2 = x^2 + y^2 + z^2 \qquad (3.14)$$

or,

$$R^2 = \left(x_r - x_s\right)^2 + \left(y_r - y_s\right)^2 + \left(z_r - z_s\right)^2 \qquad (3.15)$$

or,

$$R = \sqrt{(x_r - x_s)^2 + (y_r - y_s)^2 + (z_r - z_s)^2} \qquad (3.16)$$

where (x_s, y_s, z_s) stands for the coordinates of the satellite and (x_r, y_r, z_r) stands for the coordinates of the receiver.

Therefore, from equation 3.13, each observation made by the receiver can be parameterised as follows:

$$\rho = \sqrt{(x_r - x_s)^2 + (y_r - y_s)^2 + (z_r - z_s)^2} + c.t \qquad (3.17)$$

Therefore, for four satellites we shall have following four equations:

$$\rho_1 = \sqrt{(x_r - x_1)^2 + (y_r - y_1)^2 + (z_r - z_1)^2} + c.t \qquad (3.18)$$

$$\rho_2 = \sqrt{(x_r - x_2)^2 + (y_r - y_2)^2 + (z_r - z_2)^2} + c.t \qquad (3.19)$$

$$\rho_3 = \sqrt{(x_r - x_3)^2 + (y_r - y_3)^2 + (z_r - z_3)^2} + c.t \qquad (3.20)$$

$$\rho_4 = \sqrt{(x_r - x_4)^2 + (y_r - y_4)^2 + (z_r - z_4)^2} + c.t \qquad (3.21)$$

Where, (x_1, y_1, z_1), (x_2, y_2, z_2), (x_3, y_3, z_3), and (x_4, y_4, z_4) stand for the locations of four satellites, which are known from the ephemeris data. The ρ_1, ρ_2, ρ_3, and ρ_4 are the distances of satellites from the receiver position (pseudoranges) (Figure 3.3), which can be derived using equation 3.12. Hence, solving the last four equations, we can determine four unknowns—x_r, y_r, z_r, and t. The GNSS receiver is equipped with dedicated software for routinely solving these equations.

The aforementioned explanation is applicable for a single constellation receiver, e.g., GPS, only. In the case of a dual constellation receiver, e.g., GPS/GLONASS, combined mode, the receiver must track five satellites (representing the same four previous unknowns and at least one satellite from the other constellation) to determine the GPS/GLONASS time offset. With the Galileo system, we need to track one

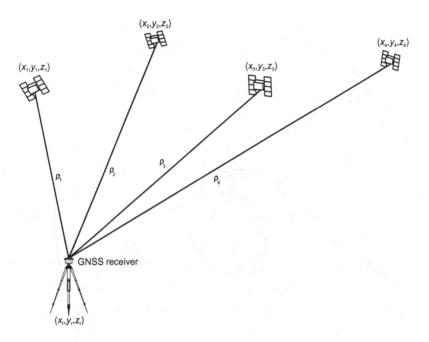

FIGURE 3.3   GNSS positioning with reference to four satellites.

more satellite. With the availability of combined GPS/GLONASS receivers, users have access to a potential 48+ satellite combined system, dramatically improving performance. A larger satellite constellation also improves real-time carrier phase differential positioning performance (refer to Chapter 6).

3.7   SATELLITE ORBIT AND LOCATION

Many of the system's characteristics can be identified from the very basic equations given in Section 3.6. The first is that there is a need for the receiver to know the satellites' locations. Second, the accuracy of the pseudorange is fundamental. If the pseudorange is wrong, then the resulting location will also be wrong. Finally, the physical constants used by the system are also very important. For example, the speed of light, introduced in the above system of equations, must be 299,792,458 m/sec (and not 3×10^8 m/sec). However, the speed of light will be affected by the atmosphere.

The satellite's location is provided by the satellite itself through the so-called ephemeris—just as it was in the early days of navigation (celestial navigation), where there was need of the celestial object's ephemeris. However, today's technology demands greater accuracy, so the ephemeris has to be much more accurate. But there still remains another difficulty—the time at which the signal departs the satellite and the time at which receiver receives it are not the same because the signal takes some time to reach from a satellite to the receiver. A new question arises from this discussion: do we need to take into account the displacement of the satellite? To

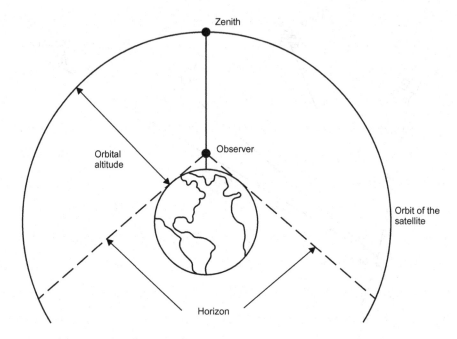

FIGURE 3.4   The distance of the satellite from a fixed point on or above the earth is shortest if the satellite is at the zenith and longest at the horizon; the satellite becomes invisible below the horizon.

give part of the answer to this question, let us just remember some elementary facts. The orbits of the satellites have been chosen in order to provide full coverage of the earth. This is therefore a compromise between the number of satellites, their height (involving the power to be transmitted and then the lifetime of the satellites), and the number of satellites visible from any terrestrial place. The distances separating the satellites from the receiver once again depend on the constellation and the location of the satellite. When the satellite is at the zenith to the location to be determined, the distance from the satellite to the receiver is shortest and is equal to the altitude of the orbit. The distance is longest if the satellite is at horizon (Figure 3.4).

NOTE

The *zenith* is the point in the sky which appears directly above any point on the earth. More precisely, it is the point in the sky with an altitude of +90°. Geometrically, it is the point on the celestial sphere intersected by a line drawn from the centre of the earth through our location. Actually, the point opposite the zenith is the nadir. If a person lies on his back and looks at the sky directly above him, he is looking at his zenith. If he moves and lies down in another place, his zenith will move with him. This moving point in the sky is useful to us because it serves as a 'point of reference', or a starting point, from which we can measure the location of objects in the sky overhead. Zenith angle is the angular distance between an object in the sky, such as the Sun, and an object directly overhead. Zenith angle is 90° minus the elevation angle.

Horizon is the circle we see when we stand in one spot and look around at the edge where the earth and the sky meet. Usually, we have to stand in a desert or in a boat at sea to see the horizon all the way around. It is a natural circle with the observer in its centre. As the observer moves, his horizon moves with him. The distance to the horizon depends on how tall we are. An ant has a horizon of only a few inches. A six-foot person sees a horizon that is 5 km away. A sailor in a ship, of about a height of 100 ft will see a horizon about 20 km. In an airplane at 130 km height, we will have a horizon of 650 km.

Considering the nearest and the farthest satellites for each constellation, one obtains 19,100 km and 24,680 km for GLONASS, 20,200 km and 25,820 km for GPS, and 23,222 km and 28,920 km for Galileo. These values lead directly to the propagation time of 64 ms (millisecond) and 82 ms for GLONASS signals, 67 ms and 86 ms for GPS, and 77 ms and 96 ms for Galileo. It is thus now possible to extract the distances travelled by the satellite during the time of transmission. This gives 252 m (respectively, 325 m) for GLONASS, 260 m (respectively, 333 m) for GPS, and 285 m (respectively, 355 m) for the smaller distance (respectively, larger distance). It is not possible to neglect this time of flight, for the consideration of satellite location, without having a direct effect on positioning accuracy. Indeed, to achieve a positioning accuracy of few meters, the accuracy of the satellite's location must be much smaller than the aforementioned values. Thus, the location of 'point of reference' needed is actually the location where the satellite was at the instant it transmitted the signal; that is, a few tens of milliseconds before the instant at which the receiver received the signal.

To solve this problem, the orbital modelling implemented in GNSS provides the receiver with parameters that enable the calculation of the location of the satellites for every moment. The signal leaves the satellite with a time tag and thus the receiver knows the location of the satellite for that instant (from the ephemeris data).

3.8   SIGNAL-RELATED PARAMETERS

Once the choices of the orbits and of the ephemeris data have been made available to the receiver, there is still the need for signals. What were the reasons for the choices of the frequencies and codes used?

The unmodulated frequencies of GNSS are called *carrier frequencies*, as their role is mainly to carry the information that is going to be the modulating data. High frequencies (above 1 GHz) were chosen for wireless applications because of their better propagation capabilities. The reservation of such frequencies for this particular application was necessary to avoid accidental matching. This was achieved under the coordination of the International Telecommunications Union (ITU) (www.itu.int).

Once the frequency band has been chosen, the structure of signal must be defined. In a satellite-based positioning system, three components are required: the identification of the satellite, the transmission of data required for the computation of the location of the satellites (typically ephemeris data), and a means to achieve the physical measurement of the time delay of the transmission from the satellite to the receiver. Different choices are available in order to fulfil these requirements, but physical limitations have to be respected, such as the allocated bandwidth. For example, the bandwidths allocated to GPS were 24 MHz on L1 and 22 MHz on L2. The future L5 will have 28 MHz. Thus, the choice was made to use a PRN code for both identification of GPS satellites (for GLONASS, satellite identification is achieved via different frequencies; refer to Chapter 4) and time measurement.

The number of frequencies that are going to be used for a given system is another consideration. For both GPS and GLONASS, two frequencies were used from the very beginning. The main reason for this was the error source, *propagation delay*, in the ionosphere. This high atmosphere layer is composed of ionised particles that have the direct effect of slowing down the information transmitted. Instead of propagating at the speed of light, it travels at a lower speed. This effect has to be taken into account and mitigated when considering the translation from time measurement into pseudorange. This is carried out by modelling the ionosphere's thickness and the proportions of the ionised particles. Unfortunately, this modelling is very complex: the thickness and the concentration of ionised particles depend on the effect of the sun, for instance. The success of the modelling is dependent on season, the sun's activity, temperature, the actual path of the wave (thus it depends on the relative location of the satellite and the receiver), and so on. The modelling takes all these aspects into account, but the remaining error, the ionosphere, still looms. Fortunately, the physics that lies behind the behaviour of the ionosphere is non-linear (the ionosphere is a dispersive medium) and having access to two measurements made at two different frequencies can help mitigate this effect. As the reader can establish for themselves, earlier the P code was available on L1 and L2 in GPS, whereas the C/A code was only available on L1. The P code-based GPS receivers had the ability to overcome the ionospheric error, but C/A receivers could not. This was a deliberate choice of the US GPS program in order to limit availability of this dual frequency to authorised users only. Introducing two signals forces the receiver to deal with two frequencies instead of one. But simple electronics cannot deal with both L1 and L2 using the same *front-end* (the part of the receiver that deals with the incoming signals). When GPS was being designed, a more complex electronic part in the receiver would have been required to accommodate a civil signal on L2, increasing the price of civilian equipment. Note that this feature is now being considered by the US government for GPS (from 2005).

NOTE

A *dispersive medium* is one in which different frequencies exhibit various behaviours. This is the case when the mathematical expression of the phenomenon observed is not a linear function of the parameter being considered. In the current case of propagation delay while passing through the ionosphere, the equation exhibits the presence of the frequency to the square.

To conclude, this chapter provided a brief discussion of how GNSS position is calculated, with emphasis on receiver position. But there are also several intrinsic matters involved which need to be addressed to understand the technological background thoroughly. In Chapters 4, 5, and 6, we shall have a closer look at these issues. However, this chapter also serves as a foundation to the next three chapters.

EXERCISES

DESCRIPTIVE QUESTIONS

1. How is triangulation used to determine a location? What is the difference between triangulation and trilateration?
2. Explain basic 2D trilateration mathematically.
3. What are the almanac and ephemeris? What are the differences between them?
4. Explain the purposes of almanac and ephemeris. How can we determine the distance with the given speed of light and propagation time?
5. Explain how pseudorange is calculated.
6. How do we determine three coordinates of the receiver's position and time bias?
7. What do you understand by time synchronisation? How can it be achieved?
8. How many satellites are necessary to determine one's position from a single and a multi constellation? Explain in brief.

SHORT NOTES/DEFINITIONS

Write short notes on the following topics

1. Almanac data
2. Ephemeris data
3. Cold and warm state of a GNSS receiver
4. Pseudorandom noise code
5. Clock bias
6. Zenith
7. Horizon
8. Carrier frequency

4 GNSS Signals and Range Determination

4.1 INTRODUCTION

We have multiple GNSS constellations at present; therefore, they can complement each other rather than compete. The dramatic increase both in the number of satellites and available signals certainly triggered the enhancement of performances and number of applications. In order to accommodate all the existing and proposed services, the number of signals from satellites is growing rapidly, promising new capabilities and potentially new applications. This chapter is devoted to the description of signals and determination of range using these signals.

GNSS signals carry two types of coded information—*ranging codes*, that are used to measure the distance (range) to the satellite, and *navigation codes* (also called *navigation messages* or *data messages*) that include the ephemeris data and information about the time and status of the satellite constellation. These codes are transferred on carrier signals. The codes or their carriers both can be used for the determination of ranges.

4.2 CONCEPTS OF RADIO WAVES

GNSS signal refers to radio waves in the microwave region on the electromagnetic spectrum. A wave is a *disturbance* that propagates through space and time, usually with transference of energy. The most familiar form of a wave for most of us is a water wave in the sea. Although we cannot see it, sound is another type of wave which travels from one place to another by means of air. An electromagnetic wave is an example of a wave that can travel through a vacuum, i.e., without the involvement of a medium. To understand the GNSS signal and its properties, let us have an overview of electromagnetic wave, especially the radio wave.

4.2.1 ELECTROMAGNETIC WAVE

In the 1860s, James Clerk Maxwell conceptualised electromagnetic radiation (EMR) as an electromagnetic energy or wave that travels through space at the speed of light that is about 3×10^8 m/s. The electromagnetic wave consists of two fluctuating fields—one electric and the other magnetic (Figure 4.1). These two fluctuating fields are at right angles (90°) to one another, both are perpendicular to the direction of travel, and have the same amplitudes (strengths) which reach their maxima–minima at the same time. Unlike other types of waves, electromagnetic waves can

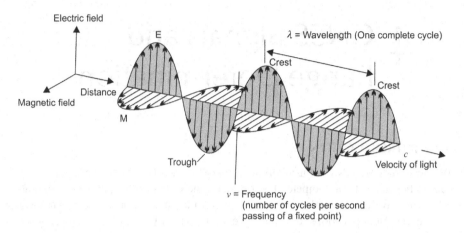

FIGURE 4.1   An electromagnetic wave composed of both electric and magnetic fields.

transmit through a vacuum. The EMR is generated whenever an electrical charge is accelerated.

Two characteristics of electromagnetic wave are particularly important to understand, *wavelength* and *frequency*. One cycle is a complete sequence of values, as from crest to crest (Figure 4.1). A *crest* is the point on a wave with the greatest positive value or upward displacement in a cycle. A *trough* is the inverse of a crest. The *wavelength* is the length of one complete wave cycle, which can be measured as the distance between two successive crests (Figure 4.1) and depends upon the length of time that the charged particle is accelerated. Wavelength is usually represented by the Greek letter *lambda* (λ), which is measured in metre (m) or some fraction of metre such as nanometre (nm, 10^{-9} m), micrometre (μm, 10^{-6} m), or centimetre (cm, 10^{-2} m). *Frequency* refers to the number of cycles of a wave passing a fixed point per unit of time. Frequency is usually represented by the Greek letter *nu* (ν) and normally measured in hertz (Hz) that is equivalent to one cycle per second. A wave that sends one crest by every second (completing one cycle) is said to have a frequency of one cycle/sec, or one hertz (1 Hz). A kilohertz (kHz) is 1000 cycles/sec. A megahertz (MHz) is 1,000,000 cycles/sec and a gigahertz (GHz) is 1,000,000,000 cycles/sec.

The relationship between the wavelength (λ) and frequency (ν) of an EMR is based on the following formula, where c is the velocity of light:

$$c = \lambda \nu \qquad\qquad (4.1)$$

or,

$$\lambda = c / \nu \qquad\qquad (4.2)$$

Note that wavelength and frequency are inversely proportional, that means the longer the wavelength, the lower the frequency; the shorter the wavelength, the higher the frequency.

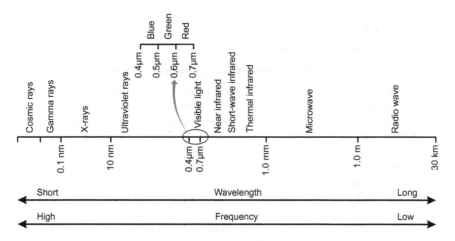

FIGURE 4.2 Electromagnetic spectrum.

4.2.2 ELECTROMAGNETIC SPECTRUM

EMR extends over a wide range of wavelengths or frequencies. A narrow range of EMR (extending from 0.4 to 0.7 μm), the interval detected by the human eye, is known as the visible region (also referred to as light, but physicists often use the term light to include radiation beyond the visible). The distribution of the continuum of all radiant energies can be plotted either as a function of wavelength or of frequency in a chart known as the *electromagnetic spectrum* (Figure 4.2). The electromagnetic spectrum ranges from the shorter wavelengths (including cosmic, gamma, and X-rays) to the longer wavelengths (including microwaves and broadcast radio waves).

Using spectroscopes and other radiation detection instruments over the years, scientists have divided the electromagnetic spectrum into several regions or intervals and applied descriptive names to them. These regions or intervals are also called *bands*. Microwave band, a part of the radio band, covers a wavelength range of 1 mm to 1 m, or frequencies between 0.3 GHz and 300 GHz. This range is again subdivided into several sub-bands, for example, P band (0.3–1.0 GHz), L band (1.0–2.0 GHz), S band (2.0–4.0 GHz), C band (4.0–8.0 GHz), X band (8.0–12.5 GHz), K band (12.5–40 GHz), and so on. Radio band ranges from a wavelength of 1 mm to 30 km (frequency range 10 kHz to 300 GHz). The radio band is divided into a number of sub-bands, as shown in Table 4.1. The GNSS carriers come from a part of the microwave region of the electromagnetic spectrum, namely L band in the ultra-high frequency range.

4.2.3 SOURCE OF RADIO WAVES

Consider electric current as a flow of electrons along a conductor (e.g., a copper wire) between points of differing potential. A direct current flows continuously in the same direction. This would occur if the polarity (the condition of being positive or negative) of the electromotive force causing the electron flow was constant, such

TABLE 4.1
Specification of Radio Bands

Band	Abbreviation	Range of Frequency	Range of Wavelength
Very low frequency	VLF	10 to 30 kHz	30,000 to 10,000 m
Low frequency	LF	30 to 300 kHz	10,000 to 1,000 m
Medium frequency	MF	300 to 3,000 kHz	1000 to 100 m
High frequency	HF	3 to 30 MHz	100 to 10 m
Very high frequency	VHF	30 to 300 MHz	10 to 1 m
Ultra-high frequency	UHF	300 to 3000 MHz	100 to 10 cm
Super high frequency	SHF	3 to 30 GHz	10 to 1 cm
Extremely high frequency	EHF	30 to 300 GHz	10 to 1 mm

as is the case with a battery. If, however, the current is induced by the relative motion between a rotating conductor and a stationary magnetic field, such as is the case in a generator, then the resulting current changes direction in the conductor as the polarity of the electromotive force changes with the rotation of the generator's rotor. This is known as *alternating current*.

The energy of the current flowing through the conductor is either dissipated as heat (an energy loss proportional to both the current flowing through the conductor and the conductor's resistance) or stored in an electromagnetic field oriented symmetrically about the conductor. The orientation of this field is a function of the polarity of the source producing the current. When the current (electron flow) is removed from the wire, this electromagnetic field will, after a finite time, collapse back into the wire.

What would happen if the polarity of the current source supplying the wire was reversed at a rate which greatly exceeds the finite amount of time required for the electromagnetic field to collapse back upon the wire? In the case of rapid pole reversal, another magnetic field, proportional in strength but exactly opposite in magnetic orientation to the initial field, will be formed upon the wire. The initial magnetic field, though the current source is gone, cannot collapse back upon the wire because of the existence of this second oriented electromagnetic field. Instead, it 'detaches' from the wire and propagates out into space. This is the basic principle of a radio antenna, which transmits a wave at a frequency proportional to the rate of pole reversal and at a velocity equal to the speed of light (Bowditch 1995).

4.2.4 STRENGTH OF RADIO WAVES

The strength of a magnetic field is directly proportional to the magnitude of the current flowing through the conductor. Recall the discussion of alternating current in the preceding section. A rotating generator produces current. That is, the magnitude of the current varies as a function of the relative position of the rotating conductor

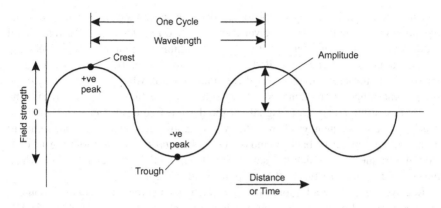

FIGURE 4.3   Relationship of current and field strength.

and the stationary magnetic field used to induce the current. The current starts at zero, increases to a maximum as the rotor completes one quarter of its revolution, and falls to zero when the rotor completes one half of its revolution. The current then approaches a negative maximum; then it once again returns to zero (Bowditch 1995).

The relationship between the current and the magnetic field strength induced in the conductor through which the current is flowing is shown in Figure 4.3. Recall from the discussion above that this field strength is proportional to the magnitude of the current; that is, if the current is represented by a sine wave function, then so too will be the magnetic field strength resulting from that current. This characteristic shape of the field strength curve has led to the use of the term 'wave' when referring to electromagnetic propagation. The maximum displacement of a peak from zero is called the *amplitude*. The forward side of any wave is called the *wave front*. For a *nondirectional antenna*, each wave proceeds outward as an expanding sphere (or hemisphere); and, in a *directional antenna*, a wave proceeds to a specific direction.

4.2.5   RADIO TRANSMITTER AND RECEIVER

A radio transmitter essentially consists of: (1) a power supply to furnish direct current, (2) an oscillator to convert direct current into radio-frequency oscillations (the carrier wave), (3) a device to control the generated signal, and (4) an amplifier to increase the output of the oscillator (Bowditch 1995). When a radio wave passes a conductor, a current is induced in the conductor. A radio receiver is a device which senses the power thus generated in an antenna and transforms it into a usable form. It is able to select signals of a single frequency (or a narrow band of frequencies) from among the many which may reach the receiving antenna. The receiver is able to demodulate the signal and provide adequate amplification. The output of a receiver may be presented audibly by earphones or loudspeaker or visually on a dial, cathode-ray tube, counter, or other display. Thus, the useful reception of radio signals requires three components: (1) an antenna, (2) a receiver, and (3) a display unit.

Unwanted signals or any distortion of the transmitted signal that impedes the reception of the signal at the receiver end are called *interferences*. The intentional production of such interference to obstruct communication is called *jamming*. Unintentional interference is called *noise*. That means interference includes both jamming and noise. Radio receivers differ mainly in (Bowditch 1995): (1) *frequency range*—the range of frequencies to which they can be tuned; (2) *selectivity*—the ability to confine reception to signals of the desired frequency and avoid others of nearly the same frequency; (3) *sensitivity*—the ability to amplify a weak signal to usable strength against a background of noise; (4) *stability*—the ability to resist drift from the frequency at which it is set; and (5) *fidelity*—the completeness with which the essential characteristics of the original signal are reproduced.

Receivers may have additional features, such as an automatic frequency control, automatic noise limiter, etc., Some of which are interrelated (Bowditch 1995). For instance, if a receiver lacks selectivity, signals of a frequency differing slightly from those to which the receiver is tuned may be received. This condition is called *spillover*, and the resulting interference is called *crosstalk*. If the selectivity is increased sufficiently to prevent spillover, it may not permit receipt of a great enough band of frequencies to obtain the full range of those of the desired signal. Thus, the fidelity may be reduced.

A *transponder* is a transmitter-receiver capable of accepting the challenge of an interrogator and automatically transmitting an appropriate reply.

4.3   GNSS SIGNALS—CARRIERS AND CODES

How does a GNSS satellite communicate all that information to a receiver? It uses *codes*. GNSS codes are binary—zeroes and ones, the language of computers. Codes are carried to GNSS receivers by *carrier* waves (Issler *et al.* 2003; Spilker 1980; Langley 1993a).

A series of waves transmitted at constant frequency and amplitude is called a *continuous wave*. When a continuous wave is modified in some manner, this is called *modulation*. When this occurs, the continuous wave serves as a carrier wave for information.

In communication technology, a carrier wave, or carrier, is a waveform (shape and form of a signal) that is modulated (modified or changed) with an input signal for the purpose of conveying information: for example, voice or data, to be transmitted, by radio wave. A carrier wave has at least one characteristic such as phase, amplitude, or frequency that may be changed or modulated (Figure 4.4), for the purpose of carrying information. For example, the information, music, or speech received from an AM radio station is placed on the carrier wave by amplitude modulation (i.e., the amplitude is altered), and the information on the signal from a FM radio station is there because of frequency modulation (i.e., the frequency is changed). Any one of the several types of modulation may be used to carry the information.

The phase of a wave is the amount by which the cycle has progressed from a specified origin. For most purposes, it is stated in circular measure, a complete cycle being considered 360° (Figure 4.6). Generally, the origin is not important—the

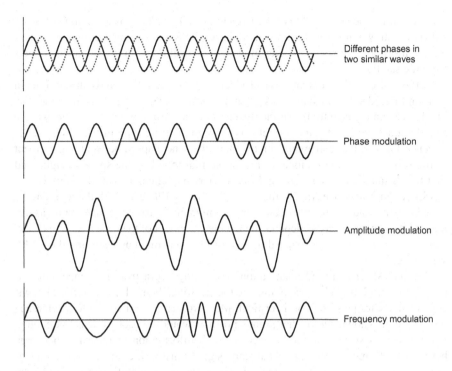

FIGURE 4.4 Different types of modulation in a wave.

principal interest being the phase relative to that of some other wave. Thus, two waves having crests 1/4 cycle apart are said to be 90° 'out of phase' (Figure 4.7). If the crest of one wave occurs at the trough of another, the two are 180° out of phase (Figure 4.8).

By means of phase modulation using radio wave, GNSS signal is carried from satellites to the receivers (Sickle 2008). We shall discuss GNSS signals later in this chapter in detail. Before that, let us discuss the information carried by these signals.

4.4 INFORMATION CARRIED BY GNSS SIGNAL

It has been mentioned earlier that GNSS is based on trilateration, i.e., on distances (ranges) measured from the satellites to the receiver. These ranges are measured with signals that are broadcast from the GNSS satellites to the GNSS receivers in the microwave part of the electromagnetic spectrum. GNSS is sometimes called a *passive system* (Sickle 2008), in the sense that only the satellites transmit signals; the users (receivers) simply receive them. As a result, there is no limit to the number of GNSS receivers that may simultaneously monitor the GNSS signals. Just as millions of television sets may be tuned to the same channel without disrupting the broadcast, millions of GNSS receivers may monitor the satellite's signals without overburdening the system. This is a distinct advantage; albeit, as a result, GNSS signals must carry a great deal of information. A GNSS receiver must be able to gather all the information it needs to determine its own position from the signals it collects from the satellites.

In the case of GNSS, distance is a product of the speed of light and elapsed time. In a ground-based survey, frequencies generated within an electronic distance measuring device can be used to determine the elapsed travel time of its signal because the signal bounces off a reflector and returns to where it started (Figure 4.5). Therefore, in general terms, the instrument can take half the time elapsed between the moment of transmission and the moment of reception, multiplied by the speed of light, and find the distance between itself and the retroprism (reflector). But the signals in a GNSS system travel one way: to the receiver. The satellite can mark the moment the signal departs and the receiver can mark the moment it arrives. Since the measurement of the ranges in GNSS depends on the measurement of the time it takes a GNSS signal to make the trip, the elapsed time must be determined by decoding the GNSS signal itself.

Both GNSS and ground trilateration positioning begin from reference (control) points. In GNSS, the reference points are the satellites themselves; therefore, knowledge of the satellite's position is critical. Measurement of a distance to a reference point without knowledge of that point's position would be useless. It is not enough that the GNSS signals provide a receiver with information to measure the range between itself and the satellite. That same signal must also communicate the position of the satellite, at that very instant. The situation is complicated somewhat by the fact that the satellite is always moving with respect to the receiver at a speed of approximately 3.5–4.0 km/s (Table 2.4).

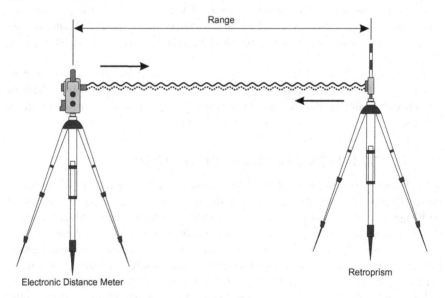

FIGURE 4.5   Two-way range measurement.

In a GNSS, as in a ground-based trilateration, the signals must travel through the atmosphere. In a ground-based trilateration, compensation for the atmospheric effects on the signal, estimated from local observations, can be applied at the signal's source for necessary corrections. This is not possible in GNSS. The GNSS signals begin in the virtual vacuum of space, but then, after hitting the earth's atmosphere, they travel through much of the atmosphere (Saastamoinen 1973). Therefore, the GNSS signals must give the receiver some information needed for atmospheric corrections.

It takes more than one measured distance to determine a new position in a ground-based trilateration or in a GNSS. For a GNSS, the minimum requirement is range measurements from at least four GNSS satellites. The GNSS receiver must be able to match each of the signals it tracks along with the origin of the signal, that is, the location of the transmitting satellite. Remember, there are several transmitting satellites in the sky, and the receiver needs to know which signal is coming from which satellite. Therefore, the GNSS signals themselves must also carry a kind of satellite identification. To be on the safe side, the signal should also tell the receiver where to find all of the other satellites as well.

Additionally, if a satellite goes out of adjustment, or if it malfunctions, the signal received from that satellite should not be used by the receiver for positioning. Therefore, the receiver should also know whether the signal received from a satellite is usable or not. This information about the 'health' of the satellite is also transmitted to the receivers. The ground control stations monitor each satellite's health and send this information to all of the satellites in a constellation, which in turn is transmitted to the receivers.

To sum up, a GNSS signal must somehow communicate to its receiver: (1) what time it is on the satellite, (2) the instantaneous position of a moving satellite, (3) information about necessary atmospheric corrections, (4) a satellite identification system to tell the receiver where it came from and where the receiver may find the other satellites, and (5) health information of the satellites.

4.5 NAVIGATION MESSAGE

The *navigation code*, or *navigation message*, is the vehicle for telling the GNSS receivers some of the most important things they need to know (GLONASS ICD 2002; Issler *et al.* 2003; Sickle 2008; Samama 2008; Zaharia 2009). The navigation code generally has a low data transfer rate, for example 50 bit/s for GPS, GLONASS, BeiDou (25 bit/s for Galileo). Navigation messages provide the information about the satellite time, satellite health information, and information for atmospheric corrections (partial), ephemeris, and almanac. These are uploaded by the uploading stations of the ground control segment of the satellites.

Unfortunately, the accuracy of some aspects of the information included in the navigation message (e.g., ephemeris) deteriorates with time. Therefore, mechanisms are in place to prevent the message from getting too old. The message is renewed frequently by upload stations. The following sections describe all the information carried by navigation message.

4.5.1 GNSS TIME

Time sensitive information is found in the navigation message. It contains information that helps a receiver in relating a standard time scale to the time of a specific GNSS constellation. Although similar, GPS, GLONASS, Galileo, and BeiDou have several differences in the way they record and report time. First of all, we have to understand that the time used by a specific GNSS constellation is different than the Coordinated Universal Time (UTC), International Atomic Time (TAI), or any other time scale we use in our daily life (Bowditch 1995). The UTC is a worldwide time scale. The rate of UTC, which is steered by timing laboratories around the world, is more stable than the rotation of the earth itself. This causes a discrepancy between UTC and the earth's actual motion. This difference is kept within 0.9 sec by the periodic introduction of *leap seconds* (one-second adjustment) in UTC. Leap seconds are necessary because time is measured using stable atomic clocks, whereas the rotation of earth slows down continually, though at a slightly variable rate. Therefore, leap seconds are used to keep UTC close to mean solar time (Bowditch 1995; Lombardi 2002). Just compare with one-day addition in a leap year.

Unlike UTC, TAI is a high-precision atomic time standard that tracks proper time on earth's geoid. The TAI, as a standard, is a weighted average of the time kept by about 300 atomic clocks worldwide. The clocks at different institutions are regularly compared against each other. The International Bureau of Weights and Measures (BIPM) combines these measurements to retrospectively calculate the weighted average that forms the most stable time scale possible. TAI is the uniform conventional time scale, but it is not distributed directly in every-day life. In case of TAI, there is no need of the leap second addition. The time in common use (broadcast by different means) is referred to as UTC. As of December 2020, TAI is exactly 38 sec ahead of UTC: 10 sec initial difference at the start in 1972, plus 28 leap seconds in UTC since 1972.

The *GPS time*, the scale for internal use in the GPS system, is similar to TAI and 19 sec behind of TAI (a constant amount). The GPS system uses the UTC time reference maintained by the United States Naval Observatory (USNO). On the internet, USNO master clock time is available at https://www.usno.navy.mil/USNO/time/display-clocks/simpletime. Since GPS time is not earthbound, leap seconds are not used. Although many leap seconds have been added (and are being added) to UTC, and none have been added to GPS time, the difference between them is increasing continuously. This complicates the relationship between UTC and GPS time. Even though their rates are virtually identical, the numbers expressing a particular instance in GPS time are different by some seconds from the numbers expressing the same instance in UTC. The GPS navigation message includes the difference (integer leap seconds and fractional difference) between GPS time and UTC, and thus they are related (Langley 1991a).

As opposed to the year, month, and day format of the Gregorian calendar, the GPS date is expressed as a week number and a day-of-week number. The week number is transmitted in the navigation messages. GPS week zero started at 00:00:00 UTC (00:00:19 TAI) on 6 January 1980 and ended on 12 January 1980. It was followed by GPS week 1, GPS week 2, and so on. However, about 19 years later, at the end of GPS week 1023—at midnight (UTC) 21 August–22 August, 1999—it was necessary to start the numbering again at 0. This is called *rollover*. The second rollover occurred at midnight 6 April–7 April 2019, when GPS Week 2047, represented as 1023 in the counter, advanced and rolled over to 0 within the counter. Many older receivers and mobile phones lost the time and stopped working during the rollover; however, they became operational after patching them.

This necessity accrued from the fact that the following week after week 1023 would have been GPS week 1024. That was a problem because the capacity of the GPS week field in the navigation message was only 10-bits, and the largest count a 10-bit field can accommodate is 2^{10} or 1024 (0 to 1023). To overcome this problem in the future, the modernised navigation message has a 13-bit field for the GPS week count; this means that the GPS week will not need to rollover again for a long time (Samama 2008).

The GLONASS uses the UTC time maintained by (the former Soviet Union (SU) and now) Russia (GLONASS ICD 2002). Unlike GPS, *GLONASS time* is directly affected by the introduction of a leap second. The GLONASS time scale is not continuous and must be adjusted for periodic leap seconds. GLONASS time is maintained within 1 ms, and typically better than 1 microsecond (µs), of UTC (SU) by the control segment with the remaining portion of the offset broadcast in the navigation message. During the leap second correction of UTC, GLONASS time is also corrected by changing enumeration of second pulses of onboard clocks of all GLONASS satellites. GLONASS users are notified in advance (at least three months before) on these planned corrections through relevant bulletins, notifications, etc. Therefore, the GLONASS satellites do not have any data concerning the UTC leap second correction within their navigation messages. Due to the leap second correction, there is no integer-second difference between GLONASS time and UTC (SU). However, there is a constant three-hour difference between these time scales due to GLONASS control segment specific features (GLONASS ICD 2002).

Galileo also establishes a reference time scale, Galileo System Time (GST), to support system operations (Stehlin 2000; Zaharia 2009). The Galileo system avoids the addition of a leap second by steering Galileo time to TAI. The GPS time is steered to a real-time representation of UTC produced by the USNO; whereas GST is steered to TAI by the BIPM. The GST time started at 0 hours UTC at midnight between 21 and 22 August 1999. The time accuracy in Galileo is maintained within 50 ns with reference to TAI.

The time reference system used in BeiDou is called *BeiDou time* (or *BeiDou System Time*, abbreviated as BDT). Similar to GPS time, BeiDou time is a continuous time scale, which does not introduce any leap seconds. BDT is derived from the atomic clock installed in the BeiDou ground control centre and can be traced to the Chinese national official time—UTC, which is kept by the National Time Service Centre (NTSC), Chinese Academy of Sciences. The BDT time started at 0h UTC on midnight between 31 December 2005 and 01 January 2006. The time accuracy in BeiDou is maintained within 100 ns with reference to UTC (Dong *et al.* 2007, 2008).

From the preceding discussion, we can understand that the time used by the GPS, GLONASS, Galileo, and BeiDou are different, and the references they use are also different. The *GNSS time* is a generic term used in this book, to refer to these time scales.

4.5.2 SATELLITE CLOCKS

Each GNSS satellite carries its own onboard clocks, also known as *time standards*, in the form of very stable and accurate atomic clocks regulated by the vibration frequencies of atoms (Langley 1991a; Lombardi 2002; Jespersen and Fitz-Randolph 1999). Since the clocks in any one satellite are completely independent from those in any other, they are allowed to drift up to one millisecond from the strictly controlled time standard of a specific GNSS. Instead of constantly tweaking the satellite's onboard clocks to keep them all in lockstep with each other and with GNSS time, their individual drifts are carefully monitored by the monitoring (or tracking) stations of the control segment. These stations record each satellite clock's information, send them to the master control station, where the satellite clock information is compared with the master control station's clock, and *clock drifts* are identified. These clock drift information are then sent to the upload stations for uploading to each satellite's navigation message, where it is known as the broadcast clock correction.

A GNSS receiver is capable of relating the satellite's clock to the GNSS time, using the correction given by the broadcast clock correction in the navigation message. This is obviously only part of the solution to the problem of directly relating the receiver's own clock to the satellite's clock. The receiver will need to rely on other aspects of the GNSS signal for a complete time correlation. The drift of each satellite's clock is neither constant nor can the broadcast clock correction be updated frequently enough to completely define the drift. Therefore, navigation message also provides a definition of the reliability of the broadcast clock correction. This is called *issue of data clock* (IODC).

4.5.3 BROADCAST EPHEMERIS

The navigation message also harbours a type of time sensitive information. It contains information about the position of the satellite with respect to time. This is called the satellite's ephemeris.

The broadcast ephemeris, however, is never perfect. In the case of GPS, Galileo, and BeiDou, the broadcast ephemeris is expressed in parameters that appear Keplerian (refer to Chapter 5, Section 5.8.1) (Samama 2008). Its elements are named after the 17th century German astronomer Johann Kepler. The GLONASS uses Geocentric Cartesian coordinates and their derivatives to compute the future orbit (GLONASS ICD 2002). But in every case, they are the result of least-squares, curve-fitting analysis of the satellite's actual orbit. Therefore, like the IODC, accuracy of the broadcast ephemeris deteriorates with time. As a result, this issue, one of the most important parts of the navigation message, is called *issue of data ephemeris* (IODE).

4.5.4 ATMOSPHERIC CORRECTION

Data provided by the navigation message offer only a partial solution to atmospheric interaction. The control segment's monitoring stations find the apparent delay of a GNSS signal caused by its trip through the ionosphere by analysing the different propagation rates of the two frequencies broadcast by all GNSS satellites, e.g., L1 and L2. Usage of two frequencies and the effects of the atmosphere on the GNSS signal will be discussed later in Chapter 5 (Section 5.4). For now, it is sufficient to know that a single-frequency receiver depends on the ionospheric correction carried by navigation message to help in removing part of the error introduced by the atmosphere (Klobuchar 1987).

4.5.5 BROADCAST ALMANAC

Navigation message tells the receiver where to find all the satellites of a particular GNSS constellation. This information is the coarser orbital information, called the *almanac*. However, these data are not complete to provide the accurate location information about the satellites for a given instant and cannot be used for positioning. Once the receiver finds its first satellite, it can look at this truncated coarser orbital information (the almanac) of that satellite's navigation message to figure the position of more satellites to track. This makes the tracking of the satellites faster.

4.5.6 SATELLITE HEALTH

Information about the health of the satellite that is being tracked by the receiver is also available from the navigation message. It allows the receiver to determine whether the satellite is operating within normal parameters or not. Each satellite transmits health data for all of the satellites in a specific constellation. GNSS satellites are vulnerable to a wide variety of breakdowns, particularly clock trouble. That is one reason why they carry at least three clocks. Health data are also periodically uploaded by the ground control station. Health data inform the receivers about any malfunctioning of the satellite before they try to use a particular signal.

4.6 RANGING CODES

In addition to the navigation message, GNSS satellites transmit a special type of code used to determine the range (distance) from the satellite to the receiver. This coded information is transmitted by the carrier signals in several frequencies. The frequencies currently in use were documented in Chapter 2 (Table 2.4). However, unlike the navigation message, the ranging codes are not vehicles for broadcasting information. They carry the raw data from which GNSS receivers derive their propagation time and distance measurements.

The ranging codes are complicated; so complicated, in fact, that they appear to be nothing but noise at first. And even though they are known as *pseudorandom noise* or PRN codes, actually, these codes have been designed carefully. This complex and specific design provides them the capability of repetition and replication (Langley 1990).

Discussion on all of these codes is beyond the scope of this book and, hence, not necessary. However, for our understanding, we may consider the P code of GPS as an example (Spilker 1980). The GPS P code, generated at a rate of 10.23 Mbps, is available on both L1 and L2. Each satellite is assigned a portion of the very long P code and it repeats its portion of the P code every seven days. The entire code is renewed every 37 weeks. All GPS satellites broadcast their codes on the same two frequencies. Still, a GPS receiver needs to distinguish transmission of one satellite from another. One method used to facilitate this satellite identification is the assignment of one particular week of the 37-week long P code to each satellite.

Note that there is a fundamental difference between GPS and GLONASS regarding the spectrum, which is induced by the addressing techniques used for the identification of the satellites. If the same codes were generated in same frequency from all of the satellites it would not be possible to identify the satellite from which a specific signal is coming. The basic difficulty, as with all radio communication systems, is to find the best way to allow multiple satellites to access the same receiver without interference; the corresponding technique is called *multiple access*. There are three main techniques available for this: Frequency Division Multiple Access (FDMA), Time Division Multiple Access (TDMA), and Code Division Multiple Access (CDMA) (Pratt *et al.* 2003).

In FDMA, specific frequencies are allocated to transmitters in order to be identified by the receiver; whereas, in TDMA, individual time slots allow identification; and in CDMA, different codes are assigned to each transmitter (Bellavista and Corradi 2007). Some systems also use a combination of these. The main advantage of CDMA is that the whole bandwidth allocated can be used by every receiver. GPS, Galileo, and BeiDou systems use CDMA, but GLONASS uses FDMA. However, GLONASS is being transformed from FDMA to CDMA. This switchover started in 2014 (with the launch of a GLONASS-M satellite) and is expected to be completed by 2040. Russia has a plan to launch the first GLONASS-K2 satellite by the year 2022. The GLONASS-K2 satellite will allow Russia to evaluate CDMA signals that are planned to broadcast on L1, L2, and L3 frequencies. However, GLONASS will also continue transmitting FDMA signals for the 'unlimited future' to provide backward capability.

For GPS, Galileo, and BeiDou, each satellite is being identified by its code, as mentioned earlier—one particular week of a very long code is assigned to each satellite. Hence, all satellites transmit on the same frequency. In case of GLONASS, each satellite is identified by its different frequencies, and all satellites have the same code. The nominal values of GLONASS L1 and L2 carrier frequencies are defined by the following expressions (GLONASS ICD 2002):

$$f_{K1} = f_{01} + K\Delta f_1 \tag{4.3}$$

$$f_{K2} = f_{02} + K\Delta f_2 \tag{4.4}$$

where K is a frequency number (frequency channel) of the signals transmitted by GLONASS satellites in the L1 and L2 sub-bands correspondingly;

f_{01} = 1602 MHz; Δf_1 = 562.5 kHz, for L1 sub-band,
f_{02} = 1246 MHz; Δf_2 = 437.5 kHz, for L2 sub-band. (These are fundamental frequency values).

Channel number (K) for any particular GLONASS satellite is provided in almanac data.

4.7 MODULATED CARRIER WAVE AND PHASE SHIFT

Since all the aforementioned codes come to a GNSS receiver on a modulated carrier, it is important to understand how a modulated carrier is generated. We know that GNSS is a passive system, meaning that the signal is transmitted one way. However, the one-way ranging used in GNSS is more complicated than the two-way ranging generally used in an electronic distance meter. A GNSS signal cannot be analysed at its point of origin. The measurement of the elapsed time between the signal's transmission by the satellite and its arrival at the receiver requires two clocks—one in the satellite and one in the receiver. This complication is compounded to correctly represent the distance between them; these two clocks would need to be perfectly synchronised with one another. Since such a perfect synchronisation is physically impossible, the problem is addressed mathematically (refer to Chapter 3, Section 3.6).

The time measurement devices used in GNSS measurements are clocks only in the most general sense; they are more correctly called *oscillators* or *frequency standards*. In other words, rather than producing a steady series of audible ticks they keep the time by chopping a continuous beam of electromagnetic energy at extremely regular intervals. The result is a steady series of wavelengths and is the foundation of a modulated carrier. GNSS carriers are the result of phase modulations as mentioned earlier.

This is illustrated with an example (Sickle 2008); the action of a shutter in a movie projector is analogous to the modulation of a coherent beam by the oscillator. Consider the visible beam of light of a specific frequency passing through a movie projector. It is interrupted by the shutter rotating at a constant rate that alternately blocks and uncovers the light. In other words, the shutter chops the continuous beam into equal segments. Each length begins with the shutter closed and the light beam entirely blocked. As the shutter rotates and opens, the light beam is gradually uncovered. It increases to its maximum intensity, and then decreases again as the shutter gradually closes. The light is not simply turned on and off; it gradually increases and decreases. In this analogy the light beam is the carrier, and it has a wavelength much shorter than the wavelength of the modulation of that carrier produced by the shutter. This modulation can be illustrated by a *sine wave*. The wavelength begins when the light is blocked by the shutter. The first minimum is called a 0° phase angle. The first maximum is called the 90° phase angle and occurs when the shutter is entirely open. It returns to minimum at the 180° phase angle when the shutter closes again.

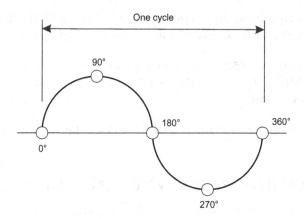

FIGURE 4.6   0° to 360° phase = one complete cycle (or one wavelength).

But the wavelength is not yet complete. It continues through a second shutter open-
ing, 270°, and closing, 360°. The 360° phase angle marks the end of one wavelength
and the beginning of the next one. The time and distance from the 0° to the 360°
phase angles is a wavelength or one complete cycle (Figure 4.6). As long as the rate
of an oscillator's operation is very stable, both the length and elapsed time between
the beginning and end of every wavelength cycle of the modulation will be the same.

The GNSS oscillators are sometimes called 'clocks' because the frequency of
a modulated carrier, measured in hertz, can indicate the elapsed time between the
beginning and end of a wavelength, which is useful information for finding the
distance covered by a wavelength. The relationship between the wavelength and
frequency has been described in equation 4.2 (Section 4.2.1). Now, consider the distance from the satellite to the receiver, which can be covered by some number of full
wavelengths and perhaps a fraction of a full wavelength. With the original *Gunter's
chain*, the surveyor could simply look at the chain and estimate the fractional part of
the last link that should be included in the measurement. Those links were tangible
(detectable with the human senses) and thus the full links could be counted and fraction of the last link could be estimated. Since the wavelengths of a modulated carrier
are not tangible, the GNSS must find the fractional part of wavelength measurement
electronically. Therefore, it compares the phase angle of the satellite signal to that of
a replica of the transmitted signal generated by the receiver to determine the *phase
shift*. That phase shift represents the fractional part of the wavelength measurement.

NOTE

Gunter's chain is a measuring device used for land surveying. It consists of a
chain formed of 100 long wire links, and it is 22 yards long, exactly 66 feet.
There are brass tags at various distances along the chain to simplify intermediate measurement.

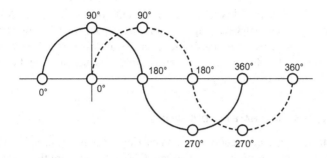

FIGURE 4.7 90° phase shift between two signals.

How does it work? First, it is important to remember that points on a modulated carrier are defined by phase angles, such as 0°, 90°, 180°, 270°, and 360° (Figure 4.6). When two modulated carrier waves reach exactly the same phase angle at exactly the same time, they are said to be *in phase*, or *coherent*, or *phase locked*. Conversely, when two waves reach the same phase angle at different times, they are *out of phase* or *phase shifted*. For example, in Figure 4.7, a sine wave shown by the dashed line and a sine wave shown by the solid line, it is out of phase by one-quarter of a wavelength cycle, i.e., 90°.

In GNSS, the measurement of the difference in the phase of the incoming signal and the phase of the internal oscillator in the receiver reveals the small distance at the end of a range. In GNSS, the process is called *carrier phase ranging*. And, as the name implies, this measurement is actually done on the carrier itself. While this technique discloses the fractional part of a wavelength, a problem still remains—determining the number of full wavelengths (cycles), often termed as *cycle ambiguity* or *integer ambiguity*. We shall discuss cycle ambiguity later in this chapter (Section 4.10.2) and in Chapter 8 (Section 8.3.3).

4.8 OBSERVABLES—PSEUDORANGE AND CARRIER PHASE

The word *observable* is used throughout the GNSS literature to indicate the signals whose measurements yield the range (distance) between the satellite and the receiver. The word is used to draw a distinction between the thing being measured (the observable) and the measurement (the observation). Observable may be thought as a basis or scale of a measurement. In GNSS there are two types of observables (Langley 1993a), the *pseudorange* and the *carrier phase*. The latter, also known as *carrier beat phase*, is the basis of the techniques used for high-precision GNSS survey. On the other hand, the pseudorange can serve applications when virtually instantaneous point positions are required or relatively low accuracy can be considered. These basic observables can also be combined in various ways to generate additional measurements that have certain advantages; it is that the pseudoranges are used in many GNSS receivers as a preliminary step toward the final determination of position by a carrier phase measurement.

The foundation of pseudoranges is the correlation of code carried on a modulated carrier wave received from a GNSS satellite with a replica of that same code generated in the receiver. This technique is called *code correlation*. Most of the GNSS receivers used for surveying applications are capable of code correlation. That is, they can determine pseudoranges. The receivers are also capable of determining carrier phase. However, first let us concentrate on the pseudorange and then the carrier phase will be discussed.

4.8.1 Encoding by Phase Modulation

Before we start discussing pseudorange and carrier phase measurements, we need to understand *phase modulation*. We know that GNSS carriers are the result of phase modulations. One consequence of this method of modulation is that the signal can occupy a broader bandwidth. This characteristic offers several advantages, including a better signal to noise ratio, more accurate ranging, less interference, and increased security. However, spreading the spectral density of the signal also reduces its power to a range of 160–166 decibels per watt (dBw). The power of GNSS signal is so low that it is often described as tantamount to a 25 watt light bulb seen from 10,000 miles away (Sickle 2008). Clearly, it is somewhat difficult to receive the GNSS signal under a dense vegetation canopy, under water, underground, or inside buildings.

The most commonly used spread spectrum modulation technique is known as *binary phase shift keying* (BPSK) (Roddy 2006; Mittal and De 2007), also called *bi-phase shift keying*. BPSK is the technique used to add a binary signal to a sine wave carrier to create the navigation message and ranging codes. All of them consist of 0s and 1s, and this binary code is imprinted on the carrier wave without altering the amplitude, frequency, or wavelength of the carrier. Rather, the BPSK modulation is accomplished by phase changes of 180°. When the value of the message is to change from 0 to 1, or from 1 to 0, the phase of the carrier wave is instantly reversed—it is flipped 180°. And each one of these flips occurs when the phase of the carrier is at the zero-crossing (Figure 4.8).

Each zero or one of the binary code is known as a *code chip*. The term 'chip' is used instead of 'bit' to show that it does not carry any pieces of information; '0' represents the normal state, and '1' represents the mirror state. Note that each shift from 0 to 1, and from 1 to 0, is accompanied by a corresponding change in the phase of the carrier (Figure 4.9). The rate of all of the components of GNSS signals are multiples of the standard rate of the oscillators, for example, 10.23 MHz for GPS P

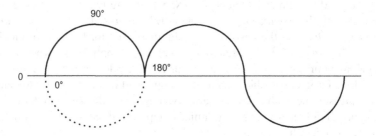

FIGURE 4.8   180° phase shift occurs at zero-crossing.

code. This rate is known as the *fundamental clock rate* and is symbolised as F_0. For example, the GPS carriers are 154 times F_0, or 1575.42 MHz, and 120 times F_0, or 1227.60 MHz, for L1 and L2, respectively (Kaplan 1996; Spilker 1980). GPS L5 will also be a multiple of the fundamental clock rate, 115 times F_0 or 1176.45 MHz.

The codes are also based on the fundamental clock rate. For example, 10.23 code chips of the GPS P code, 0s or 1s, occur in every microsecond; this is known as *chip rate*. In other words, the chipping rate of the P code is 10.23 megabits per second (Mbps), exactly the same as F_0, 10.23 MHz. This chipping rate is also called *code frequency*. From the preceding discussion, we understand that, in GPS L1 P code, 154 cycles cover a single binary code, either 0 or 1. That means the phase is changed after 154 cycles or multiples of 154 cycles (Spilker 1980; Wells 1986). Figure 4.9 shows a 2-cycle per bit code, since one bit is represented by two complete cycles; hence the phase is changed after 2 cycles or multiples of 2 cycles. The chipping rate of the GPS C/A code is 10 times slower than the P code, a tenth of F_0, 1.023 Mbps. That means, 10 P code chips occur in the time it takes to generate one C/A code chip, allowing P code-derived pseudor-anges to be much more precise. This is one reason the C/A code is known as the 'coarse acquisition' code. The code frequency of GLONASS (GLONASS ICD 2002), Galileo (Samama 2008), and BeiDou are furnished in Table 4.2.

Even though both codes, C/A and P, are broadcast on the same carrier, they are distinguishable from one another by their transmission in *quadrature* (90° phase modulation). It means that the C/A code modulation on the carrier is phase shifted 90° from the P code modulation on the same carrier.

Quadrature phase-shift keying (QPSK), another type of phase-shift keying, is also being used in GNSS. The term 'quadrature' implies that there are four possible phases which the carrier can have at a given time. These four phase shifts may occur at 0°, 90°, 180°, or 270°. In QPSK, information is conveyed through quadrature phase variations. In each time slot, the phase can change once. Since there are four possible phases, there are 2 bits of information conveyed within each time slot. The bit rate of QPSK is, thus, twice the bit rate of BPSK. Each of the four possible phase changes is assigned a specific two-bit value (or *dibit*). For example, 01 for 0°, 00 for 90°, 10 for 180°, and 11 for 270°. Figure 4.10 shows how this modulation generates a series of two-bit values.

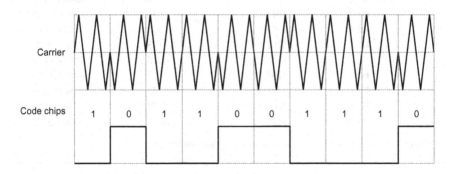

FIGURE 4.9   An example of 2-cycle per code phase modulation.

TABLE 4.2
Details of GNSS Signals*

Constellation	Signal	Carrier Frequency (MHz)	PRN Code Chip Rate (Mbps)	Signal Modulation
GPS	L1 C/A	1575.42	1.023	BPSK
	L1 P(Y)	1575.42	10.23	BPSK
	L1 M	1575.42	5.115	BOC(10,5)
	L1 C	1575.42	1.023	BOC(1,1)
	L2 P(Y)	1227.60	10.23	BPSK
	L2 M	1227.60	5.115	BOC(10,5)
	L2 CM	1227.60	0.5115	BPSK
	L2 CL	1227.60	0.5115	BPSK
	L5 C-I	1176.45	10.23	BPSK
	L5 C-Q	1176.45	10.23	BPSK
GLONASS	L1 C/A	1598.0625–1607.0625	0.511	BPSK
	L1 P	1598.0625–1607.0625	5.115	BPSK
	L2 C/A	1242.9375–1251.6875	0.511	BPSK
	L2 P	1242.9375–1251.6875	5.115	BPSK
	L3 C/A	1198–1208	4.069	BPSK
	L3 P	1198–1208	4.069	BPSK
Galileo	E1-I	1575.42	2.5575	MBOC(6,1,1/11)
	E1-Q	1575.42	2.5575	BOC(14,2)
	E2-I	1561.098	2.046	BPSK(2)
	E2-Q	1561.098	2.046	BPSK(2)
	E5a-I	1176.45	23.205	AltBOC(15,10)
	E5a-Q	1207.14	23.205	AltBOC(15,10)
	E5b-I	1207.14	2.046	BPSK(2)
	E5b-Q	1207.14	10.23	BPSK(10)
	E6-I	1278.75	10.23	BPSK(10)
	E6-Q	1278.75	10.23	BPSK(10)
BeiDou	B1-I	1561.098	2.046	BPSK(2)
	B1-Q	1561.098	2.046	BPSK(2)
	B1-C	1575.42	2.046	MBOC(6,1,1/11)
	B1-A	1575.42	2.046	BOC(14,2)
	B2-I	1207.14	2.046	BPSK(2)
	B2-Q	1207.14	10.23	BPSK(10)
	B2a	1176.45	10.23	AltBOC(15,10)
	B2b	1207.14	10.23	AltBOC(15,10)
	B3-I	1268.52	10.23	BPSK(10)
	B3-Q	1268.52	10.23	BPSK(10)
	B3-A	1268.52	10.23	BOC(15,2.5)

* Information furnished in this table are indicative only, not conclusive.

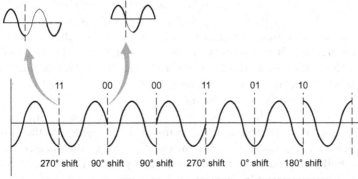

The corresponding information transmitted is therefore: 110000110110

FIGURE 4.10 Two bits of information are conveyed in the transition between time slots.

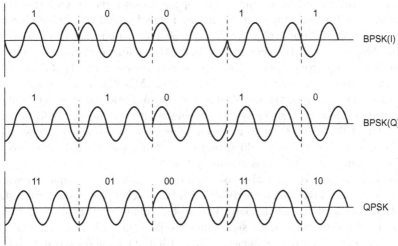

The corresponding information transmitted is therefore: 1101001110

FIGURE 4.11 QPSK is a combination of BPSK(I) and BPSK(Q).

Although QPSK can be viewed as a quaternary modulation, it is easier to see it as two independently modulated BPSK carriers. With this interpretation, the even (or odd) bits are used to modulate the in-phase (I) component of the carrier, while the odd (or respectively even) bits are used to modulate the quadrature-phase (Q) component of the carrier. BPSK is used on both carriers, and they can be independently demodulated. Figure 4.11 clearly shows how QPSK can be viewed as two independent BPSK signals. This clearly means that the QPSK carrier can be used as two BPSK carriers, e.g., as it was used in the GPS L1 signal to transmit two different codes, P(Y) and C/A, in two different BPSK carriers. However, QPSK can also be used as a single carrier to increase the data bit rate.

Modern GNSS satellites are using a new modulation process called *binary offset carrier* (BOC). With BOC, the BPSK signal undergoes a further modulation. The modulation frequency is always a multiple of the chip rate. The main reasons for creating BOC signals were, on one hand, the need to improve traditional GNSS signal's properties for better resistance to *multipath*, interferences of all kinds, and receiver noise; and, on the other hand, the need for improved spectral sharing of the allocated bandwidth with existing signals or future signals of the same class. The corresponding spectrum of BOC is shifted to each side of the central frequency, allowing many different signals on the same central frequency. The properties of this modulation are communicated in a specific way. Following the convention introduced by Betz (1999), the designation BOC(α,β) is used as an abbreviation. The subcarrier frequency is actually $\alpha \times$ 1.023 MHz, while the code chip rate is actually $\beta \times$ 1.023 Mbps (Betz 1999). For example, BOC(10,5) means that the subcarrier frequency is 10.23 MHz, and the code chip rate is 5.115 Mbps.

In 2004, the United States and the European Union signed a groundbreaking agreement to provide common, interoperable signals for civilian users of GPS and Galileo: BOC modulation with a 1.023 MHz sub-carrier frequency and a code rate of 1.023 Mbps—BOC(1,1)—centred at 1575.42 MHz. Subsequent discussions by the technical working group produced what was considered—not without dissent—to be an even better signal structure: a *multiplex BOC* (MBOC). An important part of the argument in favour of the MBOC modulation arose from the expectation that it would improve multipath suppression by adding a higher frequency BOC modulation on top of BOC(1,1), either by a method of algebraic addition (composite BOC) or by time multiplexing (time multiplexing BOC) (Simsky *et al.* 2008). Not only the Galileo system but the BeiDou is also using MBOC.

The Galileo and BeiDou also use another modulation technique known as *Alternate BOC* (AltBOC). The AltBOC modulation scheme is based on the standard BOC modulation. The standard BOC modulation is a subcarrier modulation, which splits the spectrum of the signal into two parts located at the left and right side of the central carrier frequency. The AltBOC modulation scheme aims at generating a single subcarrier signal adopting a source coding similar to the one involved in the standard BOC. The process allows for keeping the BOC implementation simplicity (Issler *et al.* 2003). The AltBOC modulation offers the advantage that the in-phase (I) and quadrature-phase (Q) component of the carrier can be processed independently, as traditional BPSK signals, or together, leading to tremendous performances in terms of tracking noise and multipath. The BOC modulation has several other modern variants as well, such as sine BOC (BOC_{Sin}), cosine BOC (BOC_{Cos}), Double BOC (DBOC), etc., and some of them have been currently selected for GNSS signals. Refer to Lohan *et al.* (2007) for further details on BOC modulation.

From the preceding discussion, it is obvious that different signals (or services) of different constellations use several modulation techniques to transfer the codes (or chips). Table 4.2 lists all of the currently available signals in detail.

4.9 PSEUDORANGE MEASUREMENT

A pseudorange observable is based on a *time shift*—the time elapsed between the instant a GNSS signal leaves a satellite and the instant it arrives at a receiver. The basic concept of time shift has been described in Chapter 3 (Section 3.4), where we tried to match our friend's count with ours. In GNSS, the elapsed time or time shift is known as the *propagation delay*, and it is used to measure the range between the satellite and the receiver. The measurement is accomplished by a combination of codes. The pseudorange is measured by a GNSS receiver using a replica of the code that has been impressed on the modulated carrier wave. The GNSS receiver generates this replica itself to compare with the code it receives from the satellite. To conceptualise the process, one can imagine two codes generated at precisely the same time and identical in every regard: one at the satellite and another at the receiver. The satellite sends its code to the receiver, but on its arrival at the receiver, the two codes do not match. The codes are identical but they do not correlate until the replica code in the receiver is time shifted.

The receiver-generated replica code is shifted relative to the received satellite code. This time-shift, to match the receiver generated replica and received satellite code, reveals the propagation delay, the time it took the signal to make the trip from the satellite to the receiver. Once the time shift is accomplished, the two codes match perfectly and the time that the satellite signal spends in transit (propagation delay) can been measured—well, almost. It would be wonderful if this time shift could simply be multiplied by the speed of light and yield the true distance between the satellite and the receiver at that instant, and it is close. But there are several physical limitations on the process that prevent such a perfect relationship, which will be described later.

4.9.1 Autocorrelation

As mentioned earlier, the almanac information from the navigation message of the first satellite a GNSS receiver acquires tells it which satellites can be expected to come into the view of the receiver. With this information the receiver can load up pieces of the ranging codes (for example C/A or P) for each of those satellites. Then the receiver tries to line up (match) the replica of ranging codes with the signals it is actually receiving from the satellites. The time required for correlation to occur is influenced by the presence and quality of the information in the almanac.

Lining up the code from the satellite with the replica from the GNSS receiver is called *autocorrelation*, and it depends on the transformation of code chips into *code states* (Borre 2001). The formula used to derive code states (+1 and –1) from code chips (0 and 1) is:

$$\text{code state} = 1 - 2x \qquad (4.5)$$

where, x is the code chip value. For example, a normal code state is +1 and corresponds to a code chip value of 0. A mirror code state is –1, and corresponds to a code chip value of 1.

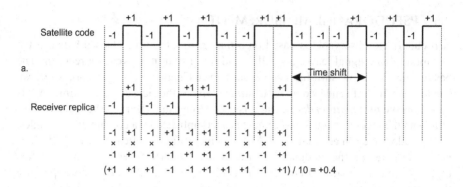

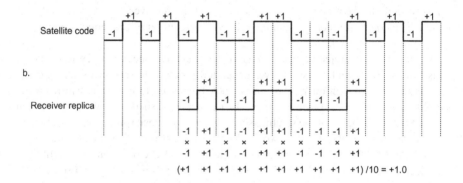

FIGURE 4.12   Concept of code correlation.

The function of these code states can be illustrated by asking three questions (Sickle 2008):

First, if a tracking loop of code states generated in a receiver does not match code states received from the satellite, how does the receiver know?

In that case, for example, the sum of the products of each of the receiver's 10 code states with each of the code states from the satellite, when divided by 10, does not result in 1 (Figure 4.12a).

Second, what does the receiver do when the code states in the receiver do not match the code states received from the satellite?

It shifts the frequency of its search a little bit from the central frequency. This is done to accommodate the inevitable *Doppler shift* (defined later) of the incoming signal since the satellite is always either moving toward or away from the receiver. The receiver also shifts its piece of code in time. These iterative small shifts in both time and frequency continue until the receiver code states do in fact match the signal from the satellite.

Third, how does the receiver know when a tracking loop of replica code states does match code states from the satellite?

In this case (illustrated in Figure 4.12b), the sum of the products of each code state of the receiver's replica 10, with each of the 10 from the satellite, divided by 10, is exactly 1.

In Figure 4.12a, the code from the satellite and the replica from the receiver are not matched, the sum of the products of the code states is not 1. Following the autocorrelation (Figure 4.12b) of two codes, the sum of the code states divided by 10 is exactly 1, and the receiver's replica code fits the code from the satellite like a key fits a lock.

4.9.2 LOCK AND TIME SHIFT

Once correlation of the two codes is achieved, it is maintained by a correlation channel within the GNSS receiver, and the receiver is sometimes said to have *achieved lock* or to be *locked on* to the satellites. If the correlation is somehow interrupted later, the receiver is said to have *lost lock*. However, as long as the lock is present, the navigation message is available to the receiver. Remember that one of the elements of navigation message is the broadcast clock correction that relates the satellite's onboard clock to the GNSS time, and a limitation of the pseudorange process comes up.

One reason the time shift, found in autocorrelation, cannot quite reveal the true range of the satellite at a particular instant is the lack of perfect synchronisation between the clock in the satellite and the clock in the receiver. Recall that the two compared codes are generated directly from the fundamental code frequency of those clocks. And since these widely separated clocks, one on the earth and one in space, cannot have perfect synchronisation with one another, therefore the codes they generate cannot be in perfect synchronisation either. Therefore, a small part of the observed time shift must always be due to the disagreement between these two clocks. In other words, the time shift not only contains the signal's propagation time from the satellite to the receiver, it also contains clock errors.

In fact, whenever satellite clocks and receiver clocks are checked against the carefully controlled GNSS time they are found to be drifting a bit. Their oscillators are never perfect. It is not surprising that they are not quite as stable as the more than a hundred atomic clocks around the world that are used to define the rate of GNSS time: they are subject to the destabilising effects of temperature, acceleration, radiation, and other inconsistencies. As a result, there are two clock offsets (one for satellite and one for receiver) that bias every pseudorange observable. That is one reason it is called a 'pseudorange'.

4.9.3 PSEUDORANGING EQUATION

Clock offsets are only one of the errors in pseudoranges. Other than the clock error, there are several errors involved (described in Chapter 5). Their relationships can be illustrated by the following equation (Wells 1986; Bossler *et al.* 2002; Gopi 2005; Sickle 2008):

$$\rho = R + d_p + c\left(dt - dT\right) + cd_{ion} + cd_{trop} + \varepsilon_{mp} + \varepsilon_p \tag{4.6}$$

where
   ρ = the pseudorange measurement
   R = the true range
   d_p = satellite orbital (ephemeris) errors
   c = the speed of light through vacuum (constant)
   dt = the satellite clock offset from GNSS time
   dT = the receiver clock offset from GNSS time
   d_{ion} = ionospheric delay
   d_{trop} = tropospheric delay
   ε_{mp} = multipath
   ε_p = receiver noise

Please note that the pseudorange, ρ, and the true range, R, cannot be made equivalent without consideration of clock offsets, atmospheric effects, and other biases that are inevitably present.

This discussion about time may make it easy to divert attention from the real objective, that is, the positioning of the receiver. Obviously, if the coordinates of the satellite and the coordinates of the receiver were known perfectly, then it would be a simple matter to determine time shift or find the true range (R) between them. In fact, receivers placed at known coordinated positions can establish time so precisely that they are used to monitor atomic clocks around the world. Several receivers by simultaneous tracking of the same satellite can achieve resolutions of 10 nanoseconds or better. Also, receivers placed at known positions can be used as base stations to establish the relative position of receivers at unknown stations, which is a fundamental principle of most GNSS surveying.

It is useful to imagine that the true range (R), also known as the *geometric range*, actually includes the coordinates of both the satellite and the receiver. However, they are hidden within the measured value, the pseudorange (ρ), and all of the other terms on the right side of the equation 4.6. The objective then is to mathematically separate and quantify these biases so that the receiver coordinates can be revealed. Needless to say, any deficiency in describing (or modelling) the biases will degrade the quality of the final determination of the receiver's position. These issues will be addressed in Chapter 5.

4.10   CARRIER PHASE MEASUREMENT

The wavelengths of the carrier waves are very short compared to the code chip lengths. For example, GPS L1 P code chip length is 154 times longer than the carrier wavelength. Carrier phase can be measured to millimetre precision compared with a few decimetres for P code measurements (and several meters for C/A

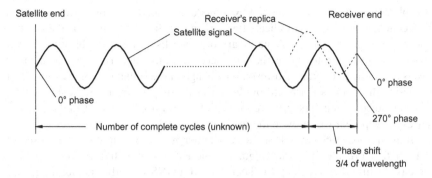

FIGURE 4.13 Phase shift and number of complete cycles (Note that receiver's replica is generated within the receiver not transmitted to the satellite; this figure is a schematic representation).

code measurements). Therefore, if we can determine the range by means of carrier instead of code, we can achieve very high accuracy in comparison to pseudorange measurement.

Understanding the carrier phase is more difficult than the pseudorange, but the foundations of the measurement have some similarities. For example, the foundation of a pseudorange measurement is the correlation of the codes received from a GNSS satellite with replicas of those codes generated within the receiver. The foundation of the carrier phase measurement is the combination of the carrier itself received from a GNSS satellite with a replica of that carrier generated within the receiver. The phase difference between the incoming signal and the receiver's internal reference reveals the fractional part of the carrier phase measurement in GNSS (Figure 4.13). The number of complete cycles to reach the signal from the satellite to the receiver (i.e., the integer part of the measurement) is not immediately known.

The carrier phase observable is sometimes called the *reconstructed carrier phase* or *carrier beat phase* observable. In this context, a beat is the pulsation resulting from the combination of two waves with different frequencies. It can occur when any pair of oscillations with different frequencies are combined. In GNSS, a *beat* is created when a carrier generated in a receiver and the carrier received from the satellite are combined. At first, that might not seem sensible. How could a beat be created by combining two absolutely identical carriers? It is known to us that there should not be any difference in frequency between a carrier generated in a satellite and a replica carrier generated in a receiver. If there is no difference in the frequencies, how can there be a beat? But there is a slight difference between these two carriers. Something happens to the frequency of the carrier on its trip from a GNSS satellite to a receiver and its frequency changes. The phenomenon is described as the *Doppler effect*.

4.10.1 DOPPLER EFFECT

The Doppler effect (or *Doppler shift*), is the change in frequency and wavelength of a wave for an observer moving relative to the source of the waves (Jones 1984).

This phenomenon can be explained by a model provided by sound. An increase in the frequency of a sound is indicated by a rising pitch; a lower pitch is the result of a decrease in the frequency. A stationary observer listening to the blasting horn of a passing train will note that as the train gets closer, the pitch rises, and as the train travels away, the pitch falls. However, the change in the sound to the observer standing beside the track is not heard by the driver driving the train. He hears only one constant, steady pitch. The relative motion of the train with respect to the observer causes the apparent variation in the frequency of the horn.

From the observer's point of view, whether it is the source, or the observer, or both that are moving, the frequency increases while they move closer and decreases while they move apart. Therefore, if a GNSS satellite is moving toward an observer (receiver), its carrier wave comes to the receiver with a higher frequency. If a GNSS satellite is moving away from the observer, its carrier wave comes into the receiver with a lower frequency. Since a GNSS satellite is virtually always moving with respect to the observer, any signal received from a GNSS satellite is Doppler-shifted.

In the case of GNSS, the receiver also moves in some applications (e.g., the receiver of a car navigation system). Therefore, the movement of the receiver should also be accounted for. This is achieved, once again, through Doppler shift analysis. In fact, when the receivers' hardware tries to 'find' the signal of a given satellite, it has to take into account two components of the Doppler shift: the motion of the satellite and that of the receiver. The Doppler results from the relative displacement of any one or both. It is essential to take the Doppler shift into account. This is achieved through the measurement of the frequency shift of the receiver oscillator (see Chapter 8, Section 8.3.1 for details).

4.10.2 CARRIER PHASE MEASUREMENT EQUATION

The carrier phase observation in cycles is often symbolised by φ (*phi*) in GNSS literature. Other conventions include the use of superscripts to indicate satellite designations and the use of subscripts to define receivers. For example, in the following equation φ_r^s is used to symbolise the carrier phase observation between satellite s and receiver r. The difference that defines the carrier beat phase observation is (Wells 1986):

$$\varphi_r^s = \varphi^s\left(t\right) - \varphi_r\left(T\right) \tag{4.7}$$

where, $\varphi^s\left(t\right)$ is the phase of the carrier broadcast from the satellite s at time t. Note that the frequency of this carrier is the same, nominally constant frequency that is generated by the receiver's oscillator. $\varphi_r\left(T\right)$ is its phase when it reaches the receiver r at time T.

A description of the use of the carrier phase observable to measure range can start with the same basis as the calculation of the pseudorange, i.e., travel time. The time elapsed between the moment the signal is broadcast, t, and the moment it is received,

T, multiplied by the speed of light, c, will yield the pseudorange between the satellite and receiver, ρ:

$$(T-t)\,c \approx \rho \qquad\qquad (4.8)$$

This expression now relates the travel time to the range. However, in fact, a carrier phase observation cannot rely on the travel time for two reasons. First, in a carrier phase, observation of the receiver has no codes with which to tag any particular instant on the incoming continuous carrier wave. Second, since the receiver cannot distinguish one cycle of the carrier from any other, it has no way of knowing the initial phase of the signal when it left the satellite. In other words, the receiver cannot know the travel time and therefore it is hard to see how it can determine the number of complete cycles between the satellite and itself. This unknown quantity is called the *cycle ambiguity* (or *integer ambiguity*). A carrier phase observation must derive the range from a measurement of phase at the receiver, not from a known travel time of the signal.

The missing information is the number of complete phase cycles between the receiver and the satellite at the instant the tracking began. The critical unknown integer, symbolised by N, is the cycle ambiguity, and it cannot be directly measured by the receiver. The receiver can count the complete phase cycles it receives from the moment it starts tracking until the moment it stops. It can also monitor the fractional phase cycles by determining the phase shift between the satellite and receiver carriers (Figure 4.13), but the cycle ambiguity, N, is unknown. However, there are several tricks and techniques to solve this 'ambiguity', which will be discussed in the subsequent chapters.

A 360° cycle in the carrier phase observable is the wavelength λ. Therefore, the cycle ambiguity included in the complete carrier phase equation is an integer number of wavelengths, symbolised by λN. So, the complete carrier phase observable equation can be stated as (Wells 1986; Bossler *et al.* 2002; Gopi 2005; Sickle 2008):

$$\varphi = R + d_p + c\left(dt - dT\right) + \lambda N - cd_{ion} + cd_{trop} + \varepsilon_{mp} + \varepsilon_p \qquad (4.9)$$

where
  φ = carrier phase observation in cycles
  R = the true range
  d_p = satellite orbital (ephemeris) errors
  c = the speed of light through vacuum (constant)
  dt = the satellite clock offset from GNSS time
  dT = the receiver clock offset from GNSS time
  λ = the carrier wavelength
  N = the integer ambiguity in cycles
  d_{ion} = ionospheric delay
  d_{trop} = tropospheric delay
  ε_{mp} = multipath
  ε_p = receiver noise

Note that, unlike the pseudoranging equation (equation 4.6), the ionospheric delay is negative here. (This is an important matter that will be discussed in Chapter 5, Section 5.4.1.) Furthermore, the errors mentioned in the equations 4.6 and 4.9 have not been defined yet. To solve these equations, we need to understand these errors and their determination. Chapter 5 will address these issues.

EXERCISES

DESCRIPTIVE QUESTIONS

1. Explain the wave model of electromagnetic energy and derive the relation between frequency, wavelength, and speed of light. Define electromagnetic spectrum.
2. How is the radio wave generated, transmitted, and received?
3. What do you understand by 'modulation of carrier wave'? Describe different types of modulations.
4. What do you understand by 'navigation code'? What information is carried by this code?
5. Explain the time standards used in GPS, GLONASS, and Galileo in comparison to UTC and TAI.
6. What do you understand by 'observable'? Explain how autocorrelation is achieved in a pseudorange observable.
7. Explain the concepts of BPSK and QPSK.
8. What do you know about BOC modulation? Explain BOC (1,1).
9. Explain pseudorange measurement equation.
10. Explain Doppler shift and its importance in GNSS.
11. Explain carrier phase measurement equation.
12. What do you understand by 'cycle ambiguity'? Why can a GNSS receiver not solve cycle ambiguity?
13. What do you understand by 'pseudorange' and 'geometric range'? How do they relate to one another?
14. How can you determine code states from the code chips? Explain the function of code states.
15. Explain autocorrelation (lock) with proper illustration.

SHORT NOTES/DEFINITIONS

Write short notes on the following topics

1. Band
2. Microwave band
3. Radio band
4. L band
5. Ultra-high frequency
6. Interference of signal

7. Phase modulation
8. TAI
9. UTC
10. Galileo system time
11. Issue of data clock
12. Multiple access
13. CDMA
14. FDMA
15. Issue of data ephemeris
16. Binary phase shift keying
17. Code chip
18. Chipping rate
19. Lock
20. Carrier beat phase

5 Errors and Accuracy Issues

5.1 INTRODUCTION

In the real world, there are several factors that influence GNSS performance and could make it less than mathematically perfect. These aspects of GNSS performance are (Petovello 2008):

- *Accuracy*, at a certain level when the appropriate hardware, software, and operational procedures are used.
- *Availability*, the extent to which the system is available to all users, anywhere on the earth, and at any time of the day.
- *Continuity*, the degree to which a certain level of accuracy is maintained on a continuous basis.
- *Reliability*, of the system and results, often evidenced by a certain 'repeatability' of the positioning accuracy.
- *Integrity*, the capacity to monitor performance and warn users when accuracy falls below a certain level.
- *Cost*, of hardware and software as well as indirect operational costs.
- *Competitive technologies*, do they exist? What do they offer in terms of superior accuracy, etc.?

However, the most important performance measure for most users is accuracy. Hence the main factors influencing positioning accuracy shall be examined in this chapter. To achieve the best possible accuracy from a system, it is necessary for a good receiver to take into account a wide variety of possible errors. The management of errors is indispensable for finding the true geometric range from a pseudorange, or carrier phase observable.

The equations for pseudoranging and carrier phase measurement (Equations 4.6 and 4.9, respectively) have been described in Chapter 4. Both include environmental and physical limitations called *range biases*, such as atmospheric errors, clock errors, receiver noise, multipath errors, orbital errors, and so on. The objective of this chapter is to separate and quantify each of these errors and address other accuracy-related issues.

5.2 IMPACTS OF ERRORS IN PSEUDORANGES

Every error leads to an incorrect estimation of the corresponding pseudoranges. Although there are several ways to represent error sources, for our understanding, the error sources can be split into three categories, depending on where they take place: (1) errors due to satellite-based uncertainties, (2) errors due to signal propagation, and (3) errors due to receiver-based uncertainties. On the satellite side, both clock synchronisation bias (with reference to constellation time) and satellite location accuracy essentially generate final positioning errors. On the receiver side, clock bias and antenna *centre of phase* location are sources of positioning inaccuracy.

In addition to these physical errors, a voluntary noise can be introduced, in some constellations, by the system management. This was the case for the GPS system with its so-called *selective availability* (SA). The main idea was to generate intentional error sources in both the time synchronisation data of the satellites and the ephemeris data. In such a way, the calculations carried out at the receiver end were affected.

All of the errors involved in GNSS may also be classified into the following categories (Samama 2008):

- *Synchronisation errors*—these occur either at the satellite end or at the receiver end and cause the transmitting time and the receiving time not to be considered in the same time reference frame.
- *Propagation errors*—these cause the signal flight time from the satellite to the receiver to appear different from what it actually is. The two effects concerned are the *atmospheric propagation* and the *multipath*. In the first case, the information transmitted is slowed down while crossing some layers, which should be taken into account. In the second case, the signal is lengthened. Note that both effects lead to an increased value of the pseudorange.
- *Location errors*—these relate to both the satellite and receiver. The precise location of satellites, which is sent through ephemeris data, is essential for the receiver to calculate its position. Therefore, the accuracy of these satellite location calculations has a direct influence. But there is another issue concerning the real physical location of the receiver, which is calculated through positioning. This may sound strange, as its location is precisely what is being looked for. Indeed, when dealing with high-precision positioning, i.e., centimetre accuracy, one should think of the real significance of this centimetre. Since the size of the receiver is much larger than a centimetre, where, exactly, is the location point to be considered? The answer is a specific point on the receiving antenna, the centre of phase. Thus, one has to be aware of the real centre of phase location of the antenna in order to use the positioning result.

In addition, one should include all the noise-like errors such as internal receiver noise, thermal noise, and so on. The following sections describe the major biases or accuracy-related problems commonly encountered by a GNSS system.

5.3   SATELLITE CLOCK ERROR

One of the most significant errors is satellite clock bias. The impact of this error can be quite large especially if the broadcast clock correction is not used by the receiver to bring the time signal acquired from a satellite's onboard clock in line with GNSS time.

5.3.1   RELATIVISTIC EFFECTS ON THE SATELLITE CLOCK

It is beyond the scope of this book to provide a comprehensive explanation of the theory of relativity. However, it has an influence on many processes, among them the proper functioning of the GNSS system. According to the theory of relativity (Einstein and Lawson 2001), due to their constant movement and height relative to the earth-centred inertial reference frame, clocks on the satellites are affected by their speed (special relativity) as well as their gravitational potential (general relativity). Since we know that time runs slower during very fast movement, we can expect that for satellites moving at a speed of, for example, 3870 m/s (Table 2.4), satellite clocks run slower when viewed from earth. This relativistic time dilation leads to an inaccuracy of time, approximately 7.2 µs per day (1 µs = 10^{-6} s) for GPS.

The theory of relativity also says that time moves slower if the field of gravitation is stronger. For an earthbound observer, the clock onboard a satellite is running faster (as the satellite at 20, 200 km height (for GPS) is exposed to a much weaker field of gravitation than the observer on the earth). And this effect is six times stronger (approximately 45.65 µs faster per day for GPS) than the time dilation explained above.

Altogether, the clocks of the satellites seem to run a little faster (approximately 38.44 µs per day for GPS). In the GPS system, the combination of these two effects leads to a clock rate offset of 4.45×10^{-10} times faster for the satellite clock in comparison to nominal 10.23 MHz (Kaplan 1996; Tapley *et al.* 2004). Rather than having to take this into account at the receiver end, the relativistic effects are resolved at the satellite end by shifting the frequency of the central oscillator in such a way that the behaviour is equivalent to that with no relativity. To ensure that the clocks will actually achieve the correct fundamental frequency of 10.23 MHz in space, their frequency is set a bit slow before launch to 10.22999999543 MHz. For GLONASS, the same problem led to the nominal value of the frequency, as observed at the satellite, to be biased by the relative value. This appears 4.36×10^{-10} times faster than the nominal standard of 5.0 MHz. This has been adjusted by setting the frequency as 4.99999999782 MHz before launch to achieve 5.0 MHz after launch. Galileo and BeiDou have made similar adjustments.

Other relativistic effects, such as the *eccentricity* (deviation) of the orbit of satellites and the *Sagnac effect*, are also considered. With an eccentricity of 0.02 this effect can be as much as 45.8 ns. Fortunately, the offset is eliminated by a calculation in the receiver itself, thereby avoiding a ranging error of about 14 meters. The *Sagnac effect* is caused by the movement of the observer on the earth's surface, who also moves with a velocity of up to 500 m/s (at the equator) due to the rotation of the earth (Tapley *et al.* 2004). The influence of this effect is very small and complicated to calculate as it depends on the direction of the movement; therefore, it is only considered in special cases.

5.3.2 SATELLITE CLOCK DRIFT

Satellite clock drift is different from relativistic effects. Onboard satellite clocks are independent of one another. And the rates of these rubidium, caesium, and hydrogen maser oscillators are more stable if they are not disturbed by frequent tweaking, so adjustment is kept to a minimum. However, while it is understood that the atomic clocks are very precise, they are not perfect. Minute discrepancies may occur, and these translate into travel time measurement error. These 'very accurate' atomic clocks can also accumulate an error of 1 ns for every 3 h. Let us try to understand this with the example of GPS.

While GPS time, for example, itself is designed to be kept within one microsecond of UTC (excepting leap seconds), satellite clocks are allowed to drift up to a millisecond from GPS time. Three kinds of time are involved here: the first is UTC as per the US Naval Observatory (USNO), the second is GPS system time, and the third is the time determined by each independent GPS satellite.

Their relationship is as follows. The master control station gathers the GPS satellites' data from monitoring stations around the world. After processing, this information is uploaded back to each satellite to become the broadcast ephemeris, broadcast clock correction, and so on. The actual specification for GPS time demands that it be within one microsecond of UTC as determined by USNO, without consideration of leap seconds. Leap seconds are used to keep UTC correlated with the actual rotation of the earth, but they are ignored in GPS time. In GPS time, it is as if no leap seconds have occurred at all in UTC since 00:00:00 h, January 6, 1980. And in practice, GPS time is much closer than the microsecond specification; it is usually within about 25 ns of UTC, except leap seconds. Each independent satellite clock should match exactly with the GPS system time, however, in practice, they are often deviated. Similar time-related problems also occur for GLONASS, Galileo, and BeiDou.

To resolve satellite clock drifts, they are continuously monitored by ground stations and compared with the master control clock systems that are combinations of more than ten very accurate atomic clocks. The errors and drifts of the satellites' clocks are calculated and included in the messages that are transmitted by the satellites. We recall that clock corrections are part of the navigation message sent to the receiver. While computing the distance to the satellites, the GNSS receiver deducts the satellite clock errors from the reported transmit time to calculate the true signal travel time.

The satellite clock error at any time *t* can be computed from the following model (Bossler *et al.* 2002):

$$dt = \alpha_0 + \alpha_1\left(t - t_{oc}\right) + \alpha_2\left(t - t_{oc}\right)^2 \tag{5.1}$$

where α_0, α_1, and α_2 are the clock offset from reference time (e.g., UTC), the clock drift-rate, and half the clock drift acceleration at the reference clock time t_{oc} (time of clock). This provides a good prediction of the satellite clock behaviour.

5.4 ATMOSPHERIC EFFECTS

While travelling from the satellite to the receiver, the signal has to cross various layers of the atmosphere. This crossing induces two effects: the deceleration of the signal and the deviation of its path (Klobuchar 1991; Prolss 2004). The first effect, however, is fundamental and must be eliminated, either by dual-frequency analysis (for ionosphere-related effects) or by appropriate modelling.

5.4.1 IONOSPHERIC DELAY

Another significant error in GNSS positioning is attributable to the atmosphere. The relatively unaffected travel of the GNSS signal through the virtual vacuum of space changes as it passes through the earth's atmosphere. The atmosphere alters the apparent speed and, to a lesser extent, the direction of the signal through both refraction and diffraction.

The ionosphere is the first part of the atmosphere that the GNSS signal encounters. This is a dispersive medium, ionised through the action of the sun's radiation. Dispersive means that the behaviour depends on the frequency of the signal. It extends from about 50 km to 1000 km above the earth's surface. The ionosphere has layers sometimes known as the *mesosphere* and *thermosphere* (Figure 5.1) that are themselves composed of D (50–90 km), E (90–1700 km), and F (170–1000 km) regions (Prolss 2004). Farthest outside of the ionosphere is the area known as the *exosphere*. All of these divisions are based on electron density which decreases as one goes further from the earth. These layers refract the electromagnetic wave coming from the satellite, resulting in an elongated runtime (longer travel time) of the signals (Figure 5.2). Travelling through this part of the atmosphere, the most troublesome effects on the GNSS signal are known as the *group delay* and the *phase delay* (Parkinson and Spilker 1996). They both alter the measured range. The magnitude of these delays is determined by the density and the thickness of various layers of the ionosphere at the moment the signal passes through it. The electron density of the ionosphere changes with the number and dispersion of free electrons released when gaseous molecules are ionised by the sun's ultraviolet radiation. This density is often described as *total electron content* (TEC), a measure of the number of free electrons per cubic meter (Figure 5.1).

The ionospheric delay changes slowly as it goes through a daily cycle. It is usually minimal between midnight and early morning, and maximal around local noon time. During daylight hours in the mid-latitudes, the ionospheric delay may grow to be as much as five times greater than it is at night, but the rate of that growth is rarely more than 8 cm per minute. Furthermore, the time of the year, seasons, and weather also have a high impact on both the height of the ionosphere layer and the density of its ionised particles. It is nearly four times greater in November, when the earth

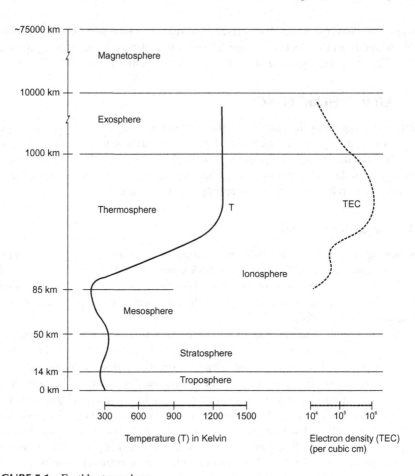

FIGURE 5.1 Earth's atmosphere.

is at its closest approach to the sun, than it is in July when the earth is at its farthest point from the sun. The effect of the ionosphere on the GNSS signal usually reaches its peak in March, about the time of the vernal equinox (the moment when the sun is positioned directly over the earth's equator).

The ionosphere is not homogeneous. It changes from layer to layer within a particular area, and its behaviour in one region of the earth is liable to be unlike its behaviour in another. For example, ionospheric disturbances are generally high in the polar regions. But the highest TEC values and the widest variations in the horizontal gradients occur in the band of about 60° of geomagnetic latitude. That band lies 30° north and 30° south of the earth's magnetic equator.

The severity of the ionospheric effect varies with the amount of time the GNSS signal spends travelling through the layer. A signal originating from a satellite near the observer's horizon must pass through a larger amount of the ionosphere to reach the receiver than does a signal from a satellite near the observer's zenith. The longer

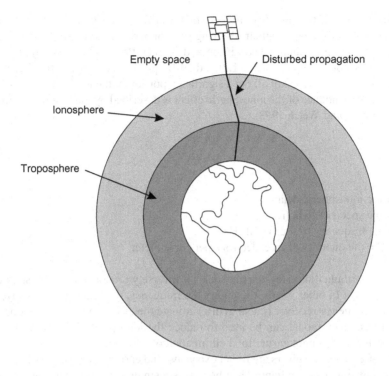

FIGURE 5.2 Influenced propagation of radio waves through the earth's atmosphere.

the signal is in the ionosphere, the greater the ionospheric effect, and the greater the impact of horizontal gradients within the layer.

In summary, the error introduced by the ionosphere may be large when the satellite is near the observer's horizon, the vernal equinox is near, and sunspot activity is at its maximum. It varies with magnetic activity, location, time of day, and even the direction of observation. Fortunately, the ionosphere has a property that can be used to minimise its effect on GNSS signals. It is dispersive, which means that the apparent time delay contributed by the ionosphere depends on the frequency of the signal. One result of this dispersive property is that during the signal's trip through the ionosphere, the codes, the modulations on the carrier wave, are affected differently than the carrier wave itself.

All the modulations (codes) on the carrier wave, ranging code and the navigation code, appear to be slowed; they are affected by the *group delay*. But the carrier wave itself appears to speed up in the ionosphere; it is affected by the *phase delay*. It may seem odd to call an increase in speed a delay, but, governed by the same properties of electron content as the group delay, phase delay just increases negatively. Please note that the algebraic sign of d_{ion} is negative in the carrier phase equation and positive in the pseudorange equation (Chapter 4, equations 4.6 and 4.9, respectively).

Another consequence of the dispersive nature of the ionosphere is that the apparent time delay for a higher frequency carrier wave is less than it is for a lower frequency wave. If the signals of higher and lower frequencies that reach a receiver are analysed with regard to their differing time of arrival, the ionospheric runtime elongation can be calculated (Parkinson and Spilker 1996). Therefore, by tracking two carriers (for example, L1 and L2), a dual-frequency receiver has the facility of modelling and removing not all but a significant portion of the ionospheric bias. The frequency dependence of the ionospheric effect is described by the following expression (Brunner and Welch 1993):

$$v = \frac{40.3}{c \cdot f^2} TEC \qquad (5.2)$$

where

v = the ionospheric delay
c = the speed of light (m/s)
f = the frequency of the signal (Hz)
TEC = the quantity of free electrons per cubic meter

As the formula illustrates, the time delay is inversely proportional to the square of the frequency. In other words, the higher the frequency, the less is the delay. Hence the dual-frequency receiver is able to discriminate the effect on L1 from that on L2. A dual-frequency model can be used to reduce the ionospheric bias. Still, it is far from perfect and cannot ensure total elimination of the effect.

Dual-frequency receivers are rather expensive and earlier two signals (L1 and L2) were not available to civilians. Then how does a single frequency receiver account for atmospheric delay? Single frequency users can partially model the effect of the ionosphere using the *Klobuchar model* (Klobuchar 1987, 1996). As mentioned in Chapter 4, an ionospheric correction is also available to the single frequency receiver in the navigation message, which can be used for atmospheric correction to some extent. Eight parameters for the Klobuchar model are transmitted with the navigation message and are used as the coefficients for two third-order polynomial expansions which are also dependent on the time of day and the geomagnetic latitude of the receiver. These polynomials result in an estimate of the vertical ionospheric delay, which is then combined with an obliquity factor dependent on satellite elevation and produces a delay for the receiver-satellite line of sight (Grewal *et al.* 2001). The final value provides an estimate within 50% of the true delay, with delays ranging from 5 m (night-time) to 30 m (daytime) for low elevation satellites and 3–5 m for high elevation satellites at mid-latitudes.

NOTE

The *mid-latitudes* (also expressed as *midlatitudes*) are the areas on earth between the tropics and the polar regions; approximately 30° to 60° north or south of the equator. The mid-latitudes are an important region in meteorology, having weather patterns which are generally distinct from weather in the tropics and the polar regions.

5.4.2 TROPOSPHERIC DELAY

The troposphere is the lowest layer of the atmosphere, and is directly in contact with the earth's surface (Figure 5.1). Its height varies from 7 to 14 km, depending on the observation point. In fact, it extends from the earth surface to about 7 km over the poles and 14 km over the equator. The space between troposphere and ionosphere is known as the *stratosphere*. In this discussion, stratosphere will be combined with the troposphere, as it has been done in much of the GNSS literature. Therefore, the following discussion of the tropospheric effect will include the layers of the earth's atmosphere up to about 50 km above the surface.

The troposphere and the ionosphere are by no means similar in their effect on the satellite's signal. The troposphere is a non-dispersive medium for frequencies under 30 GHz. In other words, the refraction of a GNSS satellite's signal is not related to its frequency in the troposphere. The troposphere is part of the electrically neutral layer of the earth's atmosphere; that means it is neither ionised nor dispersive. Therefore, all GNSS frequencies are equally refracted (bending of signal). This delay (due to refraction) depends on temperature, pressure, humidity, and elevation of the satellite. Similar to the ionosphere, the density of the troposphere also governs the severity of its effect on the GNSS signal.

Tropospheric effect due to satellite elevation is analogous to atmospheric refraction in astronomic observations; the effect increases as the energy passes through more of the atmosphere. For example, when a satellite is close to the horizon, the delay of the signal caused by the troposphere is maximised. The tropospheric delay of the signal from a satellite at zenith (directly above the receiver) is minimised. The induced effect on pseudorange of troposphere varies from 2 m (if the satellite is at the zenith) to 30 m (for a 5° elevation satellite) (Brunner and Welsch 1993).

Modelling the troposphere is one technique used to reduce the bias in GNSS data processing, and it can be up to 90%–95% effective. However, the remaining 5%–10% is quite difficult to remove. For example, surface measurements of temperature and humidity are not strong indicators of conditions on the path between the receiver and the satellite. But instruments that can provide some idea of the conditions along the line between the satellite and the receiver are somewhat more helpful in modelling the tropospheric effect. Several models have been developed, including the *Saastamoinen model* (Saastamoinen 1973; Wells 1986) and the *Hopfield model* (Hopfield 1969). They perform well at reasonably high elevation angles. The modelling currently used for GNSS is based on Saastamoinen and Hopfield; and such parameters are included at the receiver level. Other models also do exist and can be used for specific purposes, such as the *exponential model*.

The tropospheric delay is mainly affected by the altitude of the receiver and is usually considered a combination dry component/wet component. The dry component which contributes most of the delay (perhaps 80%–90%) is closely correlated to the atmospheric pressure and is more easily estimated than the wet component. The wet component depends on local atmospheric conditions and varies very quickly. It is fortunate that the dry component contributes the larger portion of range error in the troposphere since the high cost of water vapour radiometers and radiosondes generally restricts their use to only the most high-precision GNSS work. Local measurements could help in achieving better accuracy but are rather difficult to implement. Thus, models are usually also used for wet component as with the dry component. Models put forward by Hopfield (1969), Saastamoinen (1973), and Black (1978) are all successful in predicting the dry part delay to approximately 1 cm and the wet part to 5 cm.

The differential technique (refer to Chapter 6) can also be used for receivers close to each other (when the receivers are in equivalent meteorological conditions). Nevertheless, the user should check for very close altitude values of both receivers in order to achieve acceptable differential elimination of the tropospheric delays. There are other practical consequences of the atmospheric biases in respect of differential technique. For example, the character of the atmosphere is never homogeneous and the importance of atmospheric modelling increases as the distance between receivers increases. Consider a signal that is travelling from one satellite to two receivers that are very close together. That signal would be subjected to very similar atmospheric effects, and, therefore, atmospheric bias modelling would be less important to the accuracy of the measurement of the relative distance between them. But a signal travelling from the same satellite to two receivers that are far apart may pass through levels of atmosphere quite different from one another, and atmospheric bias modelling would be more important.

5.5   MULTIPATH SIGNAL

As the name implies *multipath* is the reception of the GNSS signal via multiple paths rather than from a direct line of sight (Figure 5.3). Multipath is the phenomena by which the GNSS satellite signal is reflected by some object or earth surface before being detected by the receiver antenna (Weill 1997). This increases the travel time of the signal, thereby causing errors. The signal can also be reflected off a part of the satellite (for instance, the solar panels) although this is usually ignored as there is nothing that can be done by the user to prevent this. Multipath differs from both the apparent slowing of the signal through the ionosphere and troposphere and the discrepancies caused by clock offsets. The range delay in multipath is the result of the reflection of the GNSS signal (Grewal *et al.* 2001).

While measuring the distance to each satellite, we assume that the satellite signal travels directly from the satellite to the antenna of the receiver. But in addition to the direct signal, there are reflected signals from the ground and the objects near the antenna. These reflected signals reach the antenna through indirect paths and interfere with the direct signal. The compound signal creates an uncertainty about the actual propagation time. The effect of multipath on pseudorange solutions is larger than it is in carrier phase solutions. However, multipath in carrier phase is much harder to mitigate than multipath in pseudoranges.

In the propagation domain, it is usual to distinguish between two configurations when considering the radio link between a transmitter and a receiver (Figure 5.3): the *line of sight* (LOS) configuration where there is a direct geometrical unobstructed path; and the *non line of sight* (NLOS) configuration where no such direct path exists.

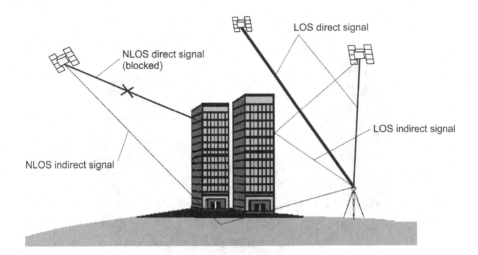

FIGURE 5.3   Interference caused by reflection of the signals.

When considering a NLOS situation in satellite-based navigation, it appears that as the receiver is fundamentally based on the hypothesis that the signal it is receiving is LOS, the resulting positioning will be erroneous. However, even when considering LOS configuration, there may also be a reflected path that is obtained by reflection of the signal on surfaces such as building facades or the earth's surface. The incident signal on the receiving antenna will in fact be the time combination of all the incoming signals, namely the LOS and NLOS both. As the receiver has not been designed to have any knowledge of its environment, it assumes that the signal received is in LOS. The real problem is in fact the time resolution of the receiver. If it is able to discriminate between two signals delayed by an amount of time dt, then the interference effect is reduced to reflected signals that reach the receiving antenna less than dt after the LOS did. Unfortunately, the discrimination capabilities of a code correlation-based receiver are not infinite. A typical value is given by the correlation spacing in *early–late-based correlations* (Samama 2008). This correlation helps to pick the signal received early by the receiver and to reject the latter signal (applicable for same signal that reaches through direct LOS as well as indirect LOS). Remember that the propagation time of direct signals is less than that of indirect signals. This leads to values in the order of 10–15 m of remaining errors for so-called *narrow correlators* (Van Dierendonck *et al.* 1992) and 100–150 m for one chip spaced correlators (known as *wide correlators*). Specific correlation configurations have been developed by almost all the receiver manufacturers to reduce multipath impact on the resulting positioning (Grewal *et al.* 2001; Raquet 2002; Fu *et al.* 2003; Samama 2008).

Although the performance of narrow correlators is better than that of wide correlators, another technique, *edge correlators*, is even better. These techniques are typical signal processing approaches, as compared to 'hardware' solutions consisting in reducing the amount of reflected signal received by the antenna. These latter methods can only be carried out when the environment is known or specific. This is

FIGURE 5.4 A choke-ring antenna.

the case for ground-based reflection that can be mitigated by the *choke-ring antenna* (Figure 5.4) which is designed to reject the waves reflected by the ground. A long-term analysis of the *residuals* can also help in eliminating multipath-induced error; this approach clearly is not possible for real-time applications and is reserved for static surveying applications only.

The case of *phase-based receivers* (carrier-based) is even more complicated because direct and reflected signals are added to the antenna. The resulting signal exhibits a phase that can be quite different from the direct LOS one. Such effects have to be carefully analysed when using phase measurements in multipath environments.

But neither the signal processing techniques nor the choke rings can remove the effect of multipath signals completely. When the GNSS signal is reflected from a distance of ~10 m or less (distance from the antenna to the reflector) that is subject to a short delay, these approaches are not significantly effective. A widely used strategy is the 15° *cutoff* or *mask angle*. This technique calls for the tracking of satellites only after they are more than 15° above the receiver's horizon (this will also reduce atmospheric bias). Careful attention in placing the antenna away from reflective surfaces, such as nearby buildings, water or vehicles, is another way to minimise the occurrence of multipath.

5.6 RECEIVER CLOCK ERROR

Another major error in GNSS positioning can be caused by the receiver clock (receiver's oscillator or frequency standard). Both a receiver's measurement of phase differences and its generation of replica codes depend on the reliability of this internal frequency standard (Misra 1996). Similar to satellite clock errors, any error in the receiver clock causes inaccuracy in distance measurements. However, it is not practical to equip receivers with very accurate atomic clocks (atomic clock weighs more than 20 kg, costs about US $50,000, and requires extensive care of temperature control).

GNSS receivers are usually equipped with *quartz crystal clocks* (often called *crystal oscillators*), which are relatively inexpensive and compact. They have low power requirements and long-life spans. Their reliability ranges from a minimum of about 1 part in 108 to a maximum of about 1 part in 1010; this means a drift of about 1 ns in every 10 s. Even at that, quartz clocks are not as stable as the atomic clocks (a drift of about 1 ns in every 3 h) and are sensitive to temperature changes, shock, and vibration. Some receivers augment their frequency standards with external timing from the augmentation stations.

The most common solution to the synchronisation problem of a receiver's clock with the satellite's clock is the use of four range measurements from four satellites. Making simultaneous measurements to four satellites, we not only compute the three coordinates of our position, but also find the error in our receiver clock with very good accuracy (described earlier in Chapter 3, Section 3.6). This unique solution is valid because the number of unknowns is not greater than the number of observations. The receiver tracks 4 satellites simultaneously; therefore, four equations can be solved simultaneously for every *epoch* of the observation. An epoch in GNSS literature is a very short period of observation time (think of an instant), and is generally just a small part of a longer measurement.

5.7   RECEIVER NOISE

Errors which are due to the measurement processes used within the receiver are typically grouped together as receiver noise (Langley 1997). These are dependent on the design of the antenna, the method used for the analogue to digital conversion, the correlation processes, and the tracking loops and bandwidths. Noise within the pseudorange measurements can be reduced by a factor of 50% by combining with the more precise carrier phase observations. Receiver noise is also an uncorrelated error like the multipath; thus, it cannot be modelled mathematically (Ward 1994). Fortunately, it is a small bias. Generally speaking, the receiver noise error is about 1% of the wavelength of the signal involved in carrier phase observation. In code solutions, the size of the error is related to chip width. For example, the receiver noise error in a C/A code solution is much higher than a P code solution. And in carrier phase solutions, the receiver noise error contributes millimetres to the overall error.

When dealing with typical accuracy values of around a few meters, the question of knowing the exact position of the calculated point does not arise. In such cases it is assumed to be 'the receiver's antenna' that is usually a few centimetres tall or wide. However, this question should be considered when dealing with high-accuracy positioning. The physical point where signals are 'collected' is the *centre of phase* (also called *phase centre*) of the antenna. This point is very difficult to define precisely and depends on the incident angle of incoming signal; thus, this point is not necessarily geometrically fixed. Remember that the phase centre is not the physical centre of the antenna; therefore, perfect calibration of the antenna is required. When centimetre accuracy is needed, this is fundamental.

5.8   ORBITAL/EPHEMERIS ERRORS

Although satellites are positioned in very precise orbits, slight shifts of the orbits are common for several reasons. The orbital motion of a satellite is not only the result of the earth's gravitational attraction; there are several other forces that act on the satellite. The primary disturbing forces are the non-spherical nature of the earth's gravity, the attractions of the sun and the moon, and solar radiation pressure. However, the sun and the moon have a weaker influence on the satellite's orbit. The three types of ephemeris error are radial, tangential, and cross track (Parkinson and Spilker 1996).

The orbits of satellites are monitored continuously from several monitoring stations around the earth and sent to the master control station. This master control station predicts the precise orbit of a satellite for the next few hours based on the monitored orbit and this predicted orbital information is transmitted to the satellites using uploading stations, which, in turn, are transmitted to the receivers by means of navigation message. Although the satellites' positions are constantly monitored, they cannot be watched and updated with correct information every second. Generally, the update rate of ephemeris for GPS is 2 h (30 min for GLONASS, 3 h for Galileo, and 1 h for BeiDou). Hence, slight ephemeris errors can sneak in between monitoring/updating times. However, the errors are usually very low (generally not more than 2 m), but must be accounted for if greater accuracy is needed. There is also the effect arising from the accuracy of the orbit computation procedure itself. Refer to Chapter 6 (Section 6.4.2.2) to learn how ephemeris errors are removed.

5.8.1   ORBITAL CHARACTERISTICS OF SATELLITES

The ephemeris is derived based on the *Keplerian elements* of the orbit for GPS, Galileo, and BeiDou; however, for GLONASS, it is based on geocentric cartesian coordinates and their derivatives (Langley 1991b; Tapley *et al.* 2004; GLONASS ICD 2002). Although it is beyond the scope of this book to discuss them in detail, a brief description of Keplerian elements may help us to understand the orbital characteristics of satellites.

Johannes Kepler (1571–1630) was a German astronomer and mathematician who developed three laws, based on the observations of Tycho Brahe (1546–1601), which describe the motion of the planets around the sun. This planetary motion, or Keplerian motion, is now being used to describe the path of an orbiting satellite around the earth with the only force acting on it being the gravitational attraction of the earth (Seeber 1993). Kepler's three laws are:

1. *The orbit is an ellipse in a plane with the centre of mass of the attracting body at one of its foci.* This indicates that a satellite orbiting the earth will not be at the same distance from the earth at all times, unless the orbit is

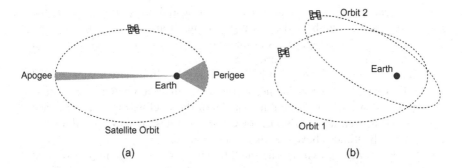

FIGURE 5.5 (a) Speed of the satellite is fastest at apogee and slowest at perigee; (b) two satellites with the same orbital semi major axis but with different eccentricities have same orbital period.

circular. The point of a satellite's orbit where the satellite is closest to the earth is known as the *perigee* and the farthest point is known as the *apogee* (Figure 5.5a).

2. *The satellite's radius vector sweeps out equal area in equal time.* This means that the speed of the satellite is not constant and will be fastest at the apogee and slowest at the perigee.
3. *The ratio between the square of the orbital period and the cube of the semi major axis of the elliptical orbit is the same for all satellites.* This indicates that two satellites with the same orbital semi major axis but with different eccentricities (the flattening of the ellipse), will take the same time to complete one revolution of their orbits (Figure 5.5b).

Just six parameters (the Keplerian elements) can be used to describe a satellite following Keplerian motion with respect to the earth. They are:

• The semi major axis of the elliptical orbit.
• The eccentricity of the elliptical orbit.
• The inclination of the orbital plane to the equatorial plane.
• The right ascension of the ascending node.
• The argument of the perigee.
• The time at which the satellite passes the perigee.

Kepler's laws are for an idealized satellite orbit where the only attracting force is a spherical gravity field. For any satellite orbiting the earth that does not follow this ideal case, its Keplerian position will be affected by the following perturbing forces:

• *Asphericity of the earth*: The earth is not a perfect sphere and has an uneven density distribution. The effect that this has on the earth's gravitational field is represented by spherical harmonic coefficients which are used to compute the disturbing potential at a particular location.

- *Lunisolar gravitational attraction*: Other celestial bodies (in particular the moon and the sun) have their own gravity fields and exert an attraction on the satellite. The magnitudes of these forces are extremely well modelled, and, therefore, their effects are greatly reduced.
- *Atmospheric drag*: The satellite is not travelling in a perfect vacuum and will experience frictional atmospheric drag. This is a function of the atmospheric density at the orbital height and of the satellite's mass and surface area. However, it is negligible for GNSS satellites, as they are orbiting at very high altitude above the earth's atmosphere.
- *Solar radiation pressure*: The satellite will experience the impact of light photons emitted by the sun both directly and indirectly (the *albedo* effect). This is known as solar radiation pressure and will be a function of the satellite's effective area (the surface area normal to the radiation), the surface reflectivity, the luminosity of the sun, and the distance to the sun. For GNSS satellites, this effect cannot be ignored, it is difficult to model, and therefore represents the largest unknown error source.

Most of the aforementioned effects can be modelled, but this is difficult for the solar radiation pressure and albedo perturbing forces. All of the perturbing forces alter the gravitational pull on the satellite and are quantified in terms of their disturbing accelerations. If ignored, these disturbing accelerations will have an effect on the GNSS positioning. The broadcast ephemeris, sent down via the GNSS navigation message, uses the Keplerian elements to represent the idealised orbit and incorporates additional terms to account for the effects of the perturbing forces. However, it is important to realise that although ephemeris data are very accurate, they are never perfect and may cause errors in meters. Furthermore, the ephemeris cannot be updated at every instant, and, as time passes, it becomes outdated until the new ephemeris is uploaded by the control segment. Therefore, the navigation message carries the information about the quality of ephemeris—*issue of data ephemeris*. This information is used by the receiver to decide which satellites should be considered in the calculation process and which should not be.

5.9 OTHER ACCURACY RELATED ISSUES

In Chapter 4, we discussed several error-related issues (Chapter 4 equations 4.6 and 4.9). All of them have been addressed in the preceding sections of this chapter as well. However, there are remaining issues involved in the accuracy of GNSS positioning that have yet to be addressed. In a strict sense, these are not considered errors but rather factors influencing the accuracy matter. The following sections describe these issues briefly.

5.9.1 NUMBER OF SATELLITES

It has been established that we need at least four satellites to derive the 3D positioning and time error—the more satellites 'seen' by the receiver, the better the accuracy.

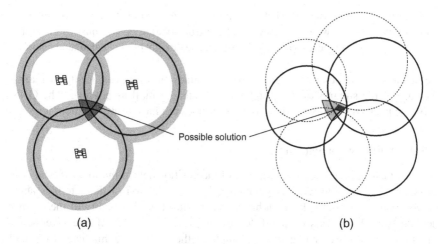

(a)   (b)

FIGURE 5.6   More satellites means more accuracy in positioning.

We also know that GNSS works on the basis of 3D trilateration. For still better accuracy, a technique called *3D multilateration* has been developed. In this is the process, the receiver considers more satellites for multiple trilateration. Ideally, the determined position from every trilateration should overlap exactly. That means all of the spheres of trilateration would intersect at exactly one point, causing there to be only one possible solution, but in reality, the intersection forms an oddly-shaped area. The device could be located within any point in that area, forcing devices to choose from many possibilities.

Figure 5.6a shows such an area created from three satellites. The current location could be any point within the dark-grey area. The signal runtime cannot be determined absolutely, and, therefore, the circles cannot be drawn precisely. The possible positions are therefore marked by the light grey donuts. This situation could be improved by adding more satellites to the fix (Figure 5.6b).

The errors from various satellites are not generally identical, and these multiple errors can compensate for each other. In other words, by using more satellites, the derived position will be more accurate than that derived by using the minimum number of satellites. This can be achieved by least-square method. The basic idea of the least-square method is to minimise the sum of the square of the residuals. If our receiver has recorded pseudorange measurements to six satellites, then how do we combine these and find our position? The solution is least-square. There are many other techniques that could be used, but least-square is usually adopted as it is known as the best estimate or the *BLUE* (Best Linear Unbiased Estimate):

Best   It produces the smallest variances. This means that the precision of a
           least square solution is higher than by any other method. It also pro-
           duces the highest precision for derived quantities, such as a derived
           heading from a GNSS position computation.

Linear   The observations relate to the parameters by some linear model.

Unbiased   On average, the least squares solution is closest to the true solution.
Estimate   Since we don't know what the true solution is, we must estimate it
              using the data available. This can be thought of as a 'calculated guess'.

Least square works by minimising a specific quadratic form, essentially by making
the sum of the squares of the weighted residuals as small as possible. The GNSS
software packages deal with this, so the user does not have to worry about it.

5.9.2 DILUTION OF PRECISION

Another factor influencing the accuracy of the position determination is the 'satellite
geometry' or 'satellite-receiver geometry'. The distribution of the satellites above
an observer's (receiver's) horizon has a direct impact on the quality of the position
derived from them. This is called *dilution of precision* (DOP). If a receiver sees 4
satellites, and all are arranged for example in the northwest, this leads to a 'bad'
geometry. In the worst case, no position determination is possible at all, when all
distance determinations point to the same direction. Even if a position is determined,
the error of the positions may be up to 100–150 m. If, on the other hand, the 4 satellites are well distributed over the whole firmament, the determined position will be
much more accurate. Let us assume that the satellites are positioned in the north,
east, south and west in 90° steps. Distances can then be measured in four different
directions, resulting in a 'good' satellite geometry.

Figure 5.7 shows this for the two-dimensional case. The signal runtime cannot
be determined precisely as explained earlier. The possible positions are therefore

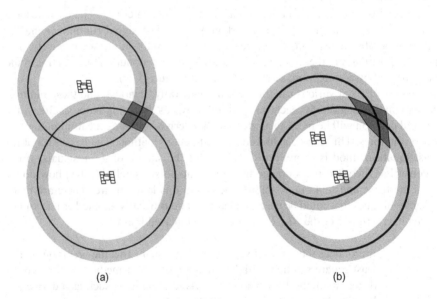

(a)   (b)

FIGURE 5.7   (a) Good geometrical alignment of two satellites; (b) bad geometrical alignment of two satellites.

marked by the grey donuts. The area of intersection of the two donuts in Figure 5.7a is rather small, more or less a quadratic field; the determined position will be more accurate. If the satellites are more or less positioned in one line from the view of the receiver, or if the satellites are very close to each other, the area of intersection of possible positions is considerably larger and elongated (Figure 5.7b). Precision of position is said to be 'diluted' when the area becomes larger.

Most GNSS receivers do not indicate only the number of locked satellites but also their position in the firmament (known as *sky plot*; Figure 5.9. This enables the user to judge whether a relevant satellite is obscured by an obstacle and if changing the position for a couple of meters might improve the accuracy. Many instruments provide a statement of the accuracy of the measured values, mostly based on a combination of different factors.

Traditionally, for navigation applications, the components of the variance-covariance matrix of the parameters are transformed into the *DOP factor* (Langley 1991c). A low DOP factor is good; a high DOP factor is bad. In other words, when the satellites are in the optimal configuration for a reliable GNSS positioning, the DOP is low.

A high DOP factor is considered as a warning that the actual errors in a receiver positioning are liable to be larger than we might expect. But remember, it is not the errors themselves that are directly increased by the DOP factor; it is the uncertainty of the GNSS positioning that is increased by the DOP factor. Here is an approximation of the effect (Langley 1991c):

$$\sigma = DOP.\sigma_{\text{UERE}} \qquad (5.3)$$

where

σ = the uncertainty of the position

DOP = the dilution of precision factor

σ_{UERE} = the uncertainty of the measurements (root-sum-square of the measurement errors)

Now since a receiver's position is derived via a three-dimensional solution, there are several DOP factors that are used to evaluate the uncertainties in the components of a receiver's position. For example, there is *horizontal dilution of precision* (HDOP) and *vertical dilution of precision* (VDOP) where the uncertainty of a solution for positioning has been isolated into its horizontal and vertical components, respectively. When both horizontal and vertical components are combined, the uncertainty of three-dimensional positions is called *position dilution of precision* (PDOP) (Siouris 2004).

$$\text{HDOP} = \sqrt{\sigma_E^2 + \sigma_N^2} = \sqrt{\sigma_x^2 + \sigma_y^2} \qquad (5.4)$$

$$\text{VDOP} = \sqrt{\sigma_h^2} = \sqrt{\sigma_z^2} \qquad (5.5)$$

$$\text{PDOP} = \sqrt{\left(\text{HDOP}^2 + \text{VDOP}^2\right)} = \sqrt{\sigma_E^2 + \sigma_N^2 + \sigma_h^2} = \sqrt{\sigma_x^2 + \sigma_y^2 + \sigma_z^2} \qquad (5.6)$$

There is also *time dilution of precision* (TDOP) that indicates the uncertainty of the receiver clock. The *geometric dilution of precision* (GDOP) is the combination of all the above (HDOP, VDOP, and TDOP).

$$\text{TDOP}=\sqrt{\sigma_T^2} \tag{5.7}$$

$$\text{GDOP}=\sqrt{\left(\text{PDOP}^2 + \text{TDOP}^2\right)} = \sqrt{\sigma_E^2 + \sigma_N^2 + \sigma_h^2 + \sigma_T^2} = \sqrt{\sigma_x^2 + \sigma_y^2 + \sigma_z^2 + \sigma_T^2} \tag{5.8}$$

where $\sigma_E^2, \sigma_N^2, \sigma_h^2$ are variances of the east, north, and height error components
$\sigma_x^2, \sigma_y^2, \sigma_z^2$ are variances of the (x, y, z) coordinate components
σ_T^2 is the variance of the error of the estimated receiver clock offset parameter

Finally, there is *relative dilution of precision* (RDOP) that includes the number of receivers, the number of satellites they can handle, the length of the observing session, as well as the geometry of the satellites' configuration (Sickle 2008).

The larger the volume of the body defined by the lines from the receiver to the satellites, the better the satellite geometry and the lower the DOP (Figure 5.8). An ideal arrangement of four satellites would be one at the zenith (directly above) of the receiver, the others 120° from one another in azimuth near the horizon. With this distribution, the DOP factor would be nearly 1, the lowest possible value. In practice, the lowest DOP factors are generally around 2. For example, if the standard

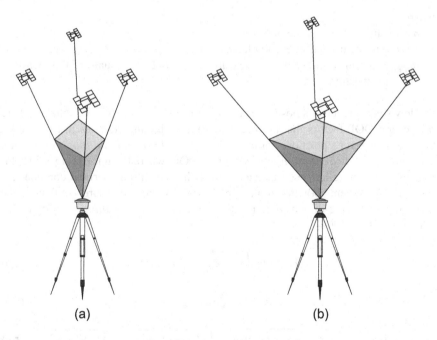

(a)   (b)

FIGURE 5.8   Dilution of precision: poor (a) and good (b).

deviation of a position was ±5 m and the DOP factor was 2, then the actual uncertainty of the position would be 2 times of ±5 m or ±10 m (refer to equation 5.3). Many GNSS receivers allow setting a PDOP mask to guarantee that data will not be logged if the PDOP goes above the set value. A typical PDOP mask is set to 6. As the PDOP increases, the accuracy of the positions generally deteriorates; and as it decreases, they improve; however, it is not always certain.

It has been observed that the DOP value for the vertical component estimate (VDOP) is consistently larger than the DOP for the horizontal positioning problem (HDOP). It is well-known that vertical error in GNSS-based position determination is larger than horizontal error by a factor of 1.5–2.0 (approximate); this is primarily due to satellite geometry. To get the most accurate altitude in the positioning calculation, the receiver should use the satellites that are located at right angles to each other (uniformly distributed), in all the directions, and one directly overhead. However, the receiver will likely choose satellites nearer the horizon in the interest of getting a more accurate horizontal position, since that is what most navigators are interested in. Furthermore, it is not possible to provide the receiver with a uniform distribution of satellites, and it is not possible to receive signals from the satellites below the horizon. As a result, the VDOP is higher than the HDOP. This problem can be overcome by using local augmentation, like pseudolites or repeaters, as it is then possible to improve the 'constellation' thus achieved. The idea is simply to locate such augmentations below the receiver's position. This has an immediate effect on vertical accuracy (Samama 2008).

Another reason for larger VDOP error is because of the strong correlation between the vertical axis and the clock state. Kline (1997) explains this phenomenon by the fact that a clock bias is very similar to moving the antenna along the vertical axis in terms of impact on the pseudorange measurements. A clock bias adds or subtracts the same amount from each pseudorange; while moving the antenna vertically it changes each pseudorange in the same direction although not equally. Hence, there is a strong correlation between the vertical accuracy and time accuracy (Kline 1997). Geodesy also influences vertical error in GNSS measurement, but that will be explained in Chapter 9, Section 9.5.1.

When a DOP factor exceeds a maximum limit in a particular location, indicating an unacceptable level of uncertainty over a period of time, the period is called an *outage*. This expression of uncertainty is useful both in interpreting the measured baselines and planning a GNSS survey. The position of the satellites above an observer's horizon is a critical consideration in planning a GNSS survey, so, most software packages provide various methods of illustrating the satellite configuration for a particular location over a specified period of time. For example, the configuration of the satellites over the entire span of the observation is important; as the satellites move, the DOP changes. Fortunately, the DOP can be worked out in advance, and the DOP factor can be predicted. And since most software allow calculation of the satellite constellation from any given position and time, they can provide the accompanying DOP factors.

A free online software from Trimble for predicting DOP for any location and many constellations (GPS, GLONASS, Galileo, BeiDou, as well as regional systems) is available at www.gnssplanning.com. This type of mission-planning software provides a number of plots to help in planning the GNSS survey or mission. One such plot, known as *sky plot*, represents the user's sky window by a series of concentric circles. Figure 5.9 shows the sky plot for *Kolkata* (formerly *Calcutta*; at 22°33′N 88°20′E) on 1 January 2009 (for GPS constellation, time span 10:00 am to 12:00 noon), produced by the Trimble software. The centre point represents the user's zenith, while the outer circle represents the user's horizon. Intermediate circles represent different elevation angles. The outer circle is also graduated from 0° to 360° to represent the satellite azimuth (direction of movement). Once the user defines his/her approximate position, desired constellation (e.g., GPS, GLONASS) or multi-constellations, desired observation period, the path of each satellite will be shown on the sky plot. This means that relative satellite locations, the satellite azimuth and elevation, can be obtained. The user may also specify a certain elevation angle, normally 10–15°, to be used as *mask* or *cutoff angle*. A mask angle is the angle below which the receiver will not track any satellite, even if the satellite is above the horizon. Other important plots include the satellite availability (or visibility) plot, which shows the total number of visible satellites above the user-specified mask angle (Figure 5.10), and the DOP plot (Figure 5.11).

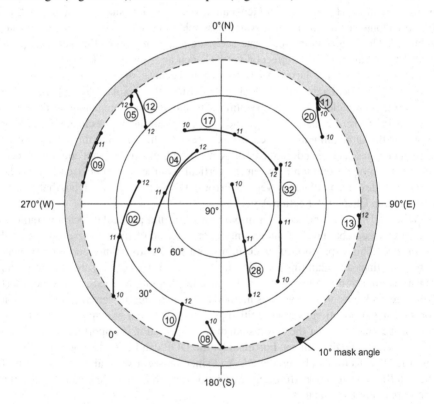

FIGURE 5.9 GPS sky plot for Kolkata on 1 January 2009 (satellite IDs are plotted inside circles and times of day are plotted in italic numerals).

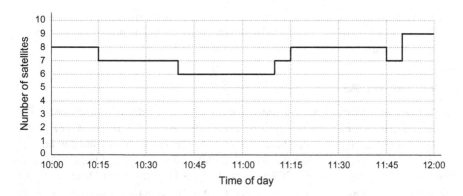

FIGURE 5.10 GPS visibility plot for Kolkata on 1 January 2009.

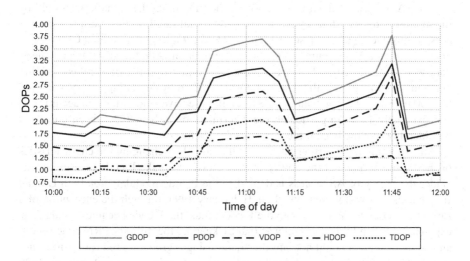

FIGURE 5.11 GPS DOP plot for Kolkata on 1 January 2009.

It is critical that DOP values are used as an indication of when GNSS is likely not to produce good positioning results, and equally should not be used as a measure that describes the quality of positioning that has actually taken place. There are a number of reasons why a DOP value, by measuring geometry, may be misleading if used to describe resultant positions:

- There may be outliers in the pseudorange observations resulting in a poor fix. This would not be picked up by a DOP value.
- Low elevation satellites will improve the geometrical configuration. However, ranges observed to these satellites will have larger atmospheric errors and, again, will lead to poor positioning.

- There was no indication of the level and rate of SA-introduced errors on all observations of GPS system. However, currently this is not an issue, since SA is currently not active (refer to Section 5.9.3).

5.9.3   SA and AS

Selective availability (SA) and *anti-spoofing* (AS) are related to GPS only. The most relevant factor for the inaccuracy of the GPS system, SA, is no longer an issue since it was shut off on 2 May 2000 at 5:05 AM. The US military's intentional degradation of the satellite signal was known as SA and was intended to prevent civilians from using the accurate GPS signals. The US Department of Defence (DOD) thought that providing a high level of precision to the general public was not in the US national interest. Therefore, they had introduced manmade intentional errors to degrade the position accuracy of GPS (Georgiadou and Doucet 1990).

The SA was an artificial falsification of the time in the L1 signal transmitted by the satellite. This was done by introducing additional satellite clock biases. This process of introducing clock biases was called *clock dithering*. For civilian GPS receivers, SA led to a less accurate position determination. Additionally, the ephemeris data were transmitted with lower accuracy, meaning that the transmitted satellite positions did not agree with the actual positions, thereby degrading accuracy to only 50–150 m. Since the deactivation of SA, civilian receivers now enjoy an accuracy of approximately 10 m; in particular, the determination of heights has been improved considerably (having been more or less useless before).

Anti-spoofing (AS) further alters the GPS signal by changing the characteristics of the P code to prevent the ability of the receiver to make P code measurements. Under the policy of AS, access is denied to the P code modulated on both L-band frequencies. AS was implemented on 31 January 1994, through the encryption of a further secret code (W code) onto the P code; thus, the P code became Y code. The explanation behind this decision was that by keeping the military PRN code secret, an enemy of the US could not jam the signal using a ground-based transmitter, nor 'spoof' (gain unauthorized access by posing as an authorized user) military GPS receivers by transmitting a false P code signal from a satellite. The AS in this context refers to a countermeasure against the malicious broadcast of a signal masquerading as the GPS signal. However, several GPS receiver manufacturers have developed proprietary techniques for making dual-frequency measurements even in the presence of AS. Navigation message tells the military receiver when a security system, SA or AS, has been activated by the GPS ground control stations. Thus, they are capable to address these issues, while the civilian receivers are not.

NOTE

Although GNSS spoofing—transmitting spurious signals to fool a receiver—has not yet emerged as a major problem for civilian users, it represents a growing risk. Certainly, the capability exists and, with ever more security-related applications coming online, the motivation for spoofing is increasing, too.

Ledvina *et al.* (2009) discussed a variety of countermeasures and demonstrated one successful method to detect GPS spoofing with a multiple antenna array.

5.10   ESTIMATION OF ERROR BUDGET

The ultimate goal of the GNSS community is to reduce the various errors or accuracy problems on the receiver's positioning. The preceding sections must give indications of the respective importance of each individual contribution in the global error budget. The assumption made here is that the budget on an individual pseudorange is directly linked to the corresponding receiver location (which is not completely true but is, nevertheless, an acceptable hypothesis). Table 5.1 gives an idea of respective values associated with different sources on a single pseudorange measurement (Samama 2008). In addition to the errors described earlier, there is also the need to draw a new line to incorporate all the non-predictable sources at the ground, space, and user segments.

Generally, each bias is expressed as a range itself, each quantity is known as *range error*. A simplified way of examining the positioning accuracy may be achieved through the introduction of the *user equivalent range error* (UERE) or *user range error*. Assuming that the measurement errors are identical and independent, a quantity known as UERE may be defined as the root-sum-square of the individual range errors listed in Table 5.1.

$$\text{UERE}=\sqrt{\sum_{i=1}^{n}\left(\text{UERE}_i\right)^2} \tag{5.9}$$

A prediction of maximum anticipated total UERE (minus ionospheric error) is provided in each satellite's navigation message as the *user range accuracy* (URA). URA is used to express the accuracy of a single range measurement.

The standard deviation of GNSS position is often referred to as *drms* (distance root mean square) value or *sigma* value. The drms can be expressed as a measurement used to describe the accuracy of a fix. Multiplying the UERE by the appropriate DOP value produces the expected precision of positioning at the one-sigma (1σ) level (Equation 5.3). Precision of positioning is also expressed as 2drms or 'two drms' or 'two sigma'. The 2drms does mean twice the drms. In practical terms, a particular 2drms value is the radius of a circle that is expected to contain from 95% to 98% of the positions a receiver collects in one occupation. For example, an absolute positioning accuracy is claimed to be 10 m 2drms, that means, approximately 95% of the horizontal position solutions will be within ±10 m of the actual location.

To obtain the precision at the 2σ level we multiply the 1σ by a factor of two. For example, assuming that the UERE is 8 m for a standalone receiver, and taking a typical value of HDOP as 1.5, then horizontal positional accuracy, for 95% of time, will be (El-Rabbany 2002):

$$2\sigma = \text{UERE} \times \text{HDOP} \times 2 = 8 \times 1.5 \times 2 = 24 \text{ m} \tag{5.10}$$

TABLE 5.1
Estimation of Error Sources on Pseudorange (for GPS Only Solution)

	Precise Positioning Service (range error in meter)	Standard Positioning Service (range error in meter)
Satellite clock drift	3.0	3.0
Receiver noise	0.2	1.5
Ionosphere delay	2.3	4.9–9.8
Troposphere delay	2.0	2.0
Multipath	1.2	2.5
Satellite ephemeris	4.2	4.2
Ground/spatial/user/others (sum)	2.9	2.9

EXERCISES

DESCRIPTIVE QUESTIONS

1. What do you understand by 'satellite clock drift'? Explain relativistic effects on a satellite clock.
2. Explain ionospheric delay for a GNSS signal.
3. Explain tropospheric delay for a GNSS signal.
4. What is multipath? How can the effects of multipath be reduced?
5. Explain 'receiver clock error' and 'receiver noise'.
6. What do you understand by 'ephemeris error'? Why are these errors introduced?
7. Explain the Keplerian elements of orbit.
8. How does the number of satellites influence positioning accuracy in GNSS?
9. Explain DOP in consideration of HDOP, VDOP, GDOP, TDOP, and RDOP.
10. Discuss atmospheric errors in brief.
11. Calculate the 2drms value with the errors listed in Table 5.1, considering the HDOP as 1.6.

SHORT NOTES/DEFINITIONS

Write short notes on the following topics

1. User equivalent range error
2. Sagnac effect
3. Relativistic time dilation
4. Ionosphere
5. Mesosphere
6. Total electron content

7. Group delay
8. Klobuchar model
9. Albedo
10. Best Linear Unbiased Estimate
11. GDOP
12. Selective availability
13. Anti-spoofing

6 Positioning Methods

6.1 INTRODUCTION

At this point of our overall understanding, some issues have to be addressed concerning the positioning methods in relation to positioning accuracy. When dealing with code-correlation functions (pseudorange measurement) that exhibit a few meters of accuracy, it is imperative to accept the errors discussed earlier in Chapter 5. It is also well understood that these errors have to be modelled carefully. However, it is totally different when dealing with carrier phase measurements, where centimetre level accuracy is desired. With this level of accuracy, some additional questions are necessary to be answered.

Concerning the representation, on a map for instance, of such an accurate position, one has to question the real accuracy of the map being used. Could it be possible this map has this level of accuracy or should this positioning only be considered as a sort of relative positioning with respect to the initial point located in a local reference frame? The other questions are relative to the errors to be considered in carrier phase measurement. First, let us recall briefly the major bias sources discussed in Chapter 5. Of course, now one has to deal with integer cycle ambiguities also. Furthermore, for clock biases, one should take into account both the satellite and the receiver clock biases, as the corrections broadcast by the navigation message are not precise enough to allow centimetre accuracy. Thus, the approach of carrier phase measurements, unlike that implemented for code phase measurements, is based on physical methods that allow most of all possible biases to be eliminated. This chapter aims to briefly discuss all of the positioning methods, either by code correlation or by carrier phase measurement. Once we go through this chapter, we shall have a clearer understanding of accuracy issues addressed by different methods.

6.2 CLASSIFICATION OF POSITIONING

Classification of GNSS positioning and navigation is rather overlapping. There are many theoretical combinations in existence. One of the classifications we have discussed earlier is pseudorange measurement (for coarser positioning) and carrier phase measurement (for more precise positioning). For both of the observations, there are two broad classes *kinematic* (or *dynamic*) and *static positioning*. In static GNSS technique, the receivers are motionless during the observation. Kinematic applications imply movement; one or more GNSS receivers are actually in periodic or continuous motion during their observations. A moving GNSS receiver on land, sea, or in air is characteristic of kinematic GNSS. However, it is worth mentioning that, in practice, the word 'kinematic' is generally used to indicate the 'carrier-phase-based kinematic relative technique' (refer to Section 6.4.2). We shall use the word 'dynamic' for code-based solutions.

Static applications, on the other hand, use observations from receivers that are stationary for the duration of their measurement. Most static applications, either code or carrier-based, can afford higher redundancy and a bit more accuracy than dynamic/kinematic method. The majority of GNSS surveying control and geodetic work still relies on carrier-based static technique.

Differential/kinematic and static applications again include two classifications, *point* (or *autonomous* or *absolute* positioning) and *relative* positioning. Point or autonomous positioning uses a single receiver for positioning, whereas relative positioning uses differential technique and more than one receiver. This relative technique can again be classified theoretically as *receiver differential, satellite differential, frequency differential*, and *time differential*. It is necessary to understand that, in practice, more than one of these techniques are often combined to achieve higher accuracy. Thus, another classification appears as *single difference, double difference*, and *triple difference*. In case of relative or differential positioning, the differential corrections can be done in real-time or at a later time after completion of the entire survey. Therefore, relative techniques can again be classified as *real-time* and *post-processed*.

Note that the aforementioned classifications are all the possible theoretical techniques for determining a position. However, in practice some of the techniques are not used due to either their complexities in implementation or one technique is similar in accuracy achievement than another. For example, we do not have any practical solution for carrier-based relative static and autonomous kinematic. As a result, a practice can also be seen to classify the GNSS positioning according to its application (Ghilani and Wolf 2008) as shown in Figure 6.1. Further note that we shall use specific terms for specific solutions for better understanding, e.g., the terms *autonomous* and *relative* are preserved for carrier-based solutions, while point and differential are used for code-based solutions. In the following sections we shall try to understand the basic concepts of widely used techniques. However, in this chapter we shall give more emphasis on code-based solutions; carrier-based solutions will be emphasised in Chapter 11.

6.3   POINT POSITIONING AND AUTONOMOUS POSITIONING

Point positioning is the most basic and common technique for the users who do not require high accuracy but require fast positioning. For example, a car driver does not require centimetre level of accuracy, rather a few meters are enough; further, they cannot wait for a long time for the positioning information—they are moving. Point positioning mode (generally dynamic) is the best solution for such type of applications. Point positioning method is also called *standalone positioning, absolute positioning, single-point positioning*, or the *navigation solution*. This method uses a single receiver for positioning solution, where the receiver handles incoming signals only from the satellites. The only current possibility of single-point positioning is to carry out code correlation because there is no solution for integer cycle ambiguity.

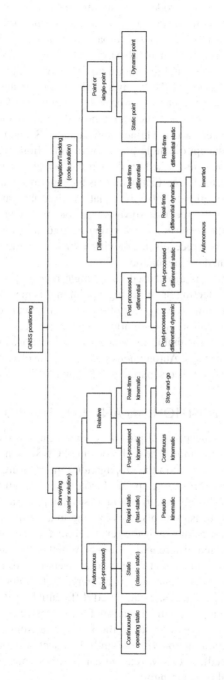

FIGURE 6.1 Classification of GNSS positioning according to applications.

The single-point positioning solution is in one sense the fulfilment of the original idea behind GNSS. It relies on a coded pseudorange measurement and can be used for quick positioning (virtually instantaneous). In this method the positions of the satellites are available from the data in their broadcast ephemeris. The satellite clock offset and the ionospheric correction are also available from the navigation message. Even if all these data are presumed to contain no errors, which they surely do, four unknowns remain: the position of the receiver in three coordinates (x, y, z) and the receiver's clock bias. Three pseudoranges provide enough data for the solution of three coordinates. And the fourth pseudorange provides the solution of the receiver's clock offset. The concept was explained in Chapter 3 (Section 3.6), and we can clearly understand the concept of single-point positioning. Both static and dynamic point positioning techniques are based on the similar principle and in fact there is no difference between them. Theoretically there is enough information in any single epoch of the navigation solution to solve the equations described in Section 3.6. The software installed within the receiver solves these four unknown variables (three coordinates along with the receiver's clock error) continuously for every epoch. In fact, the trajectory (path) of a receiver in a moving vehicle or in the hand of a moving person could be determined by this method.

Autonomous positioning is also based on a single receiver; however, it uses a carrier-based solution. Collecting or surveying a point in this method is not a single-epoch solution; rather it requires a very long observation time and hundreds (or thousands) of epochs to get the positioning solution. Since it requires a long observation time to get a solution, it is unsuitable for kinematic applications and used only for different types of static surveys. Autonomous positioning involves several technical complexities that will be discussed separately in Section 6.5.

6.4   DIFFERENTIAL POSITIONING AND RELATIVE POSITIONING

Point positioning mode is incapable of providing high accuracy because of its code correlation technique. Furthermore, the real problem of GNSS signals lies in the various error sources that are present in the propagation time measurement. In order to reduce the impact of these errors, one approach consists in carrying out differences between measurements. This allows, in certain cases, the removal of quantities that are identical in different measurements, thus leading to a significant improvement. Achieving such differences is known by the generic term *differential*. The term 'relative' is used for carrier-based differential solutions. Whether they are based on code or carrier, these techniques involve at least two receivers for positioning solutions. One receiver remains stationary and acts as a reference; the other(s) roves around to collect or record the coordinates. Code-based differential positioning can attain higher accuracy than point positioning because of the extensive correlation between observations taken to the same satellite at the same time from separate stations. Carrier-based relative technique provides slightly less accuracy than carrier-based absolute positioning. Needless to say, relative carrier phase measurement is more accurate than differential pseudoranging.

Depending on the goal one wants to achieve, there are many differential/relative approaches; however, all of them calculate differences. In the GNSS community the word *differencing* has come to represent several types of simultaneous solutions of combined measurements. The main idea is to implement differential approaches in order to physically remove certain biases by carrying out difference of quantities. The term differential GNSS or *DGNSS* (for example, DGPS—differential GPS), has come into common usage nowadays. Use of this acronym usually indicates a method of relative positioning where coded pseudorange measurements are used rather than carrier phase. The carrier phase measurement techniques are often referred to as simply *relative techniques* in order to differentiate them from code-based differential techniques. Theoretically speaking, both of these use differential techniques. Code-based technique uses mainly receiver position differential, whereas carrier-based technique uses more than one differential techniques (explained later). It is important to understand that although both code-based and carrier-based techniques use more than one receiver, the positioning calculation techniques they use are not same because they use two different observables.

Carrier-based relative technique is the standard mode for GNSS surveying, which essentially measures the *baseline components* (Δx, Δy, Δz) between simultaneously observing receivers. In case of relative positioning, one receiver is at location A, whose absolute coordinates are already known (x_A, y_A, z_A), and another receiver is at point B, whose position is to be determined. Both receivers observe the same satellites, and the observation data collected at both sites are then used to compute position B, but relative to A. As the coordinates of A are known, the absolute position of B will simply be the addition of the coordinates of A to the baseline components (Δx, Δy, Δz). Code-based DGNSS uses a different approach. Receiver at location A, whose absolute coordinates are already known, determines the gross error in signal travel time for each satellite in view. This error is then used by receiver B to correct its position.

There are several comments that can be made for the relative or differential approach:

- Receiver at point B may be stationary or moving, but receiver at point A is essentially stationary.
- There are essentially two strategies for data processing in this technique: (1) data differencing so that the mathematical model contains the baseline components (Δx, Δy, Δz) explicitly, or (2) correction of positioning at B by using the known position of A and the correction data from A. The former is usually implemented in carrier phase processing software. The latter is the normal mode for DGNSS using pseudorange data (in case of DGNSS, there are two possible implementations—see Section 6.4.1).
- If B is stationary, then data can be collected over an 'observation session' (with duration that may range from several minutes to many hours), and a more precise solution is possible.
- The accuracy of the relative position is a function of the distance between the two receivers A and B.

- The relative position can be determined in real-time if the raw measurement data (in case of carrier-based) or data-corrections (in case of code-based) from the reference receiver A are transmitted to receiver B, where they are combined with B's raw measurement data before processing. In the post-processed application (non-real-time), the raw measurement data or the data-corrections are recorded at A, and raw measurements are recorded at B; finally, the data of receiver B is compared with receiver A at a later time for necessary processing and correction of raw data recorded at B.

However, from the preceding discussion, the use of two receivers, simultaneously tracking the same satellite, is an effective means of overcoming the effect of spatially correlated biases.

6.4.1 CODE-BASED DIFFERENTIAL TECHNIQUE

Code-based differential measurement, the so-called DGNSS, is based on differencing technique (Kaplan 1996). This method can yield efficient measurements to a couple of meters (1–5 m) in moving applications and even better in stationary situations sometimes. Same principle is applicable for moving and stationary applications; however, the achieved accuracy varies.

This involves the cooperation of two receivers, one in a known location (having precise positioning information, or by seconding the average position placing the receiver for a long time) and another that roams around making position measurements (Figure 6.2). The receiver in the known location is called *base receiver, stationary receiver*, or *reference receiver* and the other receiver with unknown location is called *roving receiver* or *rover*. The stationary receiver is the key and this ties all the satellite measurements into a solid local reference. If two receivers are fairly close to each other, let us say within a few hundred kilometres, the signals that reach both of them will have travelled through virtually the same slice of atmosphere and hence, have virtually the same errors. This is the basic idea underlying this technique.

The idea is simple: placing the stationary receiver on a point that has been very accurately surveyed and allowing it to remain there. This receiver along with its position and additional facilities (such as transmission) is known as a *reference station*. This base receiver receives the same GNSS signals as the roving receiver, but instead of working like a normal GNSS receiver, it attacks the trilateration equations backwards. Instead of using time measurements to calculate its position, it uses its known position to calculate estimated propagation time. This estimated propagation time is the time which a GNSS signal should take to come from the satellite to the receiver. Simultaneously, it also measures the actual propagation time (the time which has been taken by the signal to come from the satellite to the receiver). This actual propagation time is different than the estimated time due to atmospheric interferences, satellite clock biases, satellite ephemeris, etc.

Now the reference station compares these two time measurements. The difference of these two times can be converted into pseudorange error by multiplying the time

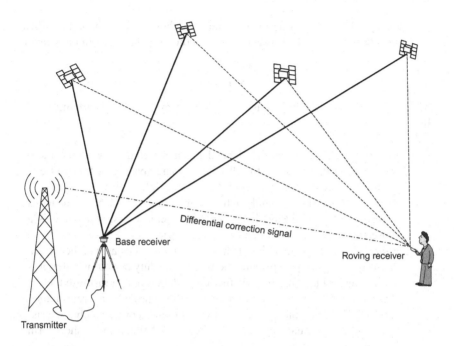

FIGURE 6.2   Code-based differential positioning.

difference with the speed of light, which is an 'error correction' factor for a given satellite. The reference station then transmits this error correction information to the rover so that it can use it to correct its measurements. Since the reference station has no way of knowing which of the many available satellites a roving receiver might be using to calculate its position, the reference station quickly runs through all the visible satellites and computes errors for each of them. Then it encodes this information into a standard format and transmits it to the roving receivers via communication links (Langley 1993b). The roving receivers obtain the complete list of errors and apply the corrections for the particular satellites being used. Corrections can be transmitted over FM radio frequencies, by satellite, or by tower transmitters. This technique is known as *real-time DGNSS*, as the error corrections are executed in real-time.

The distance between the base receiver and rover is called *baseline*. When the baseline is small, i.e., when the receivers are very close to each other, the range errors for the two receivers are nearly identical; therefore, we could use the range errors calculated by the base to correct for the rover position. As the baseline gets longer, the correlation between the range errors becomes weaker. In other words, some residual errors might occur in the computed position of the rover that depends on its proximity to base.

Finally, it is worth mentioning that the code-based differential method cannot handle the errors individually; rather the error calculation output is a 'common bias error vector', which will be removed from the resulting navigation solution of the rover (Langley 1993b). For code measurements, although other theoretical differential approaches are possible, this is the only method that has been implemented.

NOTE

The following considerations must be addressed by DGNSS communication links:

- *Coverage*: This is generally dependent on the frequency of the radio transmission that is used, the distribution and spacing of transmitters, the transmission power, susceptibility to fade, interference, etc.
- *Type of Service*: For example, whether the real-time DGNSS service is a 'restricted' one and available only to selected users, whether it is a subscriber service, or an open broadcast service.
- *Functionality*: This includes such link characteristics as whether it is a one-way or two-way communications link, the duty period (whether it is continuous or intermittent), whether other data is also transmitted, etc.
- *Reliability*: Does the communications link provide a 'reasonable' service? For example, what are the temporal coverage characteristics? Is there gradual degradation of the link? What about short-term interruptions?
- *Integrity*: This is an important consideration for critical applications, hence any errors in transmitted messages need to be detected with a high probability, and users alerted accordingly.
- *Cost*: This includes the capital as well as ongoing expenses, for both the DGNSS service provider as well as users.
- *Data rate*: In general, the faster the data rate, the higher the update rate for range corrections, and hence better positioning accuracy. Typically, a set of correction messages every few seconds is acceptable.
- *Latency*: Refers to the time lag between computation of correction messages and the reception of message at the roving receiver. Obviously, this should be kept as short as possible, and typically a latency of less than 5 sec is suggested.

6.4.1.1 Position Domain and Measurement Domain Differential Strategies

There are two possible strategic implementations of DGNSS. They are (1) *position domain* and (2) *measurement domain* differential strategies (Kaplan 1996). DGNSS positioning can be accomplished by the continuous transmission of the coordinate solution from the base station to the roving receiver. This *block shift* technique, also called *position domain differential strategy*, is the easiest to implement (although it does have certain serious limitations). In this case, the base receiver is located at a known point. It is necessary to compare the known position of the base receiver; instantaneously compute the position of the base receiver to generate corrections (δx, δy, δz); and then transmit the corrections to the roving receiver for immediate correction of 'raw' point coordinates at the rover end. In this case, it is important that both the rover and base receivers use the same satellites to generate their point solutions, otherwise severe errors can result, possibly worse than those of the (uncorrected) point positioning. This is a significant limitation, as it is rarely the case that the same satellites are simultaneously visible if the two receivers are a long distance apart or when the roving receiver is being operated in an urban canyon (urban canyons cause significant occlusion of the satellite signals).

A popular real-time DGNSS strategy is the method of *range correction* technique. Rather than making corrections to the coordinates, correction factors are calculated for pseudoranges; this is also known as the *measurement domain differential strategy*. This is achieved by a process similar in many respects to that of the block shift method. The base station is at a known position: using the known position, compute true geometric range; generate corrections to all pseudorange data by comparing 'true' to 'observed' range; and, finally, transmit correction data to the roving receiver for correction of ranges before positioning solution is carried out. This technique is far more flexible because the correction is made to the pseudoranges and hence the rover can use any combination of corrected ranges to obtain a solution.

6.4.1.2   Real-Time and Post-Processed Techniques

The DGNSS technique described earlier (in Section 6.4.1) is a real-time technique where the error-corrections are transmitted almost instantaneously from the base to the rover for immediate final positioning calculation. In some cases, real-time transmission from the stationary receiver to the rover is not required. For these cases, we could save money using the *post-processed* technique (Seeber 1993). The technology is simple. All applications are not created or used for the same purpose. Some do not need a real-time radio link because high positioning accuracy is not necessary immediately. For instance, if we want to make a map, the roving receiver just needs to record all of its measured positions and the exact time at which it made each measurement. Later, this data can be merged with corrections recorded at a reference receiver for a final clean-up (correction) of the data. Therefore, we do not require the radio link as in real-time systems. However, both of these techniques have their relative merits and demerits as furnished in Table 6.1.

An important consideration for post-processed schemes is data file formats. There are two file format options available for data to be post-processed. The first are the receiver-specific data formats, only useful if the same make of receiver is operated as both the base and roving receiver; and the universally recognised standard format known as the RINEX (Receiver Independent EXchange) format (refer to Chapter 8, Section 8.5). Privately owned and public base stations generally make data available to users in the RINEX format.

6.4.1.3   Autonomous and Inverted Techniques

Real-time DGNSS has two different implementations—*autonomous* and *inverted*. The former provides precise positioning to the rover for its own use. The latter is a means by which a roving receiver's location can be monitored at some central facility. The inverted system is useful in certain tracking applications. Let us say there

TABLE 6.1

Relative Merits and Demerits of Real-time and Post-processed DGNSS Implementations

Real-time DGNSS Implementation	Post-processed DGNSS Implementation
Advantages:	*Limitations:*
• No data archiving required and no post-processing is necessary.	• The operation requires coordination of data captured both at rover and base receivers.
• The roving receiver equipment can therefore be small and lightweight.	• The roving receiver equipment can therefore be larger and heavyweight.
• Transmission of the industry-standard DGNSS correction message format means that real-time DGNSS capability is built into all receivers at low (additional) cost.	• Cannot be used for real-time positioning, as in the case of most navigation applications.
	• Post-processing software is likely to be instrument-specific.
• When DGNSS is in broadcast (or 'open service') mode, all roving receivers operate independently.	• Roving receivers cannot operate independently; they are required to process their data with reference to the base receiver for necessary corrections.
• Can take advantage of a communication link to transmit other (non-positional) data to and from the base facility.	
Limitations:	*Advantages:*
• The requirements of a communication channel lead to greater infrastructure complexity and associated problems, such as signal coverage, fade, etc.	• No additional instrumentation (such as communications equipment) is required.
• This ultimately increases the establishment and maintenance cost.	• Lower establishment and maintenance cost.
• Real-time tracking has capacity limitations.	• Quality assurance measures can be applied.
• Real-time tracking is likely to be a restricted system available to authorised users.	
• Quality assurance within a real-time system is more difficult than in the case of post-processing.	

are a fleet of buses and we would like to pinpoint them on street maps with very high accuracy, but we do not want to buy expensive differential-ready receivers for every bus. With an inverted system, the buses would be equipped with standard receivers and a transmitter, which would transmit the standard receiver-positions back to the tracking office via the internet. Then at the tracking office the corrections would be applied to the received positions. It requires a computer to do the calculations at the central facility. It provides us a fleet of very accurate positions in real-time for the cost of one reference station, a computer, and a lot of standard receivers. It is important to realise that the inverted system can also be implemented without the differential corrections, if we do not require a high accuracy by means of differential correction. The procedure is same; however, in the case of normal inverted GNSS, the differential corrections are not made.

Real-time autonomous differential technique provides corrections to the rover from the reference station so that the rover can correct its measurements and determine a more accurate position. In this case the positioning is determined at the rover end. The use of range corrections (Section 6.4.1.1) is the preferred mode for real-time autonomous navigation. The range corrections are calculated at the base receiver and then transmitted to one or more rovers. The base and rovers operate independently of each other. We can use an unlimited number of rovers to a base. In the inverted real-time case, only a limited number of rovers can be tracked by the base station, due to information processing capacity limits. Nevertheless, such a system is ideally suited for such applications as the monitoring of precious goods while in transit. The system can be extended further by combining the tracking and navigation modes, through a two-way communications link, to provide the ideal all-round navigation/tracking system.

NOTE

In the early days of GNSS, reference stations were established by private companies who had big projects demanding high accuracy—groups like surveyors or oil drilling organisations. And that is still a very common approach. We have to buy a reference receiver and set up a communication link with our differential ready rovers. But now there are enough public agencies who are transmitting corrections that we might be able to get it for free. The United States Coast Guard and other international agencies are establishing reference stations all over the world, especially around popular harbours and waterways. These stations often transmit beacons on the radio that are already in place for radio direction finding (usually in the 300 kHz range). Many new receivers are being designed to accept corrections, and some are even equipped with built-in radio receivers. This is desirable for applications in which some actions need to be performed in the field, such as placing markers (stakeout) or moving objects to exact locations.

6.4.2 CARRIER-BASED RELATIVE TECHNIQUE

In the preceding section we have described the code-based differential technique that can give us an accuracy of 1–5 m. However, in many applications we need higher accuracy. For example, preparing a large-scale map requires centimetre level accuracy. For centimetre-accuracy positioning, the so-called *phase measurement technique* is required. Obviously, these two differential approaches are quite different from each other, other than the fact that they both calculate differences.

For the carrier phase measurements, the concept is based on three successive differences—*single difference, double difference, triple difference*, which respectively implement (i) a receiver differential technique where two receivers are used to make measurements from a single satellite, (ii) a satellite differential technique where two satellites, at the same instant, in combination with receiver differential are used, and (iii) a time differential method where a given satellite is followed while moving in combination with receiver differential and satellite differential. The frequency differential method must also be implemented in order to reduce ionosphere propagation error effects and to allow centimetre accuracy. Thus, the carrier phase measurement technique uses four differential methods together (Samama 2008).

6.4.2.1 Single Difference

Single difference, also known as a *between-receivers difference* or *simple difference*, can refer to the difference in the simultaneous carrier phase measurements from one GNSS satellite as measured by two different receivers (Figure 6.3). However, do not confuse these two receivers with the code-based differential method described earlier. In this case, we may initially think both of the receivers are at unknown locations. These two receivers are placed closer to each other compared to the code-based

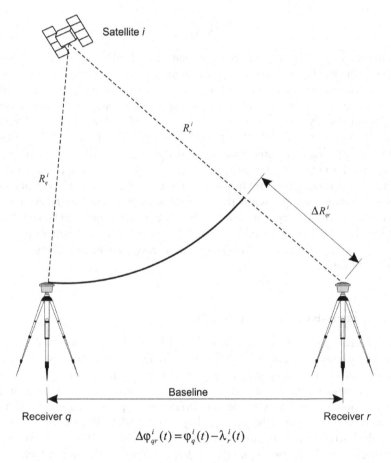

$$\Delta\varphi_{qr}^{i}(t) = \varphi_{q}^{i}(t) - \lambda_{r}^{i}(t)$$

FIGURE 6.3 Single difference measurement technique.

differential method. The baseline (distance between two receivers) is negligible if we compare it with the distance from the satellite to the receiver. Unlike the code-based differential technique, here, error calculation output is not a 'common bias error vector'; rather each individual error is addressed mathematically.

The single difference, obtained by using two receivers and a single satellite, allows the clock bias of the satellite to be completely removed by achieving the difference, at any given instant (epoch) t, of the pseudo-distances resulting from the signal processing of the two receivers. This is possible because the two receivers are not 'too far' from each other and given the fact that the same satellite is being considered for both the receivers. Since both receivers are observing the same satellite at the same time the satellite clock bias is identical for them and can be removed. The atmospheric biases and the orbital errors recorded by the two receivers in this solution are nearly identical, so they too can also be virtually eliminated (Figure 6.3).

To describe the concept, let us start with the general carrier phase observable equation (described in Chapter 4, equation 4.9). We shall have two equations for two receivers (q and r) and one satellite i at an epoch t (El-Rabbany 2002; Samama 2008; Sickle 2008).

$$\phi_q^i(t) = R_q^i(t) + d_p^{\ i} + \text{c.}\left(dt^i - dT_q\right) + \lambda N_q^i - \text{c.}d_{\text{ion}}^i + \text{c.}d_{\text{trop}}^i + \varepsilon_{\text{mp}_q} + \varepsilon_{\text{p}_q} \qquad (6.1)$$

$$\phi_r^i(t) = R_r^i(t) + d_p^{\ i} + \text{c.}\left(dt^i - dT_r\right) + \lambda N_r^i - \text{c.}d_{\text{ion}}^i + \text{c.}d_{\text{trop}}^i + \varepsilon_{\text{mp}_r} + \varepsilon_{\text{p}_r} \qquad (6.2)$$

where
ϕ = carrier phase observation in cycles
R = the true range
d_p = satellite orbital (ephemeris) errors
c = the speed of light through a vacuum
dt = the satellite clock offset from GNSS time
dT = the receiver clock offset from GNSS time
λ = the carrier wavelength
N = the integer ambiguity in cycles
d_{ion} = ionospheric delay
d_{trop} = tropospheric delay
ε_{mp} = multipath errors
ε_{p} = receiver noise

Therefore, the single difference is

$$\Delta\phi_{qr}^i(t) = \phi_q^i(t) - \phi_r^i(t) = \Delta R_{qr}^i - \text{c.}\Delta dT_{qr} + \lambda\Delta N_{qr}^i + \Delta\varepsilon_{\text{mp}_{qr}} + \Delta\varepsilon_{\text{p}_{qr}} \qquad (6.3)$$

where Δ represents the difference between receivers.

In Equation 6.3, the satellite orbital error (d_p) and satellite clock bias (dt) have been eliminated. Atmospheric delays (d_{ion} and d_{trop}) can also be eliminated by placing the receivers very close to each other; for both of the receivers their values should be same and the differences would be 0.

Unfortunately, there are still four factors in the carrier beat phase observable that are not eliminated by single differencing—the difference between the integer cycle ambiguities at each receiver, the difference between the receiver clock errors, multipath error, and the receiver noise.

6.4.2.2 Double Difference

The double difference is the difference of two single differences for two satellites (El-Rabbany 2002; Samama 2008; Sickle 2008). It allows the clock bias of the receivers (q and r) to be removed because the differences are carried out at the same instant (t) for the two satellites (i and j). It involves the addition of what might be called another kind of single difference, also known as a between-satellites difference—that is two satellites for a single receiver. Therefore, double difference is a combination of between-receivers difference and between-satellites difference (Figure 6.4). Thus, mathematically the double difference is

$$\Delta\nabla\phi_{qr}^{ij}\left(t\right) = \Delta\phi_{qr}^{i}\left(t\right) - \Delta\phi_{qr}^{j}\left(t\right) = \Delta\nabla R_{qr}^{ij} + \lambda\Delta\nabla N_{qr}^{ij} + \Delta\nabla\varepsilon_{\mathrm{mp}_{qr}}^{ij} + \Delta\nabla\varepsilon_{\mathrm{p}_{qr}}^{ij} \quad (6.4)$$

where Δ represents difference between receivers, and ∇ represents difference between satellites.

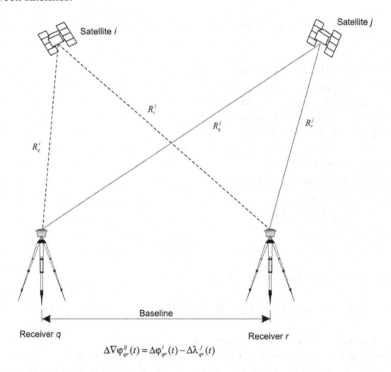

$$\Delta\nabla\varphi_{qr}^{ij}(t) = \Delta\varphi_{qr}^{i}(t) - \Delta\lambda_{qr}^{j}(t)$$

FIGURE 6.4 Double difference measurement technique.

In Equation 6.4, the receiver clock bias (dT) has been eliminated; since a receiver takes measurements for two satellites in the same instant the receiver clock error should be the same for both satellites and the difference would be 0. Therefore, for all practical purposes, the double difference does not contain receiver clock error, satellite clock error, ephemeris error, and atmospheric error. However, the disadvantage with double differencing is that the receiver noise is increased by a factor of 2 with each difference operation. Furthermore, the integer (or cycle) ambiguity, N, is still unsolved.

6.4.2.3 Triple Difference

The triple difference (Figure 6.5) allows the removal of the integer ambiguities common to the preceding measurements, where only the changing cycles are accounted for (El-Rabbany 2002; Samama 2008; Sickle 2008). This is achievable by considering the difference of two double differences, at two successive measurement instants (at two successive epochs, t_1 and t_2). That means triple difference uses between-receivers difference, between-satellites difference, and between-epochs difference; and it is also known as *receiver-satellite-time triple difference*. Mathematically the triple difference is

$$\Delta\nabla\phi_{qr}^{ij}\left(t_1 t_2\right) = \Delta\nabla\phi_{qr}^{ij}\left(t_1\right) - \Delta\nabla\phi_{qr}^{ij}\left(t_2\right) \qquad (6.5)$$

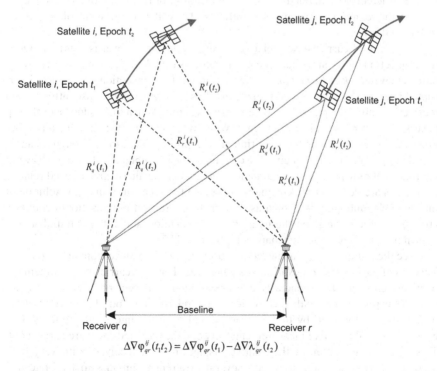

$$\Delta\nabla\varphi_{qr}^{ij}(t_1 t_2) = \Delta\nabla\varphi_{qr}^{ij}(t_1) - \Delta\nabla\lambda_{qr}^{ij}(t_2)$$

FIGURE 6.5 Triple difference measurement technique.

In the triple difference two receivers (q and r) observe the same two satellites (i and j) during two consecutive epochs (t_1 and t_2). This solution can be used to resolve the integer cycle ambiguity, N because, if all is as it should be, then N is constant over the two observed epochs. Therefore, the triple difference makes the detection and elimination of cycle slips relatively easy. Actually, a triple difference is not sufficiently accurate. It is used to resolve the integer cycle ambiguity. Once the cycle ambiguity is determined it can be used with the double difference solution to calculate the actual carrier phase measurement.

However, still there are two errors, namely, multipath error and receiver noise. These errors are local to each receiver and cannot be cancelled by the aforementioned simple differential method. Several techniques have been proposed to remove these errors by several researchers (see Raquet 2002; Fu *et al.* 2003; Samama 2008). These are some approaches based on mathematical models. It is beyond the scope of this book to explain them. Other than the model-based approaches, there are hardware approaches as well, e.g., introducing choke ring antenna for mitigating multipath error as discussed earlier (Chapter 5, Section 5.5).

Furthermore, as one knows that the propagation errors are at least two orders of magnitude higher than the accuracy being searched for (around 1 m where one is seeking centimetre accuracy), there is the need for a method to eliminate these effects thoroughly. This can only be achieved by a local differential method, that is the distance between the two receivers being close in order to avoid there being different meteorological conditions at both locations or by the use of dual-frequency method for removing the ionosphere effects. Thus, the requirement of frequency differential arises.

In the case of carrier-based relative survey, cycle slip at rover is common. Once the connection between the satellite and receiver is established, the receiver is said to have achieved *lock* or *locked on* (to the satellite). If the connection is somehow interrupted later, the receiver is said to have *lost lock*. A *cycle slip* is a discontinuity in a receiver's continuous phase lock on a satellite's signal. When lock is lost, a cycle slip occurs. A power loss, an obstruction, a very low signal-to-noise ratio, or any other event that breaks the receiver's continuous reception of the satellite's signal causes a cycle slip. That is, the receiver loses integer cycle ambiguity. There are several methods that may be used to regain a lost integer phase value. Carrier-based relative techniques are designed for kinematic survey. In the kinematic survey, resolution of integer cycle ambiguity is known as *initialisation*. Several initialisation techniques are used in practice, e.g., on-the-fly ambiguity resolution, static survey initialisation, known baseline, etc. (refer to Chapter 11, Section 11.7).

Once the initialisation is achieved, the user may wish to move from point to point. This is called *kinematic GNSS positioning*, based on the carrier phase kinematic relative technique. This generally consists of having a fixed receiver and a rover to achieve positioning with the carrier phase relative method. The carrier phase of the rover is followed by the electronics, and once the integer ambiguities have been solved, only relative measurements between the base and rover are carried out. This gives the evolution of the relative displacement of receivers with very high accuracy. However, again, there are several different techniques to facilitate this

(Hoffmann-Wellenhof *et al.* 1994); these techniques are *pseudokinematic*, *continuous kinematic*, *semikinematic*, or *stop-and-go* and *real-time kinematic*. Refer to Chapter 11 for further discussion.

Real-Time Kinematic (RTK) is the most popular among the kinematic solutions. This is a special implementation of the kinematic mode, where real-time moving resolution techniques are employed to get the corrected positions on the move. RTK is quite efficient as long as at least five or six satellites are available. In practice, the RTK system uses a single stationary receiver and one or more roving receivers (Figure 6.6). The stationary receiver rebroadcasts the phase of the carrier that it measured, and the mobile units compare their own phase measurements with the ones received from the base station to generate the double-differences, resolve the ambiguities, and perform the position calculations. This allows the units to calculate their relative position with respect to the base with an accuracy of a few millimetres. Then this relative position is converted into absolute position with reference to the position of the base receiver. Therefore, the absolute position of the roving receiver is accurate only to the same accuracy as the position of the base receiver.

In most cases, RTK considers all of the visible satellites in its computation, statistically determining the most likely solution for the position of the roving receiver. A search box is determined within which the position must lie. All possible solutions are then assessed and one is selected. The process is computationally demanding. The typical nominal accuracy for these dual-frequency systems

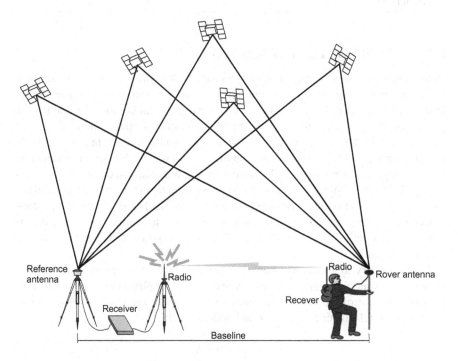

FIGURE 6.6 RTK positioning.

is 1 cm ± 2 parts-per-million (ppm) horizontally and 2 cm ± 2 ppm vertically. Although the technological complexity and costly receiver limits the use of RTK technique in terms of general navigation, it is perfectly suited to applications like high-accuracy mapping survey.

NOTE

GLONASS ambiguity resolution process is more complicated compared to GPS or Galileo or BeiDou because of its FDMA signal structure. Wavelengths of GLONASS signals are not common for all satellites. GLONASS (FDMA) double-difference carrier phase observations consist of a single-difference reference ambiguity term in addition to the usual double-difference ambiguity. Furthermore, GLONASS observations may be affected by inter-frequency biases. These two issues are not normally associated with CDMA signals and will hinder or even prevent reliable ambiguity resolution if not handled correctly. Resolving these issues in practice is challenging, especially in real-time and when different receiver brands are involved. Nevertheless, these issues are manageable, and once they have been addressed, the addition of GLONASS observations in a high-precision GNSS solution can certainly improve positioning performance compared to GPS alone. See Takac and Petovello (2009) for detailed discussion on GLONASS (FDMA) ambiguity resolution.

6.5 AUTONOMOUS POSITIONING

Autonomous positioning uses a single receiver to determine its location, and it is based on carrier solution. All autonomous solutions are static, since they need very long observation periods. Autonomous positioning techniques are classified based on the length of observation period—rapid static, classic static, and continuously operating static. These techniques will be detailed and compared in Chapter 11. In this chapter we shall discuss the solution of integer cycle ambiguity in autonomous static survey. Autonomous static (hereinafter referred as *static* in this chapter) carrier phase GNSS survey procedures allow various systematic errors to be resolved when high accuracy positioning is required (Hoffmann-Wellenhof *et al.* 1994). All static procedures record data over an extended period of predetermined time; and during this time the satellite geometry changes. This is the classical GNSS survey method for higher accuracy.

Now let us try to understand how the carrier phase measurement is achieved in static mode. Multiple components are involved in carrier phase observable. From the moment a receiver locks onto a particular satellite, there are actually three components to the total carrier phase observable (Sickle 2008).

$$\varphi = \alpha + \beta + N \tag{6.6}$$

where

φ = total carrier phase
α = fractional initial cycle
β = observed cycle count
N = cycle count at lock on (integer cycle ambiguity)

Coming back to Equation 6.6, first is the fractional initial phase, α, which occurs at the receiver at the first instant of the lock-on. The receiver starts tracking the incoming phase from the satellite. It does not yet know how to achieve a perfect synchronisation. Since the receiver lacks this knowledge, it grabs onto the satellite's signal at some fractional part of a phase. It is interesting to note that this fractional part does not change for the duration of the observation, and so it is called the *fractional initial phase*. Second one, β, is the integer number of full cycles of phase that occur from the moment of the lock to the end of the observation. It is actually the observed cycle count. This element is the receiver's consecutive counting of the full phase cycles, 1, 2, 3, 4 . . ., between the receiver and the satellite as the satellite flies over. Of the three terms, this is the only number that changes (if the observation proceeds correctly).

Third is the integer cycle ambiguity N. It represents the number of full carrier phase cycles between the receiver and the satellite at the first instant of the receiver's lock-on. It is called the *cycle count* at lock-on. The N does not change (for a given satellite) from the moment of the lock onward, unless that lock is lost. In other words, the total carrier phase observable consists of two values that do not change during the observation, the fractional phase (α), and the integer cycle ambiguity (N). Only the observed cycle count (β) changes, unless there is a cycle slip. However, in static survey cycle slip is extremely rare; rather in practice, static survey is a solution for reinitialisation in case of a cycle slip in kinematic survey.

However, in order to use the GNSS carrier phase observables for positioning purposes, the integer ambiguity (N) must be resolved turning the raw phase into a distance measurement. This is not an easy task. With one epoch of data, we expect solutions for the receiver coordinates, the clock offset, and the integer ambiguities to all of the satellites under observation. This is impossible using conventional estimation techniques since we do not have enough observables. Collecting a few epochs of data does not help matters either; although we now have enough equations (one for each epoch), and the problem is singular, or insoluble, since the satellites are still in the same relative position (roughly) with respect to the receiver. Only a significant change in satellite–receiver geometry can determine the integer ambiguity values. This requires collection of data for a long duration of time. All the data are then downloaded and processed in the post-processing software to determine coordinates of one point. It uses combination of several statistical methods to solve the ambiguity and calculating coordinates of receiver from hundreds or thousands of measurements. More discussion on integer ambiguity is laid out in Chapter 8, Section 8.3.3.

6.6   DIFFERENTIAL AND RELATIVE CORRECTION SOURCES

In the relative or differential technique, whether it is based on code or carrier, it requires a reference or base receiver. For this receiver, the use of observed carrier phase data is a mandatory operational technique that is employed to obtain accurate results from GNSS measurements. Unfortunately, the role of the reference receiver is simply to mitigate errors affecting the rover. This forces users to purchase reference receivers which are not productive in the sense that they do not occupy marks of interest. Also, the cost of a carrier-based reference receiver is high. To overcome this problem, reference station networks have been established by many countries or states and even commercial companies. Currently there are more than 6500 such stations around the world (https://www.trimble.com/trs/findtrs.asp). Data from these reference stations are available for post-processing applications and/or real-time applications. The reference network data can be obtained in the native receiver format or in RINEX format. For the post-processing, data are generally supplied via the Internet. The RINEX format enables data collected from different brands of receiver to be combined and processed.

The US Radio Technical Commission for Maritime Services (RTCM) (www .rtcm.org), is a group concerned with communication issues as they pertain to the maritime industry. Special Committee 104 was formed to draft a standard format for the correction messages necessary to ensure an open real-time differential GPS system (Langley 1994). The format has become known as RTCM 104, and has recently been updated to version 3.1. Trimble Navigation has also published the message format used by their real-time systems. The Compact Measurement Record (CMR) was developed by and initially used by Trimble in 1992. The format was developed as a method of transmitting code and carrier phase correction data in a compact format from GPS base stations to GPS rovers for RTK GPS surveying. In 2009, Trimble introduced a new broadcast observation format called CMRx, developed to support other GNSS constellations. The purpose of CMRx is to improve the initialisation time, to cover additional GNSS core constellations, to deal with new GNSS signals, and to improve performance in urban and under canopy environments.

According to RTCM 104 version 3.1 recommendations, the pseudorange correction message transmission consists of a selection from a large number of different message types. Not all message types are required to be broadcast in each transmission; some of the messages require a high update rate while others require only occasional transmission. Provision has also been made for carrier phase data transmission to support carrier-based RTK positioning using the RTCM message protocol. GLONASS differential corrections can also be transmitted within this protocol. Many message types are still undefined, providing for considerable flexibility.

The greatest consideration for the DGNSS data link is the rate of update of the range corrections. Errors due to SA (currently not active) can vary more quickly than any other bias, such as orbit error, atmospheric refraction, etc.; hence, they were the primary concern and the major constraint for early real-time DGNSS communication options. The correction to the pseudorange and the rate-of-change of this correction were determined and transmitted near-instantaneously for each satellite. If the message *latency* (or *age*, or *time lapse*) is too high, then temporal decorrelation occurs and the benefit of the differential corrections is minimised.

Most navigation-type GNSS receivers are 'RTCM-capable', meaning that they are designed to accept RTCM messages through an input port, and hence can output a differentially corrected position. RTCM is not instrument-specific; therefore, a rover of any manufacture can apply the corrections even though the corrections were generated by a base receiver of different manufacture. There are also a number of free-to-air and commercial real-time DGNSS (especially DGPS) correction services available across the countries which are made available by either radio or communication satellites. These can be very useful for navigation, mapping, and lower accuracy surveys. There are also commercial services based on transmitting differential corrections to end users. Refer to Chapter 7 for more details.

6.6.1 COMMUNICATION (RADIO) LINK

RTK and real-time DGNSS operations require a communication (or radio) link to transmit the information from the base receiver to the rover receiver. Both ground- and satellite-based communication links are used for this purpose. However, we shall restrict our discussion in this section to dedicated ground-based communication links. Information about satellite-based communication link is furnished in Chapter 7. RTK data are typically transmitted at a *baud rate* (transmission rate at which data flows between transmitter to receiver) of 9600 bps, while the DGNSS corrections are typically transmitted at 200 Kbps (El-Rabbany 2002). A variety of radio links that use different parts of the electromagnetic spectrum are available to support such operations. The spectrum parts mostly used in practice are the low/ medium frequency (LF/MF) bands (i.e., 30 kHz–3 MHz) and the very high and ultrahigh frequency (VHF/UHF) bands (i.e., 30 MHz–3 GHz) (Langley 1993b; PCC 2000). Further, GNSS users may also utilise their own dedicated radio links to transmit base station information.

Dedicated ground-based GNSS radio links are most frequently established by using the VHF/UHF band. Radio links in this band provide line-of-sight coverage with the ability to penetrate into buildings and other obstructions. One example of such a radio link is the widely used *Position Data Link* (PDL) (Figure 6.7a). PDL allows for a baud rate of 57600 bps and is characterised by low power consumption

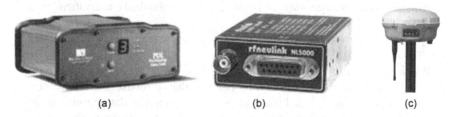

(a)   (b)   (c)

FIGURE 6.7 (a) PDL radio modem from Pacific Crest Corporation; (b) spread-spectrum radio transceiver from RfWel; (c) Trimble R8s receiver with inbuilt radio modem and antenna.

and enhanced user interface. This type of radio link requires a license to operate. Another example is the license-free spread-spectrum radio transceiver, which operates in a portion of the UHF band (Figure 6.7b). This radio link has coverage of 1–5 km and 3–15 km in urban and rural areas, respectively. More recently, some manufacturers adopted cellular technology, the digital personal communication services, and the third generation (3G) wideband digital networks, as an alternative communication link (El-Rabbany 2002). The 3G technology uses common global standards, which reduces the service cost. In addition, this technology allows the devices to be kept in the 'ON' position all the time for data transmission or reception, while the subscribers pay for the packets of data they transmit/receive. A number of GNSS receivers currently available in the market have built-in radio modems that use one or more of these technologies—e.g., Trimble R8s or R10s (Figure 6.7c)

It should be pointed out that obstructions along the propagation path, such as buildings and terrain, attenuate the transmitted signal, which leads to limited signal coverage. The transmitted signal attenuation may also be caused by ground reflection (multipath), the transmitting antenna, and other factors (Langley 1993b). To increase the coverage of a radio link, a user may employ a power amplifier or high-quality coaxial cables, or the height of the transmitting/receiving radio antenna may be increased. If a user employs a power amplifier, however, he/she should be cautioned against signal overload, which usually occurs when the transmitting and the receiving radios are very close to each other (Langley 1993b). A user may also increase the signal coverage by using one or more repeaters. In this case, it might be better to use a unidirectional (single directional) antenna at the base station and an omnidirectional (all directional) antenna at the repeater station (PCC 2000). The use of a repeater is also useful for reaching inside buildings, inside caves, or beyond obstructions (Figure 6.8).

6.7 PROCESSING ALGORITHMS, OPERATIONAL MODE AND OTHER ENHANCEMENTS

Finally, the accuracy of positioning is also dependent on a host of operational, algorithmic and other factors, such as:

Whether the user is moving or stationary. The redundant observations in a stationary condition permit an improvement in precision due to the effect of averaging down random errors over time; a moving receiver does not offer this possibility.

Whether the results are required in real-time or if post-processing of the data is possible. Real-time positioning requires that a 'robust' but less precise technique is used. The luxury of post-processing the data permits more sophisticated modelling and processing of GNSS data which minimises the magnitude and impact of residual biases and errors. However, post-processing is not an option for navigation applications.

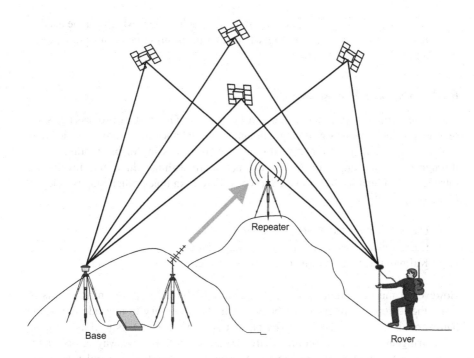

FIGURE 6.8 Use of repeaters to reach beyond obstructions.

The level of measurement noise has a considerable influence on the precision attainable with a GNSS. Low measurement noise would be expected to result in comparatively high accuracy. Hence, carrier phase measurements are the basis for high accuracy techniques, while pseudorange measurements are used for low accuracy applications. In addition, carrier phase data can be used to 'smooth' the relatively noisy pseudorange measurements prior to their use in the positioning algorithm.

The degree of redundancy in the measurements. Factors such as the number of tracked satellites (dependent upon the elevation cutoff angle, the number of receiver channels, satellites from different constellations, such as GPS and GLONASS, the use of pseudolites, etc.), the number of observations (dual-frequency carrier phase, dual-frequency pseudorange) permit more sophisticated quality control procedures to be implemented, which 'trap' (and delete or downweight) bad quality data that would otherwise introduce errors.

The algorithm type may also impact on GNSS accuracy. For example, 'exotic' data combinations are possible (carrier phase plus pseudorange), Kalman filter (Kalman 1960) solution algorithms, more sophisticated phase processing algorithms, etc.

Techniques of data enhancements and aiding may be employed. For example, the use of carrier phase smoothed pseudorange data, external data such as from inertial navigation systems (and other such devices which can be used to navigate by dead-reckoning when satellite positioning is not possible), additional constraints, etc.

6.7.1 SOFTWARE ENHANCEMENTS

There are many enhancements that can be made to GNSS data processing, some relatively minor, some significant, though most enhancements are considered in a hierarchical manner: from the straightforward processing of pseudorange data, through levels of complexity for the estimation algorithms which may involve the introduction of 'exotic' additional data types. Three examples which may be relevant to many applications are:

- Clock or height-aided position solution.
- Carrier phase smoothing of pseudorange data.
- Kalman filter algorithms for data processing.

However, it must be emphasised that to ensure high quality positioning results by using the techniques referred to above, it may be necessary to upgrade the GNSS hardware also (e.g., the use of carrier phase tracking receivers), as well as to implement stringent data collection procedures (e.g., the most effective way of improving accuracy is for the receiver antenna to be static for periods of several minutes or more).

6.7.1.1 Clock-Aiding and Height-Aiding

Clock-aiding refers to the process by which we assume that the receiver clock offset from GNSS time is not an entirely unknown parameter. *Height-aiding* refers to the technique by which we can assume that the height of the receiver is known. There are, in fact, two ways in which these enhancements can be implemented: *with* or *without* extra hardware.

With no extra hardware (and hence extra observations), information on the receiver clock offset can be included as an extra constraint within the pseudorange solution. This constraint can be considered a *pseudo-observation*: that is, the clock offset can be input as an observable (let us refer to this as 'option 1'), or it can be considered a known quantity which does not need to be estimated ('option 2') in the standard four satellite estimation procedure. The solution becomes stronger because: (a) there is one extra observation to estimate the same number of parameters (option 1), or (b) there is one less parameter to be estimated using the same number of pseudorange observations (option 2). Now, the question is where does the receiver clock error estimate come from? If the receiver clock were of good enough quality such that the clock bias were highly predictable for a short time into the future, then once the bias and bias-rate were determined in the conventional way, the estimated clock bias at some future time could be assumed accurate enough not to warrant estimation on an epoch-by-epoch basis. However, according to the investigations by Misra (1996), the standard quartz crystal clock cannot satisfy this role, but an oven-controlled crystal oscillator would.

A similar approach can be applied to height. Most navigation applications involve 2D (horizontal) positioning. If the hardware, such as a barometer, were available, an extra observation of height could be added to the solution. (This would have a similar effect to adding an extra satellite to the constellation being tracked.) Alternatively, once the height had been estimated, it could be assumed to not change in value for some time into the future, and hence removed from the estimable parameter set. In fact, many receivers have this option when less than four satellites are visible (particularly when there is significant signal shading, as would be experienced in urban canyons).

6.7.1.2 Using Carrier Phase Data to Smooth Pseudorange Data

There are two main problems associated with pseudorange data: (a) the high measurement noise and (b) the greater multipath disturbance in comparison to carrier phase data. One way to overcome these problems is to create a pseudorange and carrier phase combination which, in effect, 'smoothens' the pseudorange data. The basis of all data smoothing techniques is to derive the rate-of-change of range from the carrier phase data, and to combine this with the absolute measurement of range provided by the pseudorange data.

An early implementation of a data smoothing technique was described by Hatch (1982), making use of dual-frequency carrier phase and pseudorange data. Alternative smoothing algorithms have been developed which use Doppler data in place of carrier phase data. Many GNSS receivers nowadays use such carrier-smoothed data in the standard navigation solution.

6.7.1.3 Kalman Filter

The use of Kalman filters (Kalman 1960) for GNSS data processing is now well established. This standard least-squares estimation technique is typically used when the estimation problem is over-determined, in other words, when there are more observations than required to estimate the position parameters (Hoffmann-Wellenhof *et al.* 1994; Leick 2004; Levy 1997). In kinematic applications least-squares procedures can be applied to data on an epoch-by-epoch basis. However, the parameters of interest (the position), and the dominant system errors (e.g., the clock or atmospheric errors), are time-varying quantities. In addition, the time variation is more or less predictable. For such applications, the data processing techniques that are the most efficient and optimal, and therefore the most appropriate, are those based on the extension of the principles of least-squares to encompass the concepts of *prediction*, *filtering*, and *smoothing*.

The three concepts—prediction, filtering, and smoothing—are closely related and are best illustrated by an example, in this case a moving vehicle for which the parameters of interest are its instantaneous position at time t. The process of computing the vehicle's position in real-time (that is, observations are taken at time t_k, and position results are required at t_k) is referred to as filtering. The computation of the expected position of the vehicle at some subsequent time t_k, based on the last measurements at t_{k-1} is properly termed *prediction*, while the estimation of where the vehicle was (say at time t_k) once all the measurements are post-processed to time t_{k+1}, is referred to as *smoothing*.

Although the three procedures are separate, and can be applied independently, they may also be applied sequentially:

- *The prediction step*: Based on past positioning information together with a kinematic model, the expected position and its precision at the next epoch of measurement are computed. The kinematic model is composed, as is the measurement model, of functional and stochastic components.
- *The adjustment or filtering step*: This is a classical adjustment, except that a fairly good a priori estimate of the parameters that were already provided from the prediction step. Basically, the resulting parameter estimates are weighted combinations of predicted quantities and measurement data. The Kalman filter is a particular form of the generalised least-squares filter.
- *The smoothing step*: This is where all the measurements are reprocessed after the last measurement has been made and the filtering step has been completed.

As indicated above, the implementation of the filter requires the specification of the stochastic and mathematical models for both the measurement system and the system dynamics. Once the mathematical and stochastic models have been defined, the implementation within a Kalman filter is, in principle, relatively straightforward, although several different implementations exist which may have different advantages from the computational, numerical stability or quality control point-of-view (Gelb 1974; Minkler and Minkler 1993).

Kalman filtering techniques are particularly suited for GNSS navigation because:

- Standard least-squares procedure treats each measurement epoch independently, and hence does not use information on the system dynamics, such as the motion of the vehicle to which the receiver is attached.
- Permits the rigorous computation of precision and reliability measures such as error ellipses and marginally detectable errors.
- The Kalman filter is also central to many quality control or fault detection procedures which can be implemented in real-time in order to detect failure (e.g., where poor quality data is introduced into the process or where there is an error in the measurement or system dynamics models), then identify the source of error, and then adapt (or recover) the system to ensure that the results are not biased due to this system failure.
- Estimate small biases that affect the data over many epochs. For example, many measurement biases in modern navigation technology have the signature of drifts which are not apparent at the single epoch level, because they appear as 'noise' in standard epoch-by-epoch least-squares solutions.
- By taking into account information on system dynamics, such as the regular motion of the receiver, it is possible to carry out position estimation even if there is insufficient data—for example, when only two satellites are visible.
- A Kalman filter can accept data as and when it is measured and does not have to be 'reduced' to some specified epoch.

- The Kalman filter is well-suited to the mixing (or fusion) of various data types (including from non-GNSS sensors).

Some receivers incorporate Kalman filters as the navigation computing algorithm, but their real utility is generally only obvious when the positioning system involves several sensors such as when GNSS is integrated with *dead-reckoning* sensors (explained in Section 6.7.2). However, Kalman filters are not 'magical' procedures, because if the input data is of questionable quality, or there is an error in the assumptions regarding the model for the system dynamics, the resulted position will still be seriously biased.

6.7.2 HARDWARE ENHANCEMENTS: GNSS AND OTHER SENSORS

For many vehicle navigation applications, GNSS is insufficient as a standalone positioning system, particularly in urban environments, because of satellite signal degradation and obstructions. Therefore, many navigation systems have been developed based on a combination of several technologies. The integration of GNSS with dead-reckoning sensors would appear to be ideal for supporting vehicle positioning in navigation applications because they are complementary systems and can output continuous position information to the required accuracies of urban vehicle navigation. However, the dead-reckoning sensors do add to the overall costs of the navigation hardware.

Dead-reckoning is the process of calculating one's current position by using their previous positions—a technique of determining position by computing distance travelled on a given course (direction). Distance travelled is determined by multiplying speed by elapsed time. The principle of a dead-reckoning system is the relative position fixing method, which requires knowledge of the location of a vehicle and its subsequent speed and direction (for example, the last position and velocity determination before GNSS signal interruption) in order to calculate its present position (Bowditch 1995). A typical dead-reckoning system therefore comprises distance and heading (direction) sensors. Such a system can only give the 2D (horizontal) position of a vehicle (although more sophisticated dead-reckoning systems may include altitude sensors or inclinometers which can provide the 3D position of the vehicle). However, because of unfavourable error accumulation (a small error in heading grows over time into a large error in position), frequent calibration is required. It is in this context that GNSS is integrated with dead-reckoning systems. That is, the dead-reckoning sensors provide information on relative position (relative to a starting location), but GNSS receiver position measurements (x, y, z) are used to determine the dead-reckoning sensor errors, which may be fed back into the navigation computer.

The sensors that are favoured for dead-reckoning systems in navigation applications may consist of some or all of the following (Krakiwsky and McLellan 1995):

- An *odometer*, which is a distance sensor that may be mounted as single or in a pair onto either the wheel or the transmission of the vehicle. Odometers are prone to errors due to wheel slippage and changes in wheel circumference due to tyre pressure and velocity changes. Their accuracy is typically of the order of 0.3%–2% of the distance travelled.

- A *magnetic compass* that measures the heading of the vehicle. The most popular electronic compass technology for land vehicle applications uses the fluxgate principle. Empirical tests with fluxgate compasses in urban environments have shown that, because of their sensitivity to external magnetic field disturbances such as bridges, railway tracks, overpasses, etc., a standard error of about 2–4° is expected.

- A *tilt sensor* that gives information about the pitch and roll angles of the vehicle and may involve one or more inclinometers. Such a sensor is comparatively expensive, but its accuracy is on the order of 0.1°.

- A *gyroscope* that measures the rate-of-change of heading of the vehicle. The fibre-optic gyros have a drift rate of the order of 1–10°/hour; the lower cost (and more widely used) vibrational and solid-state gyros exhibit poor bias and scale factor stability, and hence require almost continuous calibration.

- *Digital maps* can be used to relate mathematical coordinates to locations on street segments and intersections. In turn, the stored coordinates of the map features provide a means of navigating in coordinate space, allowing the digital map to contribute to the navigation function. The accuracy requirements for digital maps must be high enough to ensure that the vehicle does not appear to be navigating a neighbouring street.

A generic vehicle navigation system architecture which incorporates GNSS and dead-reckoning sensors is illustrated in Figure 6.9.

There is a trade-off in accuracy requirements of the dead-reckoning sensors relative to the primary navigation device (the GNSS receiver). That means the more accurate the dead-reckoning sub-systems, the longer the duration of the maximum tolerable GNSS outage. The sensors that comprise the dead-reckoning sub-system

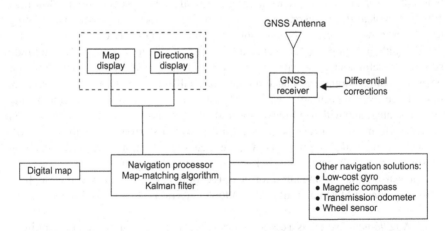

FIGURE 6.9 Generic GNSS and dead-reckoning vehicle navigation system (after Kaplan 1996).

can be classified as either sources of heading or velocity information. The vehicle's *odometer* is one attractive option for obtaining velocity information. The odometer, once calibrated, provides good long-term stability, and any effects due to tyre pressure variation can be accounted for by real-time calibration from GNSS. The only other velocity option is the use of low-cost accelerometers; however, the accelerometer bias must be removed initially and periodically recalibrated. Potential sources of heading information include magnetic compasses, accelerometers, two-wheel odometers, and gyros. Magnetic compasses are an attractive low-cost sensor but require calibration to remove the effects of local magnetic disturbances. These require that the raw heading information be filtered against other sensors, in particular GNSS. Gyros provide only heading-rate information; hence the other sensors are needed for heading initialisation. Several different types of gyros are possible, but the most common are the vibrational gyros, though fibre-optic gyros promise to become competitive as their cost decreases.

The central problem in integrating GNSS and dead-reckoning sensors is the design of the data processing algorithm. This is invariably a form of Kalman filter. There are two options: (a) a loose integration or coupling where some prior processing is carried out in sensor-specific filters, or (b) a tightly integrated implementation in which all observations are processed simultaneously. The most common filter used is the loosely coupled filter. Because of the non-homogeneous types of sensors (they all invariably come from different manufacturers) and their relatively low cost, each sensor will usually have its own filter. Fusion of the outputs from each sensor is then performed within a master filter. This means that the GNSS observations as passed to the master filter are position, velocity, and time (not the pseudorange or carrier phase measurements). This type of approach calibrates the local sensor biases and scale factors, as well as yields a globally optimal solution for the vehicle's position, velocity, time, and heading, along with accuracy and reliability measures for quality control.

However, new inertial measurement units are developed by the defence and aerospace industries. Today, dead-reckoning is rarely used in its traditional form for air navigation, but it survives in the form of inertial navigation systems. The inertial navigation system (refer to Chapter 7, Section 7.5) is used in combination with other navigation aids in order to provide reliable navigation capability under virtually any conditions. Unlike dead-reckoning (where frequent calibration is required) the advantage of an inertial navigation system is that it requires no external references in order to determine its position, orientation, or velocity once it has been initialised.

NOTE

The basic fluxgate compass is a simple electromagnetic device that employs two or more small coils of wire around a core of highly permeable magnetic material, to directly sense the direction of the horizontal component of the earth's magnetic field. The advantages of this mechanism over a magnetic compass are that the reading is in electronic form and can be digitised and transmitted easily, displayed remotely, and can be used by an electronic auto-pilot for course correction.

Fluxgate compasses and gyros complement one another. The fluxgate provides a directional reference that is stable over the long term, apart from changing magnetic disturbances, and the gyro is accurate over the short term, even against acceleration and heeling effects. At high latitudes, where the earth's magnetic field dips downward toward the magnetic poles, the gyro data can be used to correct for roll-induced heading errors in the fluxgate output. It can also be used to correct for the roll and heel-induced errors that often plague fluxgate compasses installed on steel vessels.

6.8   MISCELLANEOUS DISCUSSION

Besides the differencing methods described earlier in this chapter, developing single difference and double difference techniques with code-based measurements in order to totally remove the clock biases of the receivers and satellites could be envisaged in some specific cases, such as when initialising an initial position in a dynamic positioning using carrier phase measurements. The triple difference is of no real interest in code measurement because the removal of ambiguity concerning the code is more easily carried out by calculation. The reason behind the single difference and double difference techniques not being implemented in code-measurement is due to their increased complexity with respect to potential gain. Of course, the gain will be there, because the clock biases would be thoroughly removed, but it remains too small in comparison to the other sources of error.

The possibility of achieving code dual frequency is a substantial improvement for mass market civilian use. This has to be balanced by the fact that corresponding applications have to be found because, with current receiver architectures, the cost of the dual frequency receiver would be about 1.5 times the cost of a mono-frequency system (two radio front-ends would need to be implemented). Note that the interest in dual frequency, from a technical point of view, is both in removing the ionosphere errors and in implementing this latter correction in a single difference or double difference method.

The next point that could be of interest would be to implement a differential method taking advantage of the fact that there are four different global constellations (GPS, GLONASS, Galileo, and BeiDou) and some regional constellations (IRNSS, QZSS). Thus, one could have imagined using different signals or satellites from different constellations in order to achieve a new differential approach (known as *between-constellation differential*). Although the interoperability concept consists effectively in the possibility of computing a position using satellites from different constellations, a differential method is not achievable simply because of the various signal configurations they use. This is perhaps the next step in international cooperation for satellite-based navigation signals. To that end, an initiative has been taken to broadcast signals on at least one or two same frequencies from all of the constellations.

Remember that for the frequency differential approach, the compulsory require-ment is to receive two signals at two frequencies (e.g., L1 and L2) issuing from the same satellite. We may recall that in Section 6.2 there was an indication of hybrid techniques, such as rapid static. This consists in using specific algorithms that allow a rapid integer ambiguity resolution once at least four satellites are tracked (a few minutes are enough). This is usually achieved in dual-frequency mode, although other modes are possible.

In addition to two basic principles of measurement (code phase and carrier phase) as well as the relevant techniques described in earlier sections, some other techniques are also available. Among them one is *precise point positioning* (PPP) (Han *et al.* 2001; Colombo and Sutter 2004). This technique is based on the use of observed satellite orbit and clock data instead of predicted ephemeris. The International GNSS Service (IGS) (www.igs.org), for example, provides such data with various accura-cies. So-called 'ultra-rapid' products, available in real-time, give around 10 cm of accuracy for the orbits and 5 ns for the clocks. Better accuracies are possible, with increased latency. One advantage of PPP, compared to relative techniques, is that only a single receiver is required, although there is the need for additional correc-tion methods in order to mitigate some systematic effects that cannot be eliminated through differentiation. To achieve the highest possible point positioning accuracy, both carrier-phase and pseudorange measurements should be used. In addition, the remaining unmodelled errors, including receiver clock error, tropospheric delay, sat-ellite antenna offsets, site displacement effects, and equipment delays, must be dealt with. This approach may utilise undifferenced or between-satellite single difference measurements. Fortunately, most of these remaining errors can be modelled with sufficient accuracy (Kouba 2003). Exceptions are the receiver clock error (in the case of undifferenced measurements) and tropospheric delay, which are usually treated as additional unknowns to be estimated along with the station coordinates and ambi-guity parameters. The PPP is a low-cost solution for achieving high accuracy if one does not require the positioning information immediately (e.g., mapping); thus, it is becoming very popular. We shall elaborate on PPP in Section 6.8.1.

6.8.1 ONLINE DATA PROCESSING SERVICES

Many organisations have begun providing online GNSS data processing services—some of them are subscription-based and some are open to the world community. These services are not based on same technological principle and not meant for all applications; their performances may also vary (see Liu and Shih 2007 and Tsakiri 2008). For example, some are based on differential correction, some are based on precise observed ephemerides, and yet others are based on ambiguity resolution, etc. Some services use a combination of multiple techniques. However, they are all based on post-processing of collected observations. The main benefit of online GNSS data processing services is the use of a single GNSS receiver. Therefore, these services significantly reduce equipment and personnel costs, pre-planning, and logistics com-pared to conventional approaches (Wang *et al.* 2017). With just one dual frequency receiver, the observations taken are post-processed based on differential methods using reference stations or PPP via globally available precise satellite orbit and clock data. We shall discuss the differential correction services in Chapter 7 (Sections 7.3.1.7 and 7.3.2).

Regarding PPP, we know that the ephemeris data is downloaded by the receiver to establish the locations of the satellites; and the receiver determines its location based on these ephemerides. However, these ephemerides are actually predicted data, not the actual satellite locations (orbits are not perfectly predictable). Now think of a solution that can process the receiver collected positions with reference to the actual orbits that were followed by the satellites. This can be achieved by observing (tracking) the satellites continuously and by determining their actual ephemerides. If a GNSS receiver system stores raw observations, they can be processed later against a more accurate ephemeris than what was in the navigation messages, yielding more accurate position estimates than what would be possible with standard calculations. In addition, the actual satellite clock error, which was at the time of survey, can also be incorporated in the post-processing. This is the basic concept of PPP. Depending on the duration of receiver positioning data file, some of the systems even offer static or rapid static processing.

Other than the services for differential correction, some freely available online PPP data processing services are (Wang *et al.* 2017):

- *OPUS (Online Positioning User Service)* (https://www.ngs.noaa.gov/OPUS/index.jsp): This service is provided by NOAA with National Spatial Reference System (NSRS) coordinates. OPUS uses software which computes coordinates using the NOAA CORS Network. To use OPUS, the user needs to upload a GPS data file (collected with a survey-grade GPS receiver) to the OPUS upload page. The computed NSRS position is emailed to the user. Files that are 2 to 48 hours in duration are processed using PAGES static software. Receiver's coordinates are the average of three independent, single-baseline solutions, each computed by double-differenced carrier-phase measurements from one of three nearby CORSs. Files that are 15 min to 2 h in duration are processed using RSGPS rapid static software. Rapid static processing has stricter requirements for data continuity and geometry; in some remote areas it may not work. Under normal conditions, most positions can be computed to within a few centimetres.
- *APPS (Automatic Precise Positioning Service)* (http://apps.gdgps.net): This service is provided by NASA JPL. The users need to upload the measurement files to a special area on an FTP server, and they are processed automatically. They also offer a downloadable software GIPSY-OASIS that can be downloaded and installed in a PC for offline processing.
- *SCOUT (Scripps Coordinate Update Tool)* (http://sopac-old.ucsd.edu/scout .shtml): This is a collaborative offering by Scripps Orbit and Permanent Array Centre and California Spatial Reference Centre. SCOUT can be used to compute mean coordinates of a specific site by submitting a RINEX file of a particular day. It uses ultra-rapid orbits to allow near-real-time data processing. It supports some specific receivers and antennas (a list can be

obtained from their website). Users are limited to 10 uncompleted (queued) jobs at a time. Average run time is 30 min.

- *CSRS-PPP (Canadian Spatial Reference System Precise Point Positioning Service)* (https://webapp.geod.nrcan.gc.ca/geod/tools-outils/ppp.php?locale=en): This popular tool allows the computation of higher accuracy positions of raw GNSS data both in static and kinematic mode. CSRS-PPP online post-processing tool will use the best available ephemerides (FINAL, RAPID or ULTRA-RAPID). The FINAL can provide an accuracy of ±2 cm. It is based on combined weekly ephemerides and processed data are available 13–15 days after the end of the week. RAPID processing data are provided on the next day with an accuracy of ±5 cm. The ULTRA RAPID are available every 90 min (not available to download) with an accuracy of ±15 cm.

- *GAPS (GNSS Analysis and Positioning Software)* (http://gaps.gge.unb.ca): GAPS provides users with accurate satellite positioning using a single GNSS receiver both in static and kinematic mode. Through the use of precise orbit and clock products provided by sources such as the International GNSS Service (IGS) and Natural Resources Canada (NRCan), it is possible to achieve centimetre-level positioning in static mode and decimetre-level positioning in kinematic mode, given a sufficient convergence period.

- *AUSPOS (Geoscience Australia Online GPS Processing Service)* (https://www.ga.gov.au/scientific-topics/positioning-navigation/geodesy/auspos): The AUSPOS processing facility is provided by Geoscience Australia. It takes advantage of both the IGS Stations Network and the IGS product range. The IGS final orbit product is not available until approximately two weeks after the observation day. The rapid orbit product is available two days after observation. If both the final and rapid orbit products are unavailable, then the IGS ultra-rapid orbit product is used. The accuracy of this processing is sub-centimetre.

- *Magic GNSS PPP (magicPPP)* (https://www.gmv.com/en/Products/magicPPP): The magicPPP implements new generation PPP algorithms developed by GMV. GMV is a privately owned technological business group. The magicPPP offers four different services: (1) Post-processing service: registered users can upload, store, and manage raw data files in the magicGNSS cloud system workspace and use a number of tools for post-processing and display of results. (2) PPP webservice: this service works at the TCT/IP level, and receiver RINEX from specific users, without using the support of a graphical interface. This service can be integrated in a user mobile application for an agile interaction with the magicPPP server. (3) PPP by e-mail service (free): Users can access this free service via e-mail by sending their raw RINEX data files to magicppp@gmv.com. (4) Real-time corrections: This continuously operating infrastructure of magicPPP servers generates PPP corrections in a streaming format. Registered users can retrieve these corrections over the internet or via a satellite communication

link in case internet access is not available. GMV is currently developing magicPPP-RT real-time terminals, compatible with most commercially
available receivers.

In addition, UNB's PPPSC (Precise Point Positioning Software Centre) compares
solutions from online PPP applications (http://www2.unb.ca/gge/Resources/PPP).
All of the aforesaid systems support RINEX format, and the online processing services provide the coordinates in a recognised datum. However, it should be mentioned that while producing these coordinates, all services depend on the quality
of the data and the length of data span supplied to them by the user (Tsakiri 2008).
Generally, these services produce coordinates in the ITRF or their national geodetic
reference frame. Transformation of coordinates is possible from one reference frame
to another, either using the published transformation parameters or available online
software from governmental organisations, such as *NGS Coordinate Conversion
and Transformation Tool* (https://www.ngs.noaa.gov/NCAT).

6.9 SUMMARY OF POSITIONING METHODS

In this chapter, we have discussed so many techniques and instruments for positioning that it may confuse the reader. However, this section summarises the entire
chapter for the benefit of the beginners in our audience. The relative positioning by
carrier phase measurement is the primary observable for high-accuracy GNSS positioning. However, conditions are not always so ideal. Where obstructions threaten
to produce cycle slips, coded pseudorange measurements may offer an important
advantage over carrier phase measurement. Pseudorange measurements may also be
preferred where accuracy requirements are low and production demands are high.
There can hardly be a question that kinematic GNSS is the most productive of the
several alternative methods, under the right circumstances. However, technological
complexity and the necessity of maintaining *lock on* to four or more satellites, as the
receiver is moved, limits its application.

Differencing is an ingenious approach for minimising the effect of errors in carrier phase ranging. It is a technique that largely overcomes the impossibility of perfect time synchronisation. Double differencing is the most widely used formulation.
It is the double differenced carrier phase based fixed (static) solution that makes the
very high accuracy possible with GNSS.

However, in the discussion of errors, it is important to remember that multipath,
receiver noise, incorrect instrument heights, and a score of other errors whose effects
can be minimised or eliminated by good practice are simply not within the purview
of differencing at all. The unavoidable biases that can be managed by differencing
includes clock (satellite and receiver), atmospheric (ionospheric and tropospheric),
and orbital errors. These errors can have their effects drastically reduced by the
proper selection of baselines, the optimal length of the observation sessions, and
several other considerations included in the design of a GNSS positioning (refer to
Chapter 11). But such decisions require an understanding of the sources of these
biases and the conditions that govern their magnitudes. The management of errors
and accuracy issues cannot be relegated to mathematics alone.

EXERCISES

DESCRIPTIVE QUESTIONS

1. Why have different positioning methods been adopted? Classify positioning methods.
2. Describe the code-based point positioning method.
3. What do you understand by 'differential positioning'? Explain the working principle of code-based differential positioning.
4. What do you understand by 'position domain differential' and 'measurement domain differential'?
5. Explain autonomous and inverted DGNSS techniques.
6. Explain single difference and double difference.
7. Explain triple difference. How can it solve the integer ambiguity?
8. What do you understand by 'kinematic GNSS'? Explain real-time kinematic.
9. What are the errors that can be removed by single, double, and triple differences? List the errors that cannot be eliminated by differential approaches. Why are they not eliminated by differential approaches?
10. What do you understand by 'software and hardware enhancements' of GNSS? Briefly discuss.
11. Explain dead-reckoning. What are its advantages and disadvantages?
12. What do you understand by 'post-processed' and 'real-time' DGNSS? Explain their advantages and disadvantages.
13. What do you understand by 'clock-aiding' and 'height-aiding'? How can GNSS aid in dead-reckoning?
14. Explain the application of Kalman filtering in GNSS.
15. What do you understand by 'GNSS enhancement'? Explain ground-based radio link.
16. What is the working principle of precise point positioning (PPP)? Discuss two open online services for PPP.

SHORT NOTES/DEFINITIONS

Write short notes on the following topics

1. Receiver position differential
2. Satellite position differential
3. Frequency differential
4. Between-epochs differential
5. Base station
6. Roving receiver
7. Post-processed DGNSS
8. Inverted DGNSS
9. Common bias error vector

10. Static GNSS
11. Dual frequency receiver
12. Interoperability
13. Fix
14. Initialisation
15. Dead-reckoning
16. Precise point positioning (PPP)
17. OPUS
18. CSRS-PPP

7 GNSS Augmentations and Other Navigation Satellite Systems

7.1 INTRODUCTION

We now know that satellite navigation systems have several layers of infrastructure, such as, core GNSS, augmentation of GNSS (satellite- and ground-based), and regional navigation satellite systems. Until now, we have concentrated on core GNSSs, that is, GPS, GLONASS, Galileo, and BeiDou. Although these global systems can help in different types of applications, they have several limitations as well. Moreover, the countries who do not own a GNSS constellation are just clients of other countries who do own them. The accuracy and availability of navigation satellite signals to other countries are fully dependent on the provider countries. These uncertainties have forced the client countries to develop several augmented and regional systems. This chapter focuses on these various systems.

7.2 GNSS-1 AND GNSS-2

Civilian applications always were expected for GPS and GLONASS despite their military origins. However, non-military usage of these systems has grown far beyond what was originally envisaged. It is estimated that nine out of every ten new GNSS receivers currently sold are for civilian or commercial use. Land and offshore surveyors have been using NAVSTAR GPS for nearly three decades, and they probably have more experience and expertise than any other civil user group.

Satellite navigation has completely changed the lives of some people, and, in the near future, it will change the lives of almost everyone. Satellite navigation that provides enhanced accuracy and integrity monitoring for civilian navigation are classified as GNSS-1 and GNSS-2 (Figure 7.1) (IFATCA 1999). GNSS-1 is a first-generation system, the combination of existing satellite GPS and GLONASS with Satellite-Based Augmentation Systems (SBAS) or Ground-Based Augmentation Systems (GBAS). GNSS-2 is a second-generation system that independently provides a full civilian satellite navigation system, exemplified by the European Galileo, modernized GPS and GLONASS, Chinese BeiDou, and other regional satellite navigation systems. In this chapter, we shall first discuss the augmented systems and then regional satellite systems.

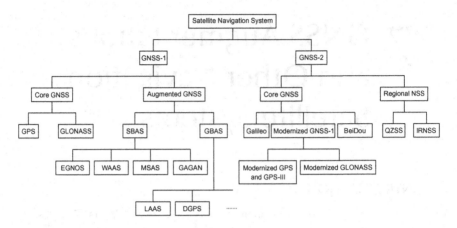

FIGURE 7.1   Satellite-based navigation systems (revised after IFATCA 1999).

7.3   GNSS AUGMENTATIONS

Augmentation of GNSS is a method of improving the navigation and positioning system's attributes, such as *accuracy, reliability,* and *availability* (refer to Chapter 5, Section 5.1), through the integration of external information into the calculation process. There are many such systems in place (IFATCA 1999) and they are generally named or described based on how the GNSS sensor receives the external information. Some systems transmit additional information about sources of error (e.g., clock drift, ephemeris error, or ionospheric delay), others provide direct measurements of how much the signal was off in the past, while a third group provides additional observable information to be integrated into the calculation process. The augmentations can be either satellite- or ground-based.

7.3.1   SATELLITE-BASED AUGMENTATION SYSTEMS

A Satellite-Based Augmentation System (SBAS) is a system that supports wide-area or regional augmentation through the use of additional satellite-broadcast messages. Such systems are commonly composed of multiple ground stations, located at accurately surveyed locations. The ground stations take measurements of one or more of the GNSS satellites, the satellite signals, or other environmental factors which may impact the signal received by the user's receiver. Using these measurements, information messages are created and sent to one or more satellites for broadcast to the end users (receivers) (Andrade 2001; Rycroft 2003).

There are many converging factors that encouraged the European Union and other countries to enter into the satellite navigation community; most importantly, the civilian GPS signal was intentionally degraded by the US government and the GLONASS signal was not available at all to civilians. Some user communities therefore developed new techniques to overcome these limitations, the so-called 'differential techniques' (code-based). The basic idea was that these errors are not random and so are identical for two receivers close to each other. Thus, it is certainly possible to remove the effects of these errors by positioning a receiver at a perfectly known fixed location and observing the difference between the computed location and the actual one. Removing the positioning error vector thus obtained from the positioning computed by another receiver placed at an unknown location, improved the resultant positioning accuracy (as explained in Chapter 6, Section 6.4.1). Differential approaches gained the popularity since they allowed an accuracy of 10–20 m (instead of the 100 m allowed when selective availability (SA) was active.)

Unfortunately, this technique requires a data transmission link. The normalized Radio Technical Commission for Maritime Services (RTCM) American format existed for such differential transmissions, but it dealt only with the format and not with the radio link. As the European Union was not yet involved in satellite navigation, this was the moment to join. The GPS system is totally driven by the US government and is thus under no 'impartial' control. This means that if one satellite were to exhibit difficulty in any way, there would be no possibility of knowing about it, unless the US authorities agreed. This has been the case on some occasions, leading to positioning errors that sometimes lasted quite a long time (a few hours) without any information from anyone. All these reasons led the European Union to decide to participate in satellite-based navigation, under the form of a 'satellite-based differential station' and 'integrity signals'. In order to achieve both integrity and differential data transmission, the main requirement was a large ground infrastructure for gathering all the necessary information. This information was then uploaded to the satellites and the satellites in turn transmitted it to the ground receivers over a large area. This is known as SBAS, since the additional information was transmitted via satellites.

Besides the European Geostationary Navigation Overlay Service (EGNOS) (European Space Agency), three other well-known SBAS programs are the Wide Area Augmentation System (WAAS) (United States), the Multi-functional Satellite-based Augmentation System (MSAS) (Japan), and GPS Aided Geo Augmented Navigation (GAGAN) (India). The coverage areas of these programs (Figure 7.2) are dictated by the ground infrastructure, not by the coverage of the transmitting geostationary satellites. Some other systems also exist and will be discussed.

There is a need for ground stations to collect the data required in order to prepare the differential and integrity messages. Discussions are in progress with some countries for the development of the EGNOS coverage, in particular in the Mediterranean region. Theoretically, an SBAS may have regional or even global coverage as well. However, all of the above are regional systems, and, to date, there is no SBAS with true global coverage. There are also commercial services based on transmitting differential corrections to end users, to improve availability, integrity, and accuracy in some regions. Some of them are also trying to offer the services over the entire globe (Section 7.3.1.7).

7.3.1.1 EGNOS

EGNOS, a satellite-based augmentation system, is effectively an overlay system that improves GPS accuracy (IFATCA 1999; Rycroft 2003; Samama 2008). The

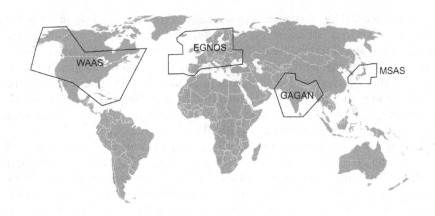

FIGURE 7.2   Coverage of different SBAS.

approach consists simply in gathering information, just like a standard local differential station, but over a large area (typically the European area). Once this has been
achieved, the differential and integrity data are sent to a geostationary satellite for
transmission over a large area.

The transmission of error correction data has some constraints. Indeed, a transmission radio data link requires the use of a frequency band that will certainly increase
the cost of a receiver. However, EGNOS uses the same band as GPS, and, furthermore, a similar code to the GPS satellites' C/A codes. This means that no hardware
changes are required in order for every GPS receiver to access the differential mode.
The only difference lies in the fact that the navigation message of EGNOS is in fact
specific to the differential mode and has to be decoded in accordance with EGNOS
message definition rather than GPS navigation message definition. Of course, as the
best C/A codes are reserved for the GPS satellites (C/A codes with the best correlation characteristics are thus reserved for the constellation), EGNOS uses other
Gold codes, exhibiting slightly degraded correlation performances. This quite clever
approach has led to the fact that 'EGNOS-enabled' receivers were available as soon
as the program was technically defined; only a software modification was required,
and all manufacturers carried out this modification early on.

The EGNOS integrity concept relies on the transmission of various corrections
(Samama 2008). In order to achieve this goal, EGNOS is capable of making corrections to orbital and clock errors of each satellite (through the so-called 'slow corrections') and to ionospheric time delays through the use of a separated set of corrections.
At the time of activation of SA, 'fast corrections' were also allocated to rapidly varying
clock errors; these corrections are still transmitted even though the SA is now inactive.

The EGNOS ground segment consists of a network of 40 Ranging Integrity
Monitoring Stations (RIMS), 2 Mission Control Centres (MCC), 6 Navigation Land
Earth Stations (NLES), and the EGNOS Wide Area Network (EWAN), which pro-
vides the communication network for all the components of the ground segment.
The main function of the RIMS is to collect measurements from GPS satellites and
transmit these raw data each second to the Central Processing Facilities (CPF) of

each MCC. The configuration used for the initial EGNOS open service includes 40 RIMS sites located over a wide geographical area. The MCCs receive the information from the RIMS and generate correction messages to improve satellite signal accuracy and information messages on the status of the satellites (integrity). The NLESs transmit the EGNOS message received from the central processing facility to the geosynchronous satellites for broadcasting to users and to ensure the synchronisation with the GPS signal. In addition to the stations/centres, the system has other ground support installations that perform the activities of system operations planning and performance assessment, namely the Performance Assessment and Checkout Facility (PACF) and the Application Specific Qualification Facility (ASQF) which are operated by the EGNOS service provider. PACF provides support to EGNOS management in such area as performance analysis, troubleshooting and operational procedures, as well as upgrades of specification and validation, and maintenance support. ASQF provides civil aviation and aeronautical certification authorities with the tools to qualify, validate, and certify the different EGNOS applications. The space segment of EGNOS is composed of three geostationary satellites broadcasting corrections and integrity information for GPS satellites in the L1 frequency band (1575.42 MHz). This space segment configuration provides a high level of redundancy over the whole service area in case of a geosynchronous satellite link failure. EGNOS operations are handled in such a way that, at any point in time, at least two of the three GEOs broadcast an operational signal. The EGNOS User segment is comprised of EGNOS receivers that enable their users to accurately compute their positions with integrity. To receive EGNOS signals, the end user must use an EGNOS-compatible receiver.

EGNOS typically offers three services intended for different user communities:

1. *Open Service*: freely available, it allows accuracy with GPS that can reach 1–3 m horizontally and 2–4 m vertically.
2. *Commercial Data Distribution Service*: offered through the Internet or through cellular phones on a controlled access basis for professional users needing enhanced performance.
3. *Safety-of-Life*: offers enhanced and guaranteed performance to the transportation sector.

EGNOS was developed under a tripartite agreement whose members are the European Space Agency (ESA), the European Commission (EC), and Eurocontrol (the European Organization for the Safety of Air Navigation). Latest information on EGNOS is available at www.esa.int/esaNA/egnos.html.

7.3.1.2 WAAS

WAAS is an air navigation aid developed by the US Federal Aviation Administration (FAA) to augment the GPS, with the goal of improving its accuracy, integrity, and availability. Essentially, WAAS is intended to enable aircraft to rely on GPS for all phases of flight, including precision approaches to any airport within its coverage area (Bowditch 1995; Samama 2008).

WAAS uses a network of ground-based reference stations, located within spaces protected from the public inside airports in North America and Hawaii, to measure small variations in the GPS satellites' signals. Measurements from the reference stations are routed to master stations, which queue the received information and send the correction messages to geostationary WAAS satellites in a timely manner (at least every 5.2 sec or sooner). These satellites broadcast the correction messages back to the earth, where WAAS-enabled GPS receiver uses the corrections while computing its position to improve accuracy. The WAAS specification requires it to provide a position accuracy of 7.6 m or better (for both horizontal and vertical measurements), at least 95% of the time. Actual performance measurements of system at specific locations have shown that it typically provides better than 1.0 m horizontally and 1.5 m vertically throughout most of the contiguous United States and large parts of Canada and Alaska.

Integrity of a navigation system includes the ability to provide timely warnings when its signal is providing misleading data that could potentially create hazards. The WAAS specification requires the system to detect errors in the GPS or WAAS network and notify users within 5.2 sec. Latest updates on WAAS are available at www.nstb.tc.faa.gov.

7.3.1.3   MSAS

MSAS is a Japanese SBAS and similar to WAAS and EGNOS. The Japan Civil Aviation Bureau (JCAB) has implemented the MSAS to utilise GPS for aviation. MSAS provides GPS augmentation information to aircraft through MTSAT (Multi-functional Transport Satellite) located at geostationary earth orbit. MSAS generates GPS augmentation information by analysing signals from GPS satellites received by monitor stations on the ground. This augmentation information consists of GPS-like ranging signal and correction information on GPS errors caused by the satellites themselves or by the ionosphere. Currently MSAS has two master control stations, four monitoring stations, and two monitoring and ranging stations.

MSAS signal provides accurate, stable, and reliable GPS position solutions to aircraft. This leads to a considerable improvement in the safety and reliability of GPS positioning, which, therefore, enables the aviation users who are under very strict safety regulations to use GPS positioning as a primary means of navigation system.

7.3.1.4   GAGAN

GAGAN is an implementation of SBAS by the Indian government. It is a system to improve the accuracy of a GNSS receiver by providing reference signals. It was developed jointly by the Airports Authority of India (AAI) and the Indian Space Research Organization (ISRO). As the name implies, this SBAS is compatible with GPS, however, Galileo is also being considered. Interoperability with other SBASs like WAAS, EGNOS, and MSAS has also been taken into account. The GAGAN was implemented to help pilots navigate in the Indian airspace and land aircraft in marginal weather and difficult approaches like Mangalore and Leh airports.

Although primarily meant for civil aviation, it is also beneficial for other government agencies. The space segment of GAGAN is designed to have three operational geosynchronous satellites. On the ground, the GPS data is received and processed in the 15 Indian reference stations (INRES), located at Ahmedabad, Bengaluru, Bhubaneswar, Kolkata, Delhi, Dibrugarh, Gaya, Goa, Guwahati, Jaisalmer, Jammu, Nagpur, Porbandar, Portblair, Trivandrum. The Indian Master Control Centre (INMCC), composed of two sites and located in Bangalore, processes the data from the INRESs to compute the differential corrections and estimate its level of integrity. The SBAS message generated by the two INMCC is uplinked to the satellites through its corresponding Indian Land Uplink Station (INLUS). GAGAN-enabled GPS receivers, with the same technology as WAAS receivers, are able to use the GAGAN signal. GAGAN uses GPS type modulation on L2 and L5 frequencies. One essential component of the GAGAN project was the study of the ionospheric behaviour over the Indian region. This was specially taken up in view of the rather uncertain behaviour of the ionosphere in the equatorial region. However, ISRO has successfully modelled the ionospheric correction. GAGAN ionospheric algorithm, known as ISRO GIVE Model-Multi-Layer Data Fusion (IGM-MLDF), is operational in the implemented GAGAN system. The accuracy standard of GAGAN is less than 7.6 m both horizontally and vertically 100% of the time. However, the position standard deviation of latitude, longitude, and altitude were found to be less than 4 meters, which indicates that the position accuracies of the GAGAN are well within the 7.6 m requirement. These position accuracies can be further enhanced with ground-based augmentation systems.

7.3.1.5   SDCM

The System for Differential Corrections and Monitoring (SDCM) is the SBAS of the Russian Federation for integrity monitoring of both GPS and GLONASS satellites. The SDCM network of reference stations is composed of 19 stations in Russia and 5 stations abroad. The central processing facilities are located in Moscow. The SDCM space segment is designed to have 3 operational and 1 backup geosynchronous satellites. The positioning accuracy is 1–1.5 m horizontally and 2–3 m vertically. In addition, it is expected to offer a centimetre-level positioning service for users at a range of 200 km of the reference stations. One may find further details at http://www. sdcm.ru/index_eng.html.

7.3.1.6   Other Government SBAS Systems

Several other countries have also initiated development of SBAS systems, e.g., Satellite Navigation Augmentation System (SNAS) (China), SACCSA (South/ Central America and the Caribbean region), Malaysian SBAS, African SBAS, etc. The main purpose is to provide contiguous SBAS coverage for the entire world. Canada also initiated an SBAS system called CDGPS, however, it was discontinued in 2011.

7.3.1.7   Commercial SBAS Systems

There are also commercial services based on transmitting differential corrections to end users, to improve availability, integrity, and accuracy of the navigation satellite systems, such as *OmniSTAR* and *StarFire*. These systems generally transmit correction data via satellite communication links accessible through user subscription. Single receivers fitted with modulation devices capable of receiving the satellite correction signal can obtain metre-level accuracy using such providers. These services provide corrections suitable for pseudorange processing and therefore may not meet the accuracy requirements specified for precise surveying applications. Important to mention: to date, these systems are based on NAVSTAR GPS.

The OmniSTAR and StarFire systems are comparable concepts; differential corrections are provided through a geostationary satellite transmission link to the users. They both send correction data to the user's receiver, which computes the positioning. Various techniques allow accuracy in the range from 10 cm to around 1 m. These companies are providing reliable differential corrections real-time for precise positioning on land and in the air in several countries, with a series of reference stations throughout the world.

There are also some commercial tracking services based on the use of geostationary telecommunication satellites for fleet management, such as *OmniTRACS* and *EutelTRACS*. The EutelTRACS system is a positioning and communication satellite system for automobiles. It provides message and report features in real-time between fleet and master control centres. The various messages use satellites to link the central station to distribution centres all over Europe. The OmniTRACS system is also a wireless communication and satellite positioning system that supports fleet management. Both are based on NAVSTAR GPS. Apart from the above, some other systems are also in the queue. Additional information on these systems and their costing details can be found at their respective websites:

OmniSTAR : www.omnistar.com
StarFire : www.navcomtech.com/StarFire
OmniTRACS : www.qualcomm.com
EutelTRACS : www.eutelsat.org

7.3.2   GROUND-BASED AUGMENTATION SYSTEMS

Ground-Based Augmentation System (GBAS) describes a system that supports augmentation through the use of terrestrial radio messages (Grewal *et al.* 2001; Hofmann-Wellenhof 2003). As with the satellite-based augmentation systems explained earlier, ground-based augmentation systems are commonly composed of one or more accurately surveyed ground stations, which take measurements concerning the GNSS, and one or more ground-based radio transmitters, which transmit the information directly to the end user. GBASs can have local and regional coverage. Generally, GBAS networks are considered localized, supporting receivers within a few tens of kilometres and transmitting in the very high frequency (VHF) or ultra-high frequency (UHF) bands. Ground-based Regional Augmentation System (GRAS), is applied to systems that support a larger, regional area, and also transmit in the VHF bands. One example of local GBAS is the US Local Area Augmentation System (LAAS), and one GRAS is Differential GPS (DGPS).

7.3.2.1 LAAS

The Local Area Augmentation System (LAAS) is a local ground-based augmentation to GPS that focuses its service on airport areas (approximately a 30–50 km radius) for precision approach, departure procedures, and terminal area operations. This system is now called GBAS instead of LAAS. In this system local reference receivers send data to a central location at the airport. This data is used to formulate a correction message, which is then transmitted to users via a VHF data link. A receiver on an aircraft uses this information to correct GPS signals. US FAA is implementing this system at major airports in the USA. This system is focusing on the resolution of outstanding integrity and safety issues to reduce the risks in aviation.

7.3.2.2 DGPS

Differential Global Positioning System (DGPS) is an enhancement to GPS that uses a network of fixed, ground-based reference stations to broadcast the difference between the positions indicated by the satellite systems and the known fixed positions. These stations broadcast the difference between the measured satellite pseudoranges and actual pseudoranges, and rovers may correct their pseudoranges by the same amount. The term DGPS can refer both to the generalized technique as well as specific implementations using it. It is often used to refer specifically to systems that rebroadcast the corrections from ground-based transmitters. These systems generally have regional coverage. For instance, the United States Coast Guard maintained one such system in the US (known as Nationwide DGPS or NDGPS; website: www.navcen.uscg.gov/dgps/Default.htm) since 1999. Canadians also had a similar system operated by the Canadian Coast Guard. Australia developed two DGPS systems, one for marine navigation, and the other for land surveys and land navigation. Due to the increased accuracy of core GNSS constellations and availability of different SBASs, all of the DGPS services have been discontinued from 2020.

7.3.2.3 Augmentation Services from Trimble, Leica, and Others

Other than the government agencies, several private organizations are also offering augmentation services. One such popular service is *CentrePoint RTX* from Trimble. Its subscription-based service offers both post-processing and real-time corrections. Trimble's RTX network is available throughout most of the world with around 6500 reference stations. It uses multiple technologies such as PPP and differential/RTK corrections, and provides centimetre-level accuracy in real-time without establishing a base station. Trimble RTX provides a satellite-delivered correction source. In addition, corrections are available via an Internet or cellular connection, adding to its versatility.

HxGN SmartNet is an integrated RTK and DGNSS correction service similar to Trimble RTX. It is an open-standard correction service, supported by any GNSS device, and is constantly monitored for integrity, availability, and accuracy. With more than 4000 reference stations, one can achieve centimetre-level accuracy. HxGN SmartNet was built to provide high-precision, high-availability RTK corrections for any application, using any constellation, while at the same time being open to all. However, this system is currently subscription-based.

Some other companies/brands that offer similar services are QXWZ (https://www.qxwz.com/en), Sapcorda (https://www.sapcorda.com), NovAtelCORRECT (https://www.novatel.com), Atlas from Hemisphere GNSS (https://www.hemispheregnss.com), MAGNET Relay from Topcon (https://www.topconpositioning.com), and proposed Melco Geo++ from Mitsubishi Electrical.

7.4 REGIONAL NAVIGATION SATELLITE SYSTEMS

As we have learned, the ability to supply signals from satellites also means the ability to deny their availability. The operator of a GNSS potentially has the ability to degrade or eliminate satellite navigation services over any territory it desires. Thus, as satellite navigation became an essential service, countries without their own satellite navigation systems effectively became client states of those countries who supply these services. Mainly for this reason, some countries sought to develop their own regional satellite navigation systems. These are similar to the core GNSS but only have a regional coverage rather than global. The early stage of BeiDou (namely, BeiDou I) was deployed by China as a regional system, and later it was converted to a global system (BeiDou III). Two other regional systems are QZSS (Japan) and IRNSS (India)

7.4.1 QUASI-ZENITH SATELLITE SYSTEM

Quasi-Zenith Satellite System (QZSS, Japanese name is Michibiki) is a Japanese regional system, which is basically a combination of augmented GPS and a regional navigation satellite system (https://qzss.go.jp/en). The QZSS uses multiple satellites that have the same orbital period as geostationary satellites with some orbital inclinations known as quasi-zenith orbits (QZOs) as well as one geostationary satellite in the constellation. The QZO satellites are placed in multiple orbital planes, so that one satellite always appears near the zenith above the region of Japan. The system makes it possible to provide high accuracy satellite positioning service covering close to 100% of Japan, including urban canyon and mountainous terrain. However, users in many other Asia-Pacific areas benefit from QZSS. The system was declared operational in the year 2018 with 4 QZO satellites. However, Japan plans to increase the number of satellites to 7 by 2024.

QZSS is compatible with GPS satellites and can be utilised with them in an integrated manner. QZS transmits the same positioning signals as GPS (L1C/A, L1C, L2C, and L5) and has clocks that are synchronised with GPS, so they can be used as if they were additional GPS satellites. The QZSS will be fully interoperable with GPS while not dependent on it. The system will also be designed to accommodate Galileo signals. Positioning accuracies will be improved by increasing the number of QZS in combination with GPS (as a greater number of visible satellites definitely increases the positioning accuracy). Further, one satellite at the zenith substantially increases the availability and positioning accuracy in the urban canyon or mountainous areas where fields of vision are often obstructed. QZS satellites also transmit augmentation signals for GPS thus enhancing the accuracy further to sub-metre level. QZSS offers several services with different levels of accuracy—from centimetre to metre.

7.4.2 Indian Regional Navigational Satellite System (NavIC)

The ISRO has deployed a project to implement Indian Regional Navigational Satellite System (IRNSS) over the Indian region. The operational name of this system is *NavIC* (Navigation with Indian Constellation). It covers the entire Indian subcontinent as its primary service area and extends service to some of its neighbouring countries. The extended service area lies between primary service area and the area enclosed by the rectangle from Latitude 30°S to 50°N and Longitude 30°E to 130°E (Figure 7.3). The entire IRNSS system (space segment, control segment, and user segment) has been built in India and is being controlled by ISRO. As of year 2019, the system consists of eight satellites in orbit; namely IRNSS-1A (02 July 2013), 1B (04 April 2014), 1C (16 October 2014), 1D (28 March 2015), 1E (20 January 2016), 1F (10 March 2016), 1G (28 April 2016), and 1I (12 April 2018). IRNSS-1H was unsuccessful—could not reach orbit. The space segment of IRNSS is designed to have seven satellites at an altitude of 35,786 km. Three of these satellites are in geostationary orbit (located at 34° E, 83° E, and 131.5° E) while the remaining are in geosynchronous orbits that maintain an inclination of 29° to the equatorial plane. Such an arrangement ensures all satellites have continuous radio visibility with Indian control stations. There are plans to expand the NavIC system by increasing the number of satellites to 11 for increased positional accuracy. The satellite payloads consist of atomic clocks and electronic equipment for generating navigation signals.

The IRNSS ground segment includes the major systems for controlling the satellite constellation and will consist of the IRNSS Spacecraft Control Facility (IRSCF), ISRO Navigation Centre, IRNSS Range and Integrity Monitoring Stations, ranging stations, a timing centre, IRNSS Telemetry Tracking & Command (TT&C) and

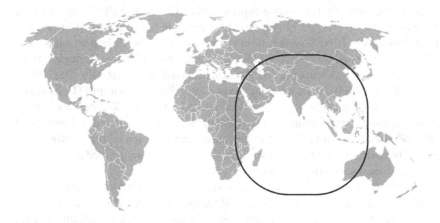

FIGURE 7.3 Coverage of IRNSS.

uplink stations, and the IRNSS Data Communication Network. IRNSS Spacecraft Control Facility is responsible for controlling the space segment through TT&C. In addition to the regular TT&C operations, IRNSS Spacecraft Control Facility also uplinks the navigation parameters generated by the ISRO Navigation Centre. IRNSS range and integrity monitoring stations perform continuous one-way ranging of the IRNSS satellites and are also used for integrity determination of the IRNSS constellation. ISRO Navigation Centre generates navigation parameters. IRNSS Data Communication Network provides the required digital communication backbone to the IRNSS network. Seventeen IRNSS Range and Integrity Monitoring Stations are distributed across the country for orbit determination and ionospheric modelling. Four ranging stations, separated by wide and long baselines, provide two-way CDMA ranging. The IRNSS timing centre consists of highly stable atomic clocks. The ISRO Navigation Centre receives all this data through communication links, then processes and transmits the information to the satellites. IRNSS is designed to have a network of 21 ranging stations geographically distributed primarily across India. They provide data for the orbit determination of IRNSS satellites and monitoring of the navigation signals. The data from the ranging/monitoring stations is sent to the data processing facility at ISRO Navigation Centre (located at Byalalu, near Bengaluru) where it is processed to generate the navigation messages. The navigation messages are then transmitted to the IRNSS satellites through the spacecraft control facility centres at Hassan and Bhopal. The state of the art data processing and storage facilities at ISRO Navigation Centre enable swift processing of data and support its systematic storage.

The IRNSS user receiver calculates its position using the timing information embedded in the navigation signal and transmitted from the IRNSS satellites. The timing information being broadcast in the navigation signal is derived from the atomic clock onboard the IRNSS satellite. The IRNSS system time is given as 27-bit binary number composed of two parameters as follows (ISRO 2017): The week number is an integer counter that gives the sequential week number from the origin of the IRNSS time. This parameter is coded on 10 bits that covers 1024 weeks (about 19 years). The time of week count is represented in 17 bits. The IRNSS system time start epoch is 00:00 UTC on 22 August 1999 (midnight between August 21 and 22). At the start epoch, IRNSS system time is ahead of UTC by 13 leap seconds. (i.e., IRNSS time 22 August 1999, 00:00:00 corresponds to UTC time 21 August 1999, 23:59:47). The first rollover, after 1023 weeks, occurred on 6 April 2019. The IRNSS Network Time (IRNWT) is a weighted mean average time similar to UTC; however, IRNWT is a continuous time without leap seconds. The IRNSS satellites carry a rubidium atomic clock onboard. These onboard clocks are monitored and controlled at the ISRO Navigation Centre. The deviation between each of the satellites and IRNWT is modelled, and the parameters of this model are transmitted as part of the IRNSS broadcast navigation messages.

The IRNSS system is designed to provide two types of services (ISRO 2017): Standard Positioning Service (SPS) and Restricted Service (RS). The SPS is an open service and has been released to the public to provide the essential information on the IRNSS signal-in-space, to facilitate research and development and aid the commercial use of the IRNSS signals for navigation-based applications. The SPS currently provides a position accuracy of better than 20 m in the primary service area, which is expected to be around 5 m in future. The RS service is an encrypted service provided only to authorised users (including the Indian military). The accuracy is also very high for the RS service. The IRNSS services are transmitted on L5 (1164.45–1188.45 MHz) and S (2483.5–2500 MHz) bands. The use of two frequencies helps in atmospheric error correction. Both of the services are based on CDMA. The navigation message is sent at a rate of 50 bps. The SPS PRN code is 1023 chips long and chipping rate is 1.023 Mcps (RS is not disclosed). Each carrier (L5 and S) is modulated by three signals namely, BPSK(1), Data channel BOC(5, 2) and Pilot channel BOC(5, 2). The SPS is based on BPSK(1) signal; RS uses all three signals. Both the services—SPS and RS—are available in L5 and S bands. Therefore, both the services are available to single as well as double frequency receivers.

IRNSS navigation system uses WGS 1984 coordinate system for the computation of user's position. The navigation message (ISRO says navigation data) includes IRNSS satellite ephemeris, IRNSS time, satellite clock correction parameters, status messages, and other secondary information. Navigation data are modulated on top of the ranging codes and can be identified as primary and secondary navigation parameters. The primary navigation parameters include satellite ephemeris, satellite clock correction parameters, satellite and signal health status, user range accuracy, and total group delay. The secondary navigation parameters are satellite almanac, ionospheric delays and confidence, IRNSS time offsets with respect to UTC and other GNSS, ionospheric delay correction coefficients, differential corrections, and earth orientation parameters. Other technical details of IRNSS can be obtained on ISRO website at https://www.isro.gov.in/sites/default/files/irnss_sps_icd_version1.1-2017.pdf.

The ISRO is also studying and analysing a new initiative to convert this regional system to a global system. This proposed global system is expected to be named Global Indian Navigation System (GINS). The GINS constellation will have 24 satellites at an altitude of around 24000 km to cover the entire earth (perhaps similar to GPS).

7.5 INERTIAL NAVIGATION SYSTEM

An Inertial Navigation System (INS) is a navigation aid that uses a computer and inertial (motion and rotation) sensors to continuously track the position, orientation, and velocity (direction and speed of movement) of a vehicle without the need for external references. Thus it is not explicitly dependent on any GNSS (Grewal *et al.* 2001; Farrell and Barth 1998; Titterton and Weston 2004).

An inertial navigation system includes at least a computer and a platform or module containing typical accelerometers, gyroscopes, magnetometers, or other motion-sensing devices. The INS is initially provided with its position and velocity from another source (a human operator, a GNSS satellite receiver, etc.), and thereafter computes its own updated position and velocity by integrating information received from the motion sensors. Unlike dead-reckoning (where frequent calibration is required) the advantage of an INS is that it requires no external references in order to determine its position, orientation, or velocity once it has been initialised. However, for the initialisation it requires positional information from any other sources, which is often obtained from a GNSS. Inertial navigation systems are now usually combined with GNSS through a digital filtering system. The inertial system provides short-term data, while the satellite system corrects accumulated errors of the inertial system time-to-time.

7.6 PSEUDOLITE

Pseudolites are ground-based radio transmitters that transmit a signal of a GNSS. They act like a navigation satellite, but from the earth's surface. Pseudolite is a contraction of the terms *pseudo* and *satellite*, used to refer to something that is not a satellite but which performs a function commonly in the domain of satellites. Pseudolites are most often small *transceivers* that are used to create a local, ground-based GNSS alternative. A *transceiver* is a device that has both a transmitter and a receiver in a combined form and share common circuitry or a single housing. The range of each transceiver's signal is dependent on the power available to the unit.

Pseudolites are intended to complement the GNSS. Being able to deploy one's own positioning system, independent of the GNSS, can be useful in situations where the normal GNSS signals are either blocked/jammed (military conflicts), or simply not available (exploration of other planets or indoors or underground). Pseudolites can aid in GNSS positioning in three main ways (Edward 2003; Samama 2008). First, they can be used to augment the GNSS satellite constellation by providing additional ranging sources when the natural satellite coverage is inadequate. Applications for which this additional coverage is beneficial include navigation in places with limited sky visibility, such as urban canyons and open pit mines, as well as some on-orbit applications involving relative positioning of spacecraft.

Second, pseudolites can be used as an aid to carrier-cycle ambiguity resolution when using carrier-phase differential GNSS for precise positioning. Motion-based integer resolution using only satellites is often a slow process because of the slow rate of change in the line-of-sight vectors to the satellites. In contrast, a receiver in the presence of a local pseudolite can see a large change in the line-of-sight vector with relatively little absolute motion on the part of the receiver. This can yield integer resolution in a matter of seconds.

Third, pseudolites can be used to replace the GNSS constellation completely. This is generally done to emulate GNSS positioning indoors, although more exotic locales such as underground or on the surface of other planets have also been explored. Most of this final class of applications uses carrier phase measurements rather than code-based pseudorange because of the high accuracy requirement within a relatively small area.

Users need to have a slightly modified GNSS receiver to make use of pseudolite signals. This is because receivers are designed to capture weak signals from distant satellites. The receiver is not designed to cope with the relatively high-powered transmissions of the pseudolite.

NOTE

A repeater is an electronic device that receives a signal and retransmits it at a higher level and/or higher power, or onto the other side of an obstruction, so that the signal can cover longer distances without degradation. Perhaps the best solution for infrastructure-based indoor positioning could be a mix of pseudolite and repeaters, using the benefits of each (Samama 2008).

7.7   INTEROPERABILITY AND INTEGRITY OF GNSS

The position calculation methods described in Chapter 6 and in this chapter are applicable to all the GNSS constellations. It is therefore possible to imagine using satellites from different constellations in order to carry out positioning. For example, one can think of two GPS satellites with two Galileo satellites, or two GPS satellites with one GLONASS and one IRNSS, or any combination. This is called *interoperability* of GNSS and is much stronger than just a superposition of constellations (Rycroft 2003; Samama 2008). Of course, it is possible to carry out multiple positioning from each constellation and then make some comparisons; but interoperability is indeed an embedded method that will allow, for example, a three-dimensional positioning in an urban canyon where no positioning is possible with any single constellation. There are GPS-only and GLONASS-only receivers available in the market. However, there is a distinct trend to develop receivers that can track and process signals from both the GPS and GLONASS satellites. One of the first commercial GPS+GLONASS systems is the Ashtech GG24 receiver. A combined GPS+GLONASS receiver can track signals from a 48+ satellite constellation, twice as many as the GPS-only or GLONASS-only constellation and therefore significantly improving availability. For example, simulation studies have shown that with a 45° obstruction to half the sky (as would be caused by tall buildings), five or more GPS satellites are only available for about 33% of the day, and four or more satellites for about 85% of the day. However, there is 100% availability of five or more satellites when both GPS and GLONASS satellites are considered (Samama 2008).

The equations to be used for a given satellite remain the pseudorange one (Equation 4.6) or the carrier phase one (Equation 4.9) as discussed earlier in Chapter 4. However, the only point common to the different constellations is the fact that one wants to get the position of the receiver. All the other variables are different; the locations of the satellites are given in different reference coordinate frames, namely WGS84 for GPS/IRNSS, PZ90 for GLONASS, GTRF for Galileo, and CGCS 2000 for BeiDou. This is not a real difficulty, as it is possible to define transformation matrices to achieve the conversions (refer to Chapter 9). This latter assertion is true when considering that all the satellites, from all the constellations, are of an identical 'quality'. This also leads to the problem of the choice of satellites to use in the case where there are more than four. The least-square method is usually based on identical hypotheses for each satellite, which is not obvious within a constellation, and not even realistic for different constellations. Nevertheless, the method can certainly be extended by individual measurement weights in order to account for different qualities.

In addition, there is also the problem of receiver clock bias. Of course, as this is a physical bias, it is identical for all constellations, but unfortunately, the time reference frames are not. Therefore, it is quite clear that the clock biases are not the same. Two approaches are then possible: either to take into account the clock corrections provided by the navigation message of a constellation (e.g., GLONASS or Galileo or BeiDou), which should be able to characterize the bias between the considered constellation time with respect to the GPS time; or to use additional satellites in order to remove these biases (by implementing some kind of differential approach) (Samama 2008). The first method is much simpler but needs to be qualified in terms of induced errors. Of course, when considering the pseudoranges, usual corrections should be applied in order to take into account the various corrections, different for each constellation; but one can consider that once implemented in a receiver, these corrections are not a real difficulty. Navigation messages have to be dealt with in different ways depending on the constellation (this is just a processing complexity, not an issue).

Another issue regarding interoperability is the dilution of precision (DOP). In some cases, like urban canyons, positioning will be achievable while using interoperability because two satellites of each constellation will be available. That means, no positioning is possible with any single constellation, but is possible with a combination of constellations. One still has to remember that the positioning, while possible, is bound to be of reduced accuracy because of the poor DOP available.

The *integrity* of GNSS can be defined as a reliability indicator of the quality of positioning (Rycroft 2003; Grewal *et al.* 2001). Given the various error sources, it is obvious that the estimated user range error, even taking into account the DOP factor (refer to Chapter 5), is a 'passive' indicator. Let us imagine that the signals transmitted from one satellite are totally wrong, then the user range error will still provide the user with the same value. In order to make such a quality factor available, there is a need for other mechanisms. The three we are going to briefly discuss in this section are: (1) receiver autonomous integrity monitoring (RAIM), (2) the SBAS integrity concept, and (3) the Galileo and other constellations' integrity concepts.

As already described, there are very often more than four satellites available in the sky for positioning purposes. The main idea of RAIM is to conduct independent position computations with various sets of chosen satellites (Parkinson and Spilker 1996; Grewal *et al.* 2001). The need for at least one supplementary satellite is obvious to determine the time difference between two constellations, and the actual availability of more than five allows the erroneous signal, if any, to be defined in addition to the detection of the problem. The basic mechanism consists of computing six positions (in the case of five satellites), the first one with all five satellites (using the least-square method for instance), and the five others considering the possible five sets of four satellites. Then, comparisons are carried out between the various positions found, and an analysis of the dispersion leads to the detection of problems relative to one satellite, if any. The problem could be linked to a navigation message error or even specific propagation conditions. Thus, having an additional satellite allows any faulty satellite to be determined. This technique was first implemented for applications requiring integrity usually linked with safety, as in air or rail transportation systems.

The SBAS have been developed with two main goals: accuracy improvements and integrity requirements. The second is clearly identified as the most important and has been the main guide for system definition. The concept has also been supported by the International Civil Aviation Organization (ICAO) and is thus clearly applicable to civil aviation. The definition of integrity is the ability of a system (infrastructure and user) to provide positioning with an associated level of confidence. Thus, it is not an intrinsic characteristic of the system, but related to a specific application and context (Samama 2008). That means the most important consideration about integrity is to remember that it is closely associated with an application and hence a context. The integrity concept implemented in EGNOS, WAAS, MSAS, and GAGAN is associated with the civil aviation context and should be very carefully extended to other applications, contexts, and environments. A typical example of a situation where care should be taken is when stating that urban applications will benefit from the integrity provided by the SBAS. However, the errors associated with multipath, which are certainly one of the main problems in urban canyons, are not at all taken into account in the SBAS integrity concept currently implemented.

NOTE

A new term, *Assisted GNSS* (AGNSS), has emerged recently (Tekinay 2000). AGNSS is a technology that enables faster position determination in an AGNSS-enabled receiver than could be achieved using the broadcast GNSS satellite data only. A standalone GNSS needs navigation models for ephemerides and clock corrections, reference location, ionosphere models, reference time, and optionally differential corrections to calculate its position. The data rate of the satellite signal is typically 50 bps; therefore, downloading orbital information, like ephemerides and the almanac, directly from satellites typically takes a long time, and if the satellite signals are lost during the acquisition of this information, it is discarded and the standalone system has to start from scratch. Further, the ephemeris data has an expiration period of 4 hours, after which the ephemeris data cannot be used and the satellite position will be lost. Therefore, the ephemeris data must be updated in fixed periods of every 2–4 hours. In the case of AGNSS, the network operator deploys an AGPS server. These AGNSS servers download the orbital information from the satellites and store it in the database. An AGNSS-enabled receiver can connect to these servers and download this information using mobile-network radio (such as GSM, CDMA, WCDMA, LTE, or even other radio bearers, such as Wi-Fi). Usually the data rate of these bearers is high; hence downloading orbital information takes less time. When the assisted data is delivered over a telecom system, the typical position fix times are in the order of 10–20 sec compared to 40–60 sec using autonomous methods or even longer fix times in weak signal conditions. Optimally, the receiver already has the assisted data in its memory, for example, in the form of extended ephemerides—and, as a result, the position fix time can be as short as few seconds.

EXERCISES

DESCRIPTIVE QUESTIONS

1. What do you understand by 'GNSS augmentation'? Classify it with examples.
2. What is SBAS? Explain in brief.
3. Describe the European Geostationary Navigation Overlay Service.
4. What do you know about WAAS and MSAS?
5. Explain the concept adopted for GAGAN.
6. What is GBAS? Give two examples of GBAS and their applications.
7. What is Regional Navigation Satellite System? Why are these systems important? Describe QZSS briefly.
8. What do you know about IRNSS? What is the coverage of IRNSS?
9. Explain the services provided by IRNSS. What is IRNSS time?
10. What is 'pseudolite'? What are its advantages over the conventional GNSS?
11. What do you understand by interoperability and integrity? What are the problems associated with these concepts and how can they be overcome?
12. What do you know about 'inertial navigation system' and AGNSS?
13. How can interoperability identify a satellite with an erroneous signal? Can any SBAS system solve the accuracy problem in urban areas? Explain logically.

SHORT NOTES/DEFINITIONS

Write short notes on the following topics

1. GNSS-1
2. GNSS-2
3. WAAS
4. OmniSTAR
5. LAAS
6. QZSS
7. Inertial Navigation System
8. Receiver Autonomous Integrity Monitoring
9. Beidou I
10. Ground-based Regional Augmentation System
11. Trimble CenterPoint RTX
12. Leica HxGN SmartNet

8 GNSS Receivers

8.1 INTRODUCTION

A GNSS receiver must collect and then convert signals from GNSS satellites into measurements. Their characteristics and capabilities influence the techniques available to the user throughout the work, from the initial planning to the processing. To understand the GNSS technology better, we need to understand GNSS receivers—how they work, what are the issues involved in terms of achieving positioning and navigation accuracy, what are the types of receivers, and how to choose an appropriate receiver for a specific application.

In 1980, only one commercial receiver was in the market, at a price of several hundred thousand US dollars (El-Rabbany 2002). This scenario, however, has changed very rapidly in the last four decades. Now every person possesses one or more GNSS receivers in their smartphones and smartwatches; that is, there is steady growth in the use of GNSS in new and existing markets. The increasing use of GNSS brings an increasing reliance on the technology. Individuals, businesses, and organisations are relying on the technology for personal pleasure, safety, and commercial advantages. Now, there are literally hundreds of different types of GNSS receivers in the market. They are generally capable of achieving accuracies from several meters to sub-centimetre. They are capable of pseudoranging and/or carrier phase measurement, with/without differential capabilities, real-time capabilities, static-mode operational, and several other hybrid techniques. They can also be accompanied by post-processing software and network adjustment software. Further, some are equipped with capacity for extra batteries, external data collectors, internal or external antennas, radio modem, etc. It is not possible to describe each of them in detail within the confines of this book. However, this chapter briefly provides an overview of basic GNSS receivers.

8.2 RECEIVER ARCHITECTURE

GNSS receiver architecture is rather complex and differs widely. However, the basic architecture is simple and composed of an internal or external antenna, a radio frequency section, a microprocessor, a control and display unit, and a storage unit with necessary power supply (Figure 8.1). The following sections describe these components briefly (for more details please refer to Langley 1991d; Van Dierendonck 1995; Kaplan 1996; Spilker and Parkinson 1995; Tekinay 2000; Sickle 2008).

8.2.1 RECEIVER ANTENNA

An antenna is required to collect the electromagnetic signals transmitted by the satellites and to convert them into electric signals that can be utilised by the receiver

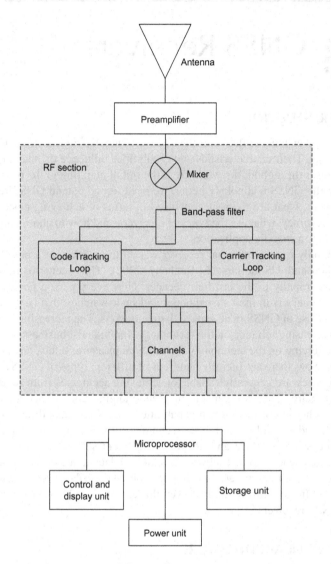

FIGURE 8.1 Block diagram of a typical GNSS receiver (Sickle 2008).

electronics. The main function of the antenna is to convert the electromagnetic waves into electric currents sensible to the *radio frequency* (RF) section of the receiver (refer to Section 8.2.2 for more about the RF section). Several designs are possible for a GNSS antenna, but the satellite's signal has such a low power density, especially after propagating through the atmosphere, that antenna efficiency is critical (Langley 1998a; Enrico and Marina 2008). The signal is generated by the satellite antenna at about 20,000 km away from the earth's surface (the altitude varies for different constellations). Therefore, GNSS antennas must have high sensitivity (also known as *high gain*). They can be designed to collect only one frequency (e.g., L1) or more than one frequency (commonly two).

The antenna may be internal or external to the receiver. Most receivers have an antenna built-in, but many can accommodate a separate tripod-mounted or range pole-mounted antenna as well. These separate antennas with their connecting coaxial (or other type) cables in standard lengths are usually available from the receiver manufacturer. Antenna cables generally come in standard lengths ranging from 3–30 m. The cables are an important element. The longer the cable, the more of the GNSS signal is lost travelling through it. In instances where an antenna is to be mounted permanently for use as a reference receiver, antenna cable lengths which are greater than the standard lengths may be required. In this situation, an in-line amplifier can be purchased which provides additional amplification of the signal on its way to the receiver. The signal from the antenna is transferred to the receiver via a coaxial antenna cable and can be distorted if the antenna cable is excessively bent, coiled, or otherwise damaged. The antenna cable also serves as the mechanism by which power is transferred to the preamplifier. Therefore, the role of the antenna cable is as important as the antenna and receiver. Users should take care of antenna cables by minimising the coiling and wear on the cable. It is recommended that antenna cables be replaced periodically, especially if the surveys being performed expose the equipment to harsh treatment (RMITU 2006). Nowadays, receiver and antenna are generally housed in a single unit to overcome these limitations. However, a separate controller is used which can be connected to the receiver via Bluetooth.

An antenna ought to have a bandwidth corresponding to its application. In general, better performance can be achieved with wider bandwidth; however, increased bandwidth degrades the signal to noise ratio by including more interference. One very important feature of a GNSS receiver antenna is the stability of its *phase centre* (RMITU 2006). The *electrical centre* of a receiver antenna is the point whose location we compute and is commonly called the *phase centre*. To determine the coordinates of a point, we must co-locate the phase centre of the antenna with that point. But usually the 'physical centre' of the antenna is positioned exactly above the point with a vertical offset (due to tripod height) and the vertical offset is accurately measured. This vertical offset will then be taken into account when calculating the coordinates of the intended point. It is therefore assumed that the 'physical centre' of the antenna is the same as its phase centre. This is true only in special advanced antennae designed for precision applications. The phase centre of a typical antenna can shift by several centimetres as the satellites move. 'Phase centre stability' is the main characteristic of an antenna for precision applications. Fortunately, the phase centre of most antennas currently being manufactured is quite stable and the magnitude of offsets is only a few millimetres.

Another feature of the antenna that requires consideration is the *antenna gain pattern*. The gain pattern describes the ability of the receiver to track signals at certain azimuth and elevations. An ideal antenna can track signals low to the horizon, at all azimuths. In general, the tracking performance of microstrip antennas is suitable for surveying. The GNSS signals become very weak by the time they reach the antenna. As a result, most antennas attach a built-in *preamplifier* which boosts the level of the signal before it is passed on to the receiver.

8.2.2   RF SECTION

The RF section is the first section of the receiver that performs operations on the signal after its amplification by the antenna preamplifier. This section is often termed as the *front end* of the receiver. Different receiver types use different techniques to process the GNSS signals, but they go through substantially the same steps that are explained in this section.

The preamplifier increases the signal's power, but it is important that the gain in the signal coming out of the preamplifier is considerably higher than the noise (Ward 1994). Since signal processing is easier if the signals arriving from the antenna are in a common frequency band, the incoming frequency is combined with a signal at a harmonic frequency. This latter, pure sinusoidal signal is the previously mentioned reference signal generated by the receiver's oscillator. The two frequencies are combined together in a device known as a *mixer*. Two frequencies emerge—one of them is the sum of the two that went in, and the other is the difference between them (Sickle 2008).

The sum and difference frequencies then go through a *band-pass filter*, an electronic filter that removes the unwanted high frequencies and selects the lower of the two (Williams and Taylor 1998). It also eliminates some of the noise from the signal. The resulting signal after filtering is known as the *intermediate frequency* (IF), or *beat frequency* signal. This beat frequency is the difference between the Doppler-shifted carrier frequency that came from the satellite and the frequency generated by the receiver's own oscillator. There are usually several IF stages before copies of it are sent into separate channels, each of which extract the code and carrier information from a particular satellite.

As mentioned earlier, a replica of the C/A or P code is generated by the receiver's oscillator and now that is correlated with the IF signal. At this point the pseudorange is measured. Remember, the pseudorange is the time shift required to align the receiver's internally generated code with the IF signal multiplied by the speed of light. The receiver also generates another replica—a replica of the carrier. That carrier is correlated with the IF signal and the shift in phase can be measured. The continuous phase observable, or observed cycle count, is obtained by counting the elapsed cycles since *lock-on* and by measuring the fractional part of the phase of the receiver generated carrier.

The antenna itself does not sort the information it gathers. The signals from several satellites enter the receiver simultaneously. But the undifferentiated signals are identified and segregated from one another in the channels of the RF section. A *channel* in a GNSS receiver is similar to a channel in a television set. It is hardware, or a combination of hardware and software, designed to separate one signal from all the others. At any given moment, only one frequency from one satellite can be on one channel at a time. A receiver may have as few as 1 or as many as several hundreds of physical channels; for example, Trimble R10 and R8s receivers have 672 and 440 channels, respectively. Although earlier we have used single channel receivers, nowadays a minimum of 12 channels is the norm. A receiver with 12 channels is also known as a *12-channel parallel receiver.* Such a receiver may actually have 24 channels if it is a dual frequency receiver. The first available receivers were based on a sequential approach; the satellites were acquired one at a time in a row. However, later the receiver's architecture made progress, and *parallel channel receivers* allowed multiple simultaneous acquisition and tracking capabilities.

Pseudorange measurement (matching the satellite code with the receiver's replica) is carried out within the *code tracking loop* (also known as *delay locking loop*) of the RF section. However, pseudoranges alone are not adequate for some of the applications. Therefore, the next step in signal processing for most receivers involves the carrier phase observable. As stated earlier, just as they produce a replica of the incoming code, receivers also produce a replica of the incoming carrier wave. And the foundation of carrier phase measurement is the combination of these two frequencies of carrier wave. It is important to remember that the incoming signal from the satellite is subject to an ever-changing Doppler shift, while the replica within the receiver is nominally constant.

The process begins after the PRN code has done its job and the code tracking loop is locked. By mixing the satellite's signal with the replica carrier, this process eliminates all the phase modulations, strips the codes from the incoming carrier, and simultaneously creates two intermediate or beat-frequencies—one is the sum of the combined frequencies, and the other is the difference. The receiver selects the latter, the difference, with a device known as a band-pass filter. Then this signal is sent on to the *carrier tracking loop*, also known as *phase locking loop* (PLL) (RMITU 2006), where the voltage-controlled oscillator is continuously adjusted to follow the beat frequency exactly.

As the satellite passes overhead, the range between the receiver and the satellite changes. This steady change is reflected in a smooth and continuous movement of the phase of the signal coming into the receiver. The rate of change is reflected in the constant variation of the signal's Doppler shift (refer to Chapter 4, Section 4.10.1). But if the receiver's oscillator frequency matches these variations exactly as they happen, it will duplicate the incoming signal's Doppler shift and phase. This strategy of taking measurements using the carrier beat phase observable is a matter of counting the elapsed cycles and adding the fractional phase of the receiver's own oscillator.

Doppler information has broad applications in signal processing. It can be used to discriminate between the signals from various GNSS satellites, to determine integer ambiguities in kinematic surveying, as a help in the detection of cycle slips, and as an additional independent observable for autonomous point positioning. But perhaps the most important application of Doppler data is the determination of the *range rate* between a receiver and a satellite. Range rate is a term used to refer to the rate at which the range between a satellite and a receiver changes over a particular period of time. Doppler shift will be discussed further in Section 8.3.1.

NOTE

The operation of a GNSS receiver can be severely limited or completely disrupted in the presence of in-band or out-of-band interference and jamming signals (Ward 1994). Most receivers filter out-of-band noise, but few suppress interference within the band. The threat of the in-band interference and jamming signals increases daily as new communication systems are put in place and the radio frequency spectrum becomes more populated. The threat is not only from the interfering signals themselves but also from their harmonics that fall inside the GNSS band.

8.2.3 MICROPROCESSOR

The microprocessor controls the entire receiver, managing its collection of data (Enrico and Marina 2008). Many of the functions of the GNSS receiver are digital, rather than analogue, operations. The control of operations, such as initially acquiring satellites, tracking the code and carrier, interpreting the broadcast navigation message, extracting the ephemeris and other information from the navigation message, and mitigating multipath and noise, among other things, are performed by a microprocessor.

Microprocessor does datum (refer to Chapter 9) conversion and produces final positions instantaneously. Then they serve up the position through the control and display unit. There is a two-way connection between the microprocessor and the control and display unit. The microprocessor and the control and display unit each can receive information from or send information to the other.

The microprocessor runs a program which is stored on a memory chip within the receiver. When the receiver firmware is upgraded to a new version, it is this program which is being updated. The power of the microprocessor defines a number of related characteristics including computing ability (whether a simple positioning solution or a real-time carrier phase survey is supported), speed of signal acquisition and calculation, receiver size, and power requirements. In general, the more powerful the processor is, the greater the functionality that can be supported by the receiver.

8.2.4 CONTROL AND DISPLAY UNIT

Generally, the GNSS receiver has a control and display unit (also called a *user interface*). From handheld keyboards to soft keys around a screen to digital map displays, and interfaces to other instrumentation, there are a variety of configurations. Nevertheless, they all have the same fundamental purpose, facilitation of the interaction between the operator (user) and the receiver's microprocessor. This unit may be used to select different surveying methods and/or set its parameters, that is, epoch interval, mask angle, and antenna height. The control and display unit may offer any combination of help menus, prompts, datum conversions, readouts of positioning, navigation results (e.g., latitude, longitude, altitude, speed, north direction, route, etc.), and so forth. Nowadays, the control and display units come separately in most cases and can be connected with a receiver via Bluetooth.

The information available in the control and display unit varies from receiver to receiver. But when four or more satellites are available, they can generally be expected to display the PRN numbers of the satellites being tracked, the receiver's position in three dimensions, sky plot, north direction, velocity information, and estimated error. Some of them also display the DOP values and GNSS time.

8.2.5 STORAGE UNIT

For almost all surveying and navigation applications, some information is required to be stored. This information may comprise pseudorange and carrier measurements for processing, station identifiers and antenna height details, position estimates determined by the receiver microprocessor, maps (used for navigation), software (to perform several operations within the receiver), etc. Most GNSS receivers have internal data logging (e.g., flash memory cards). In some cases, external data logger is also used to store the GNSS measurements and/or information supplemental to the positioning measurements. Modern receivers normally use solid state memory cards. Most allow the option of connecting to a computer and having the data downloaded directly to the computer hard drive.

The amount of storage required for a particular positioning session depends on several things: the length of the session, the number of satellites above the horizon, the epoch interval, and so forth. For example, presuming the amount of data received from a single GNSS satellite is 100 bytes per epoch, a typical 12-channel dual-frequency receiver that is observing 6 satellites and using a 1 sec epoch interval over the course of a 1 h session would require approximately 2 MB of storage capacity for that session.

$$3600 \text{ epochs/h} \times 1 \text{ h} \times 6 \text{ satellites} \times 100 \text{ bytes per satellite} \approx 2 \text{ MB}$$

A large number of range measurements and other pertinent data are sent to the receiver's storage unit during observation sessions. These data are subsequently downloaded through a serial or USB port to a personal computer to provide the user with the option of post-processing or application.

8.2.6 POWER UNIT

Since most receivers in the field operate on battery power, batteries and their characteristics are fundamental to GNSS surveying. A variety of batteries are used and there are various configurations (RMITU 2006). For example, some GNSS units are powered by camcorder batteries, and handheld recreational GNSS units often use disposable alkaline AA or even AAA size batteries. However, in surveying applications, rechargeable batteries are the norm. Lithium-ion polymer, Nickel Cadmium, and Nickel Metal-Hydride may be the most common categories, but lead-acid car batteries still have an application as well. It is fortunate that GNSS receivers operate at low power; 3–36 volts DC is generally required.

Receiver architecture may vary in terms of their construction. In low-cost receivers all components (antenna, RF section, display/control, power, and storage) are housed in a single unit, serving all the purposes in this case. Earlier, in some high-precision receivers, the antenna (with preamplifier) was a separate unit. The RF section and the display/control were housed in another unit. These two units were connected through coaxial cable. The power (from receiver to the antenna) and data (from antenna to the receiver) were transferred via this coaxial cable. The modern high-precision receivers house the antenna and RF section in one unit (commonly called receiver); the control/display is housed in another unit (known as the controller). A receiver may also house a radio modem in it (refer to Section 8.2.7). Both of these units have separate microprocessor, storage, and power units. The data can be logged either in the receiver or in the controller.

8.2.7 RADIO MODEM

As discussed earlier, in real-time DGNSS and RTK data is transmitted from the base receiver situated in a known position to the roving receiver. The rover then takes the base data into account in order to compute its own position accurately. Radio modems provide wireless communication between the base receiver and the rover (Javad and Nedda 1998). A radio modem transmitter with base receiver and a radio modem receiver with rover is required, so that when a base receiver broadcasts data via its radio modem transmitter, an unlimited number of rovers can pick up the data via their radio modem receivers.

A radio modem transmitter consists of a radio modulator, an amplifier, and an antenna. The radio modulator takes the GNSS data to be transmitted from the base receiver and converts it to a radio signal that can be transmitted. The amplifier raises the power of the signal to a level that can reach the rover (the farther the rover, the more amplification is needed). The transmitter antenna then sends the amplified signal. The power of the amplifier directly affects the distance that the signal can travel and the reliability of the communication. The range also depends on the terrain and the radio antenna setup.

The connection between the radio modem transmitter and the base receiver is generally established via the serial ports of the receiver and the radio modem transmitter. If the radio transmitter is integrated within the receiver electronics, then the connection is done internally. A radio modem receiver consists of a radio receiver antenna and a radio demodulator. The radio receiver antenna picks up the radio signal from the air and delivers it to the radio demodulator that converts the signal back to the form that can be delivered to the serial port of the rover. Modern radio modem receivers are internal to the rover. The main characteristic of a radio modem is the form to which the data is converted for its transmission. UHF, VHF, and spread spectrum (frequency-hopping or direct-sequence) are some examples. There are some advantages and some disadvantages to each form (see Abidi et al. 1999).

Recent introductions in GNSS radio modems include the use of accurate timing of the GNSS for data synchronisation between base and rover(s) in order to enhance data integrity and the use of direct-sequence and frequency-hopping combination in spread spectrum radio to enhance communication reliability. Government authorisation may be required for using certain types of radio modems, as there are international and national bodies that allocate frequency bands and issue authorisation to transmit signals. In some countries, there are bands allocated for public use without the need for any special authorisation. This is an important factor to consider when selecting a radio modem, since getting authorisation is often not an easy task. The 900 MHz band in the United States and 2.4 GHz in most European countries are allowed for spread spectrum communication without any special authorisation (but there are limitations on the power that can be used to transmit signals).

UHF and spread spectrum radio modems are the most popular for DGNSS and RTK applications. Spread spectrum radios (900 MHz and 2.4 GHz) have a range of about 20 km (unless the antenna is installed at a very high location). UHF has a longer range; with a 35-watt amplifier, a UHF radio can have a range of up to 45 km, depending on terrain and antenna setup.

NOTE

Spread-spectrum techniques are methods by which energy generated in a particular bandwidth is deliberately spread in the frequency domain, resulting in a signal with a wider bandwidth (Dixon 1984; Freeman 2005). These techniques are used for a variety of reasons, including the establishment of secure communications, increasing resistance to natural interference and jamming, and to prevent detection. *Spread-spectrum telecommunications* is a signal structuring technique that employs *direct-sequence, frequency-hopping* or a hybrid of the two, which can be used for multiple access and/or multiple functions. This technique decreases the potential interference to other receivers while achieving privacy. Spread spectrum generally makes use of a sequential noise-like signal structure to spread the normally narrowband information signal over a relatively wideband (radio) band of frequencies. The receiver correlates the received signals to retrieve the original information signal.

Frequency-hopping spread spectrum is a method of transmitting radio signals by rapidly switching a carrier among many frequency channels, using a pseudorandom sequence known to both transmitter and receiver. *Direct-sequence spread spectrum* is a modulation technique by which the transmitted signal takes up more bandwidth than the information signal that is being modulated.

8.3   SIGNAL ACQUISITION AND POSITIONING

In this section, we discuss the sometimes-difficult problems of acquisition and tracking of the GNSS signals and their solutions. The main difficulty in GNSS positioning is making the best possible measurements, i.e., the most accurate ones, in order to provide positioning with the highest accuracy. Indeed, as long as the measurements are totally accurate, the positioning will be perfect. Unfortunately, as already described in Chapter 5, a lot of errors lead to inaccuracies. Other possible errors are caused by the receiver's own electronic processing unit, which could give rise to inadequate correlations because the signal is not as pure as one would expect. Multipath, low power levels, or cross channels (different satellites) may also cause imperfect correlations. In addition, the signal has to be found in a two-dimensional area that copes with Doppler shifts, due to both the satellite's motion and the receiver's displacement on the one hand, and time shift, due to the propagation delay from satellite to receiver, on the other. Of course, both searches (i.e., Doppler and time) are fundamental to carrying out position calculation. In addition to these, the receiver is also required to solve the integer ambiguity for precise carrier phase measurement. Different approaches have been implemented throughout the years, leading to real improvements in the resultant quality of the positioning. It is worth mentioning that these problems are addressed by the RF section of the receiver.

The basic pseudorange measurement is clearly time, which is required for distance estimations from the satellites to the receiver. This is achieved through the use of a code sequence that is also the identifier of the satellite (for GPS, Galileo, BeiDou and IRNSS, as for GLONASS identification is achieved through different frequencies, but a code is still used). We know that the code correlation technique allows the pseudorange to be estimated (refer to Chapter 4). Unfortunately, the signal is distorted due to the Doppler effect (Hatch 1982; Jones 1984). The Doppler effect is a physical compression or dilation (becoming wider) of the signal due to the transmitter's and/or receiver's relative motion. As this distortion is a physical one, it applies to the carrier and also to the code, thus also leading to a distortion in the code, which must be taken into account. The effect is that the code can be either reduced in length or elongated. If this effect is not taken into account, the correlation might be non-optimal and the propagation time will not be measured accurately. Thus, the basic architecture of a receiver must implement both a frequency search (to cope with Doppler) and a time search (to cope with propagation time). Additionally, if the measurement is based on the carrier phase, instead of code, cycle ambiguity must also be solved.

8.3.1 DOPPLER SHIFT

The Doppler of a satellite (Gill 1965; Bowditch 1995; Parkinson and Spilker 1996; Sickle 2008), from the moment it appears on the horizon to the moment it disappears on the other side, is given in Figure 8.2. This typical curve depicts Doppler shift: when the satellite appears, it gets closer to the receiver (positive Doppler), then passes over the receiver exhibiting a zero Doppler, and finally goes away from the receiver, thus exhibiting a negative Doppler. Figure 8.2 also shows the value of the *Doppler frequency* or *Doppler shift*.

A typical interval of Doppler shift to be considered can be obtained by a simple calculation of the projection of the vector of maximum velocity. In such a case, the larger value of Doppler is found to be around +5 kHz. In addition, the corresponding value of the receiver's displacement-induced Doppler must be added, as well as the effect of the drift of the receiver's local oscillator. Typical cumulative resulting

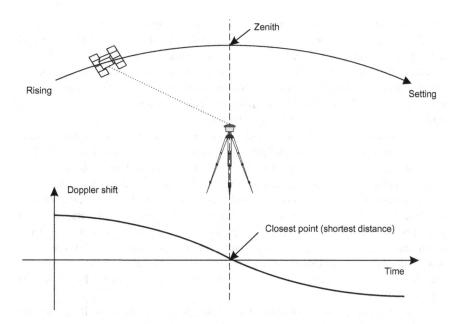

FIGURE 8.2   A typical satellite Doppler shift curve obtained at a fixed location.

values are +10 kHz. The Doppler-shift and the carrier phase are measured by first combining the received frequencies with the nominally constant reference frequency created by the receiver's oscillator. The difference between the two is the often mentioned *beat frequency* (or *intermediate frequency*), and the number of beats over a given time interval is known as the *Doppler count* for that interval. Since the beats can be counted much more precisely than their continuously changing frequency, most GNSS receivers just keep track of the accumulated cycles, the Doppler count. The sum of consecutive Doppler counts from an entire satellite pass is often stored, and the data can then be treated like a sequential series of biased range differences. The rate of the change in the continuously integrated Doppler shift of the incoming signal is the same as that of the reconstructed carrier phase. Integration of the Doppler frequency offset results in an accurate measurement of the advance in carrier phase between epochs.

8.3.2   TIME SHIFT

The time aspect is little different from the Doppler, because it relies on code structure, length, and rate. A first raw approach consists in considering that there are as many different time shifts as there are chips in the code. That means in the case of GPS L1 C/A, there are 1023 possible time shifts. For GLONASS C/A, there are 511 possible time shifts. Another important point lies in the time duration corresponding to the code lengths. For GPS C/A, the code lasts 1 ms, the same as GLONASS C/A. The direct impact is that the search domain is 1 ms for both GPS and GLONASS, for, respectively, 1023 and 511 time slots. This simple approach shows that the time search will take different values for different codes. In all cases, this time search is long. The real complete signal search is the combination of the Doppler and time searches. The difficulty of the signal search is that the two-dimensional search area is huge. When the signal has a sharp aspect it is rather a good configuration, but noise sources are bound to produce many different peaks that would lead to a much more difficult determination of the true signal peak.

The first available receivers in earlier days (obviously GPS receivers) were based on a sequential approach; the satellites were acquired one at a time in a row. Indeed, the first satellite was acquired and tracked, the corresponding pseudorange was stored, and then the second satellite acquired and tracked, then the third one, and so on. Once enough satellites were acquired and data stored, the calculations of position, velocity, and time were carried out. In 1990s, the receiver's architecture made some progress, and parallel channel receivers allowed multiple simultaneous acquisition and tracking capabilities. By channel, one has to understand a structure that deals with the time and Doppler search for one particular satellite. Multiple channel receivers are used, nowadays, in order to reduce the acquisition time by implementing parallelism.

Another approach to the complex two-dimensional search, while in the acquisition phase of the GNSS signal, is to implement heavy signal processing such as *Fast Fourier Transform* (Van Loan 1992). Here the idea is to transfer the signal into the frequency domain in order to quickly find the Doppler shift of the signal; however, this is quite disadvantageous in terms of power consumption (Corazza 2007; Samama 2008).

8.3.3 INTEGER AMBIGUITY

As mentioned earlier in Chapter 6, the solution of the integer ambiguity, the number of whole cycles on the path from satellite to receiver, is a complex matter. But it would be much harder if it was not preceded by pseudoranges, or code-phase measurements in most receivers. This allows the centering of the subsequent double-difference solution.

After the code-phase measurements, three effective methods are used to solve the integer ambiguity (Parkinson and Spilker 1996; Sickle 2008). The first is a sort of geometric method applied in static survey. The carrier phase data from multiple epochs is processed and the constantly changing satellite geometry is used to find an estimate of the actual position of the receiver. It works pretty well, but depends on a significant amount of satellite motion and, therefore, takes time to converge on a solution. The second approach uses filtering. Here independent measurements are averaged to find the estimated position with the lowest noise level. The third uses a search through the range of possible integer combinations and then calculates the one with the lowest residual. The search and filtering methods depend on heuristic calculations, in other words, trial and error. These approaches cannot assess the correctness of a particular answer, but can calculate the probability, given certain conditions, that the answer is within a specified set of limits. Most GNSS processing software use some combination of all three ideas. All of these methods are guided by the initial position estimate provided by code-based measurements.

There is another method that does not use the codes carried by the satellite's signal. It is called *codeless tracking*, or *signal squaring*. It was first used in the earliest civilian GNSS receivers. It makes no use of pseudoranging and relies exclusively on the carrier phase observable. Like other methods, it also depends on the creation of an intermediate or beat frequency. But with signal squaring, the beat frequency is created by multiplying the incoming carrier by itself. The result has double the frequency and half the wavelength of the original—it is squared. There are some drawbacks to this signal squaring method. For example, as discussed in Chapter 4, the signals broadcast by the satellites have phase shifts called code states that change from +1 to –1 and vice versa, but squaring the carrier converts them all to +1. The result is that the codes themselves are wiped out. Therefore, this method must acquire information such as almanac data and clock corrections from other sources. Other drawbacks of squaring the carrier include the deterioration of the signal-to-noise ratio because when the carrier is squared, the background noise is squared too, and cycle slips occur at twice the original carrier frequency. But signal squaring has its advantages as well. It reduces susceptibility to multipath, it has no dependence on PRN codes, and it is not hindered by the encryption of the GPS P code.

8.4 CLASSIFICATION OF GNSS RECEIVERS

GNSS receivers can be classified in various ways, such as by architecture, by method of operation, according to receiving capabilities, and so on (Gopi 2005). This section describes some of the important classifications. Commercial GNSS receivers may be divided into four main types, according to their receiving capabilities: *single-frequency code receivers*, *single-frequency carrier smoothed code receivers*, *single-frequency code and carrier receivers*, and *dual-frequency receivers*.

GNSS receivers can also be categorised according to their number of tracking *channels*, which varies from 1 to 48 (or more). Twelve-channel receivers are now very common, and there are also professional receivers having hundreds of channels. According to channel architecture, GNSS receivers can be classified as: *sequential receiver, continuous-tracking receiver*, and *multiplexing receiver*. The sequential receiver uses one or two hardware radio channels to sequentially provide individual satellite observations. These receivers are among the cheapest available due to the limited circuitry needed. However, they take a long time to lock-on a satellite and cannot track satellites while moving at high speeds. The continuous-tracking receiver (also called *multichannel* or *parallel channel* receiver) has sufficiently dedicated hardware radio channels (parallel channels) to provide continuous satellite observations. This type of receiver has the best performance of all the receiver architectures. Four hardware radio channels at minimum are required for continuous operation. A five channel receiver can view four satellites and read the navigation message from the fifth, thus continuously keeping the receiver's database of satellite orbital parameters up-to-date. A six-channel receiver can read navigation messages, track four satellites and keep a fifth satellite in reserve in case one of the four is lost for any reason.

An all-in-view receiver has sufficient hardware radio channels to lock onto all the satellites that happen to be in view at any time. The multiplexing receiver acts like a sequential receiver in that it switches between satellites being tracked; however, it does this at a fast sample rate (approximately 50 Hz) and can track more satellites than a sequential receiver. It uses the *multiplexing* (also known as *fast-switching*, *fast-sequencing*, or *fast-multiplexing*) channel instead of the parallel channel in a continuous-tracking receiver. Performance is still lower than a continuous-tracking receiver, because it can't integrate all of the satellites in view.

By method of operation, GNSS receivers can be divided into two classes—*code-based*, and *carrier phase based*. Code-based receivers determine position by processing the information found in the code (code correlation). The advantage to this method is low cost; the only drawback is moderate accuracy on the order of ±5 m or worse. Carrier phase based receivers determine position by processing measurements of the carrier phase of the satellite signals over time. They do not need to decode the information being transmitted except for locating the satellites. Such receivers can provide centimetre level accuracy in real-time when used with differential correction. The drawback is its high cost.

Finally, by application, receivers come in a wide range of variety. *General-purpose handheld receivers* are small, portable, battery-powered, and have a built-in display and keypad. Some of these receivers may have the ability to display aeronautical or marine charts from data cards. The antenna is generally placed inside the receiver unit; however, in some of the cases, the antenna may be detachable for mounting outside a vehicle.

Aviation-purpose receivers are optimised for aviation navigation and can display aeronautical charts. Accuracy varies depending on the class of aircraft in which the device is to be used. Receivers intended for use by general aviation may not use any correction, thus are limited to 10 m 95% accuracy (95% chance of recording the point within 10 m of the actual location). Receivers integrated into the navigation suite of a commercial passenger aircraft may be capable of using augmentation broadcasts, increasing the precision of the aircraft's navigation system to the point that the aircraft can land automatically. Unmanned aerial vehicles (drones) require the use of GNSS receiver to provide them with a high degree of positioning accuracy. A GNSS antenna is mounted somewhere on the drone that receives location and time data from a GNSS constellation. This data is then usually fed into the navigation system of the vehicle for its autonomous takeoff/movement/landing. *Marine receivers* are intended for marine navigation, including the ability to display marine charts and connect to other navigation equipment.

Automobile navigation receivers (Figure 2.10b, Chapter 2) are mounted in cars, trucks, and trains. The purpose of the receiver may vary depending on the application, but the characteristics will be similar. Receivers used in cars are generally intended for driver navigation or for sending the position of a car to an emergency response centre in case of accidents. GNSS receivers used on buses, trucks, and trains are generally intended for fleet tracking and navigation.

Mapping and data collection receivers (Figure 2.10a, Chapter 2) are optimised for collecting data to be exported to an external database. They often have moderate to good autonomous accuracy with differentially corrected accuracies as good as 1 meter. Often they will have an attached computer dedicated to data collection. Such data collection computers can be preloaded with feature libraries so that the operator has preset classes of items to choose from. These receivers are handheld with extra batteries and a GNSS antenna fixed to a backpack (Figure 8.3).

OEM (Original Equipment Manufacturer) receivers (Figure 8.4) are intended to be integrated into other equipment. They come from the manufacturer as a bare board (card) or module, with no display or built-in power supply. Technical characteristics of OEM receivers can vary widely depending on the market for the unit. These are able to output typical GNSS data streams (messages) that allow developers to design specific applications. For example, mobile phone companies procure these receivers to integrate into their products.

Space receivers are used on satellites both for navigation and for attitude determination. They may be radiation hardened and have special programming allowing them to operate at the high-relative velocities experienced by orbital spacecrafts.

Survey grade receivers (Figure 2.10c, Chapter 2) (also called *geodetic receiver*) are intended for high-accuracy measurements necessary for land surveying (Ghilani and Wolf 2008; Sickle 2008). Such receivers will have external tripod-mounted antennas

FIGURE 8.3 Handheld GNSS receiver with extra batteries and antenna fixed to a backpack.

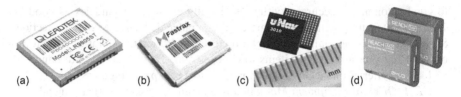

FIGURE 8.4   OEM bare-board GNSS receivers (a, b, c) and receiver module (d).

FIGURE 8.5   Survey receivers with external tripod-mounted antenna (a) and receiver with integrated antenna (b).

or integrated antenna-receiver with separate controller (Figure 8.5). Measurements of these receivers are generally based on carrier phase observation. Even though some manufacturers and users make extraordinary claims for their handheld pseudorange receivers, and the positioning outputs from such receivers have improved remarkably, such code-based point solutions are not accurate by surveying standards. However, code-based pseudoranges using DGNSS can achieve a moderate accuracy in real-time or post-processed, mapping results. For example, DGNSS is often used in collecting data for Geographical Information Systems (GIS). But, for precise surveying, carrier phase observation is the only option; they provide a level of accuracy acceptable for most surveying applications. They have multiple independent channels that track the satellites continuously, and they begin acquiring satellites' signals from a few seconds to less than a minute from the moment they are switched on. Most acquire all the satellites above them in a few minutes, with the time usually lessened by a warm start, and most provide some sort of audible tone to alert the user that data is being logged (recorded), and so forth. Most of them can have their sessions pre-programmed in the office before going to the field. Certainly, GNSS control surveying often employs several static receivers that simultaneously collect and store data from the same satellites for a period of time, known as a *session*. After all the sessions for a day are completed, their data are usually downloaded in a general binary format to the hard disk of a personal computer for post-processing. However, not all GNSS surveying is handled this way; for example, real-time kinematic surveying uses radio links to provide corrected data instantaneously. Refer to Chapter 11 for further details on survey grade receivers.

Timing receivers are intended to act as a time and frequency reference. Position is secondary information to these receivers and is often ignored by the user. The primary benefits of GNSS-derived time and frequency are long-term stability and coordination with the worldwide time network via the GNSS time standard.

Other than the preceding, GNSS receivers are now available in mobile phones (Figure 8.6a), personal digital assistants (PDAs) (Figure 8.6b), watches (Figure 8.6c), etc. *Software-aided GNSS receivers* are of another kind (Figure 8.7a).

Software techniques for GNSS are paving the way for positioning capability in a wide range of communications and navigation systems. These software-based GNSS receivers, with powerful central processing units and digital signal processing, can receive and decode GNSS signals in real-time with the help of software. They offer considerable flexibility in modifying settings to accommodate new applications without hardware redesign, using the same board design for different frequency plans, and implementing future upgrades (Borre *et al.* 2007).

Traditional GNSS receivers implement acquisition, tracking, and bit-synchronisation operations in an application-specific integrated circuit (ASIC), but a software

FIGURE 8.6 (a) GNSS (GPS) receiver in mobile phones; (b) PDAs with built-in GPS receiver; and (c) GPS receiver in wristwatches.

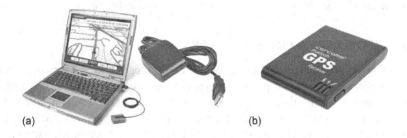

FIGURE 8.7 Software-aided GNSS receiver with USB cable (a) and Bluetooth wireless receiver (b).

GNSS receiver provides flexibility by implementing those blocks in software rather than hardware. By simplifying the hardware architecture, software makes the receiver smaller, cheaper, and more power-efficient. Software can be written in C/C++, MATLAB, and other languages and ported into all operating systems. Thus, software GNSS receivers offer the greatest flexibility for mobile handsets, PDAs, and similar applications. These receivers can be connected with a computer via data cable or wireless Bluetooth technology (Figure 8.7b).

8.5   RECEIVER INDEPENDENT EXCHANGE FORMAT

An important consideration for post-processed GNSS schemes is data file formats. There are two file format options available for data to be post-processed. The first are the *receiver-specific* data formats, only useful if the same make of receiver is operated as both the base and rover receiver; and the universally recognised standard *Receiver Independent EXchange* (RINEX) format. Each survey receiver type has its own binary data format, and the observables are defined following the manufacturers' individual concepts. Time tags may be defined in transmission time, or in fractional parts of cycles; code and carrier phase may have different or identical time tags, and satellites may be observed simultaneously or at different epochs.

As a consequence, data of different receiver types cannot easily be processed simultaneously with one particular data processing software package. To solve this problem, either all manufacturers have to use the same data output format, or a common data exchange format has to be defined that can be used as a data interface between all receiver types and software systems. The first has not been realised to date. However, a successful attempt has been made to define and accept a common data format for international data exchange. Based on the development at the University of Berne, Switzerland, the RINEX format was proposed by Gurtner *et al.* (1989) at the Fifth International Geodetic Symposium on Satellite Positioning in Las Cruzes, New Mexico. The proposal was discussed and modified during a workshop at this symposium and recommended for international use. The RINEX has been accepted by the international user community and by the community of receiver manufacturers. For most of the geodetic receivers, translator-software is provided by the manufacturers, which converts the receiver-dependent data into the RINEX format. On the other hand, all major GNSS data processing software offers the option for RINEX data input. RINEX hence serves as a general interface between receivers and multi-purpose data processing software.

8.6   CHOOSING A GNSS RECEIVER

The number of different brands and models of GNSS receiver units is rapidly expanding at present. Prices are dropping, new capabilities are being offered, and smaller units are now rivalling their larger predecessors, offering the same advanced features in more convenient sizes. GNSS units are designed for several quite different types of end use. For choosing a receiver, the first question is for which purpose it is going to be used. Next is what observable is to be tracked (code or carrier). General handheld GNSS receivers are code-based; exceptions do exist. Nevertheless, their use is growing and, while most are not capable of tracking the carrier phase observable, some have differential capability. Still these receivers were developed with the needs of navigation in mind. In fact, they are sometimes categorised by the number of *waypoints* and *track points* they can store.

Waypoint is a term that grew out of military usage. It means the coordinate of an intermediate position that a person, vehicle, or airplane must pass to reach a desired destination. With such a receiver, a navigator may call up a distance and direction from his present location to the next waypoint. Waypoints also define specific points that are recorded by the receiver during navigation (or a coarser survey). Our house, dock, airport, parked car, a great fishing/hunting spot etc.—any point which we want to mark as a point feature—is a waypoint. During travel with a GNSS receiver, the receiver by default starts recording the values of the points in between the waypoints at regular intervals. These are the track points. As we travel along, our GNSS receiver unit automatically records our journey in a *tracklog*.

These receivers do not have substantial memory because it is actually not needed for their designed applications. However, considerable memory is required for DGNSS base receivers. It is still a receiver that tracks only the code. Carrier phase on both frequencies plus the code receiver is capable of observing the carrier phase of both frequencies and requires much higher memory.

Still choosing the right instrument for a particular application is not easy. Receivers are generally categorised by their physical characteristics, the elements of the GNSS signal they can use with advantage, and by the claims about their accuracy. But the effects of these features on a receiver's actual productivity are not always obvious. For example, it is true that the more aspects of the GNSS signal a receiver can employ, the greater its flexibility, but so too the greater is its cost. And separating the capabilities a user needs, to do a particular job, from those that are really unnecessary is more complicated. The user must first have some information about how these features relate to a receiver's performance in particular GNSS positioning/navigation methods. Typical essential concerns should be:

- What observable (code or carrier) is used for positioning?
- Can this receiver give us the accuracy we need for our work?
- What is the likely rate of productivity on a project like this?
- What is the actual cost per point (position) to be surveyed?
- Can the receiver process differential corrections?
- Can it serve all of the purposes for which it is being chosen?
- What is the battery life?
- Is the memory of the unit sufficient for the chosen application?

In addition to these essential questions, there are numerous other issues as well (Bossler *et al.* 2002; Broida 2004; Gopi 2005; Sickle 2008). Most of them depend on the application. Let us consider an example of portable devices we can use in an automobile. Table 8.1 gives an idea on how to evaluate such types of receivers. Although this example is for the portable automobile navigation receiver, these discussions are essentially helpful and relevant for other types of receivers as well.

TABLE 8.1

Issues to Consider When Choosing a GNSS Receiver

Feature	What it Means and What to Look for
Model	Make sure that we are accurately comparing one model with the same model from another manufacturer when checking prices.
Price	Understand what it includes—sometimes the price is the cost of a 'bundle' or kit complete with extra components; sometimes the price is for the unit alone.
Review date/details	If using review information to help in making a decision, keep in mind that units often have software upgrades over their life, and these upgrades can improve shortcomings and add new features.
Warranty	The longer the better.
Support	Toll free support and preferably for extended hours.
Inclusions	Ideally we want the unit, documentation, a mounting adapter, main power supply and a car power adapter, mapping software and map data, a removable data card, and a data connector between the unit and our computer (usually USB). A carry case is also helpful.
Runs out of the box	Some units are ready to go almost instantly, others require the user to first copy data from a DVD to some type of memory card and possibly to register the software, etc. Clearly, units that are ready to go right out of the box are preferable.
Size	The smaller, the better, although screen size (the bigger, the better) is a limiting factor. Remember, larger screens have higher power requirement.
Weight	The lighter, the better. Consider however the total weight of what we travel with—the unit itself, plus the mount, car power charger, and any other necessary accessories.
Mounting accessories	Ideally, we want to be able to mount the unit either on the windshield or on the dash of the car, with some type of removable/reusable mount so we can move it easily from car to car.
Screen size	The bigger, the better; but note (below) that the pixel count is as important as the screen size. A considerable screen size is 3.5" diagonal, a good screen size is 5" or more, but when screens start to get to 6" or larger, we start to have compromises in terms of where it can be conveniently located in our car; and, of course, the unit becomes bigger, heavier, and more difficult to travel with. We consider screens smaller than 3.5" to be unacceptably small.
Screen pixels	More is better and translates to a clearer, crisper display. Screen pixels are more important than screen size—more pixels on a slightly smaller screen can sometimes be more readable than fewer pixels on a comparably larger screen. A minimum considerable screen resolution is 240 × 320 pixels.
Screen colours	The more, the better. Most mapping software seems to have a very limited number of colours (16 or so), but more colours provide extra information and helps us to instantly see and recognise what is being displayed on the screen. Avoid black & white units.

(Continued)

TABLE 8.1 (CONTINUED)
Issues to Consider When Choosing a GNSS Receiver

Feature	What it Means and What to Look for
Screen visibility	Important that it can be clearly read even when the sun is shining directly on it.
Screen backlighting	We want multiple levels of backlighting so as to adjust for optimum comfort when driving at night.
Day/Night mode	Some units will automatically switch between day and night colours and brightness so as to make the map display most visible in low and full light conditions. They can do this because they know, based on where they are and the time of day/time of year when the sun rises and sets. Others allow us to manually switch between these modes. Some do not have any options at all.
Controls	Does the unit use a touch screen or buttons (or both)? Because we may be driving and also attempting to adjust our receiver, we want big buttons or touch screen areas that are easy to quickly touch/push rather than small fiddly controls that are too distracting.
Interactive help files available	It is very helpful to be able to get an immediate explanation of what options mean while actually on the page in the unit.
Limited functionality when moving	Some units have the option to limit the things we can do with it while in motion, and some units have a hardwired refusal to allow us to do many things while moving—allegedly for our safety, but more commonly at great inconvenience to us, and to anyone else in the car who might be designated as the navigator. This is a feature we probably don't want.
Graphics processor speed	How quickly does it refresh the map image while driving? One that responsively refreshes as we are moving and turning is preferrable to one that updates slowly.
Max number of satellites simultaneously tracked	Although some units are promising an ability to track as many as 16 satellites simultaneously, it is very rare to have more than 12 visible from one constellation in the sky simultaneously. And even if more are visible, the extra precision from extra satellites is negligible with respect to cost.
Augmentation enhanced	This significantly improves the accuracy of the unit.
Dead-reckoning capability	This can be very helpful in areas with poor signal, such as downtown surrounded by high-rise buildings or inside a tunnel.
Satellite display	Does it show how many satellites it is receiving data from (ideally in a sky plot)? This helps us to understand how good the location data calculation may be, and also shows us where the antenna is most sensitive and where it is blocked.
Accuracy calculation	Does the unit show its calculated error estimate?
Can the unit show current latitude and longitude and compass heading	This can be useful, especially if we are going off-road.

(Continued)

TABLE 8.1 (CONTINUED)

Issues to Consider When Choosing a GNSS Receiver

Feature	What it Means and What to Look for
Can the unit show our current altitude	Of little practical value in driving, but it can be very interesting to see our varying height as we travel through the mountains.
Can the unit show the exact time	Because GNSS units can synchronise their clocks to the incredibly accurate time clocks on the satellites, there is no other accurate source of time available to us.
External antenna capability	Modern units seem to perform amazingly well without an external antenna; earlier units definitely needed them. However, having the ability to add an external antenna, if needed, adds to the overall functionality of the unit.
CPU processor speed	This is most apparent when asking the unit to calculate or recalculate a route.
Trip computer functions	Does the unit offer 'trip computer' functions such as average, current, and maximum speed, distance travelled, time spent driving (and time spent stopped)? Can it save the route that was followed?
Battery type	Even if we are planning on mainly using our unit in the car, there may be times when we cannot use the car's power supply. Does the unit use rechargeable batteries or regular batteries? How many does it use? If they are rechargeable, are they standard size batteries we can buy anywhere, or are they unique to this unit (i.e., harder and more expensive to replace)?
Battery life	How long does a set of batteries last with medium backlighting? If the batteries are not rechargeable, we can divide the cost of a new set of batteries by their expected life to get a cost per hour of operation.
Power input	Ideally the unit should accept external power via a USB port, making it compatible with a wide range of chargers and also allowing it to be recharged from our laptop or regular computer.
Auto power ON/OFF	Some units, when connected to, for example, a car's power supply, will switch ON; and OFF when the car power goes OFF. Some units will automatically switch OFF if there has been no movement for an extended period of time.
Map provider	Several sources are: Navteq, Tele Atlas, MapmyIndia, Google, Here Map. Some are subscription based, e.g., MapmyIndia.
Countries provided in map	The more, the better.
Update policy, frequency, and cost	Annual updates are desirable, but beware of cumbersome upgrade methods with copy-protection that require us to validate with our original disk, and beware of updates that cost ridiculous amounts of money.
How is map data loaded into the receiver	SD cards seem to be the lowest cost and most convenient format for data to be loaded into the receiver. Other media exist, but may not have as much capacity or may be higher priced.
Speaks directions	A voice that speaks directions is very helpful because it means we can keep our eyes on the road and don't need to look at the receiver to know when and where to take the turn.

(Continued)

TABLE 8.1 (CONTINUED)
Issues to Consider When Choosing a GNSS Receiver

Feature	What it Means and What to Look for
Speaks street names	If the unit can pronounce street names, this is even more helpful—it is the difference between a generic type instruction such as 'take the next road on the right' and a more specific instruction like 'take the next right on Park Street'.
Languages spoken	Some units give us a choice of English accents, and a choice of male or female voices, as well as a variety of different languages.
2D/3D	Some units offer a so-called 3D view of the map—this is a sort of a bird's eye view looking down from behind our location on the map, looking forward to where we are and the roads ahead, with an exaggerated perspective. Some can even show the buildings, bridges, and statues in 3D. It looks nice for a while. However, most people may find the classic 2D 'map style' view more helpful.
Can we choose between north up or direction of travel up (course up)	Typically most people will want the unit to be oriented so the screen points up in the same direction as we are travelling. That way, left turns in real life appear as left turns on the map. Sometimes though it is helpful to have a 'north up' orientation, like on a traditional map. This is particularly helpful when zoomed out a long way or when planning a route—when north is up, we are looking at the map the same way we would view a regular printed map.
Split screen mode	Some units with larger screens allow us to split the screen, each with different information. For example, one screen is a highly zoomed in detail of the nearby area, with direction of travel up, and the other screen is a zoomed out overview of the district area, with north up.
Map scale shown	What is the use of a map if it doesn't show us the scale, allowing us to understand 'this many cm on the map equals so many km on the road'? This is even more essential with receiver units that automatically 'zoom' the scale in and out without telling us.
Number of POIs provided	Units can offer prodigious numbers of points of interest (POIs), with some offering as many as 5 million, 10 million, or even more, POIs.
Number of user POIs that can be added	We definitely want to have the ability to add as many extra POIs (and destinations) as possible.
POI information includes phone number	This is very helpful, e.g., we can call a restaurant to check their hours or to make a reservation.
POI proximity alert	This has an obvious benefit and a more subtle one too. The more subtle benefit is that if we are in a place with traffic cameras, we can program their locations into our unit and have it caution us any time we are approaching a traffic camera.
Speed limit alert	This is of little value unless the unit also 'knows' the speed limit on the road we are travelling.
Does it show both miles and kilometres	If it shows both, we can conveniently switch between units for our convenience.

(Continued)

TABLE 8.1 (CONTINUED)
Issues to Consider When Choosing a GNSS Receiver

Feature	What it Means and What to Look for
Can we build a multi-stop journey with waypoints	This can be helpful if, e.g., we are planning an itinerary such as 'first we want to go to the supermarket, then to the mall, then to the bank, then to the car dealer, then back home.
Will it solve the 'travelling salesman' puzzle	This allows us to enter a series of destinations and the unit computes the best order to visit all of them as quickly as possible. This is a rare feature offered only by a very few units.
Can we program assumed speeds for different road types, and if so, how many different road types?	This information helps the unit make the best choices for routing the quickest way, and will also help it more accurately predict our time of arrival at our destination.
Can we program preferences for road/route types	Some units allow us to specify preferences for things like 'avoid/ prefer freeways' and similarly for ferries, narrow roads, and various other route options.
Can we choose different settings for different types of vehicles	Some units allow us to choose between cars, trucks, bicycles, and possibly other types of transportation, with different settings (e.g., for typical speeds on different road types) and preferences for each.
Does the unit present multiple route choices to choose from	Some units show us different routes on the screen to choose between, allowing us to visually see the different choices. This can be helpful if we have some familiarity with the route, allowing us to do a reality check on the route(s) being proposed.
Will it show breadcrumb trails?	Some units can leave a row of dots on the map to show where we have been. This can be helpful in some situations (especially in case of off-road travel).
Bluetooth phone integration	Some units can connect to a phone via Bluetooth. This can be helpful if, for example, we select a restaurant from the map and then want to dial its number.
Export data to laptop	Some units allow us to export the real-time GNSS data to a laptop mapping/navigation program. This is helpful if we have a second person as our navigator who is able to use a laptop while we are driving.
Can it play MP3/MP4 or other digital audio	Some units can do double duty as an MP3/MP4 player. This is of very little value to most people, however; it is much better to have a dedicated MP3/MP4 player for our music/video.
Can it display pictures	Some units allow us to store digital pictures, which we can then display on the unit's screen. This is another low-value gimmick application for most people.
Integrated with real-time traffic reporting	Some units can receive extra data through a radio receiver that gives us instantaneous location-specific information on traffic jams, road construction, accidents, etc.
Integrated with other location services	Some units offer additional data services such as local movie showtimes, local fuel station prices, etc.

* This table was prepared just as fun research and may miss some points. However, this gives some idea of our considerations when purchasing a GNSS receiver.

TABLE 8.2
Major Manufacturers of GNSS Receivers

Company	Website
Air Data Inc	www.airdata.ca
Allen Osborne Associates Inc	www.aoa-gps.com
American GNC Corporation	www.americangnc.com
Canadian Marconi Company, GPS-OEM Group	www.marconi.ca
Corvallis Microtechnology Inc	www.cmtinc.com
EMLID	www.emlid.com
Furuno USA Inc	www.furunousa.com
GARMIN International	www.garmin.com
GEC Plessey Semiconductors	www.gpsemi.com
Interstate Electronics Corporation	www.iechome.com
Leica Geosystems	www.leica-geosystems.com
Lowrance Electronics Inc	www.lowrance.com
Magellan Navigation Inc	www.magellangps.com
Motorola Space and Systems Technology	www.mot.com
NovAtel Inc	www.novatel.ca
Silva Sweden AB	www.silva.se
SiRF Technology Inc	www.sirf.com
SiTex Marine Electronics Inc	www.si-tex.com
Skyforce Avionics	www.skyforce.co.uk
Sokkia Topcon Co Ltd	www.sokkia.com
Spectra Precision	www.spectraprecision.com
Starlink Incorporated	www.starlinkdgps.com
Symmetricom Inc	www.symmetricom.com
TomTom International	www.tomtom.com
TRAK Microwave Corporation	www.trak.com
Trimble Navigation Limited	www.trimble.com
u-blox AG	www.u-blox.ch
Universal Avionics Systems Corporation	www.uasc.com

8.7 GNSS RECEIVER MANUFACTURERS

There are a huge number of receiver manufacturers in the international market. However, it is worth mentioning that every one of them does not manufacture all types of receivers, and all manufacturers do not provide equal quality positioning. Before purchasing a receiver we need to understand our needs, choose a receiver, and compare the manufacturers for a specific type of receiver. Table 8.2 provides a list of major GNSS receiver manufacturers; although the list is not exhaustive.

8.8 SMARTPHONE FOR SURVEY

Nowadays every smartphone carries a GNSS receiver that supports one or more GNSS constellations. The inclusion of a GNSS receiver in the smartphone is primarily targeted for navigation. However, the applications have been extended further for tracking, emergency response, and even for survey. The smartphone GNSS receivers, although not very precise, can become a low-cost solution for GIS survey. Several apps have been developed to support GIS data collection and tracking in mobile phones. Most of them are proprietary or subscription-based. One such subscription-based popular app from ESRI (www.esri.com) is *Collector for ArcGIS*. Another app from the same company is *Survey 123 for ArcGIS*. The Survey123 is a form-based data collection app—it is all about questions and answers about the facts. With Survey123, one can certainly capture geographic information, but it is just one more question in the questionnaire. This app is more useful for capturing attribute information (textual description of geographic objects) rather than geospatial information. Collector for ArcGIS is a map-based app; it can be used to capture geographic information. With Collector, one can also capture attributes associated with those features, but the map comes first—the attribute database capabilities are secondary. Data captured in Collector is stored in the cloud and therefore is accessible in other devices instantly.

A free but very efficient Android-based app is *SW Maps* from Nepal (https://sw-maps-mobile-gis.soft112.com). It supports GPS, GLONASS, Galileo, BeiDou, and QZSS (in future it will also support IRNSS). One can record points, lines, polygons, and even photos by using this app; these recorded data can be displayed over a choice of background map/satellite image (e.g., Google Maps or OpenStreetMap) and attach custom attribute data to any feature. Attribute types include text, numbers, an option from a predefined set of choices, photos, audio clips, and videos. The app can even connect to external receivers (RTK-enabled receivers as well) via Bluetooth or USB. It supports import/export of several file formats such as KMZ, Shapefile, GeoJSON, GeoPackage, and many other formats. It can define and record multiple feature layers with different styling. One can even manually draw point/line/polygon features in this app. Once we create a project in this app, it can be exported as a template for other projects. This is useful when many surveyors and instruments are deployed to collect the same type of data. Templates can be created on a Windows PC using the *SW Maps Template Builder* tool, or exported from any existing project contained in a smartphone. It can process RTK data via internet. In spotty internet or no internet situations, it can log raw kinematic GNSS observations when using an external receiver. These raw observations can then be used for post-processing. An automatic tool has been developed for correcting recorded data points using CORS station in Kathmandu, Nepal. A tutorial of SW Maps can be downloaded from http://swmaps. softwel.com.np/assets/resources/manual.pdf.

Other similar apps are GPS Essentials, Polaris GPS Navigation, GNSS Surveyor, LandMap, TcpGPS, Apglos Survey Wizard, Geo++ RINEX Logger, etc. Geo++ RINEX Logger uses the most recent Android API services to log device's raw GNSS measurement data into a RINEX file including pseudoranges, accumulated delta ranges, doppler frequencies and noise values. So far, it supports GPS/GLONASS/ GALILEO/BDS/QZSS for L1/L5/E1B/E1C/E5A (as supported by the device) and has been successfully tested on many devices. Trimble DL is an app that can be used with Trimble receivers for static and rapid static survey. This app eliminates the requirement of a costly controller.

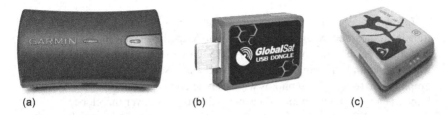

(a)   (b)   (c)

FIGURE 8.8 External GNSS receivers for mobile phones from (a) Garmin, (b) GlobalSat, and (c) GPS Precision.

We know that GNSS receivers in mobile phones are not of survey grade. Therefore, for high accuracy, it is necessary to attach external receivers via Bluetooth or USB as shown in Figure 8.7. There are several manufacturers who sell these products. Some of the receivers even support RTK. However, before purchasing such receivers it is necessary to confirm that the receiver is supported by the software to be used. Figure 8.8 shows some external receivers that can be connected to a mobile phone. The Arrow 200 from GPS Precision, shown in Figure 8.8, claims to have 1 cm RTK accuracy.

NOTE

We can edit or build our GNSS routes in the Worldwide Web and also can download programs to view topographic maps and aerial photographs without spending anything. Here are a few to give the reader an idea of what we can do with the loads of GNSS software packages available online that can be downloaded for free. There are so many other things that can be done with these downloadable GNSS software. A simple online search will produce a variety of GNSS software packages that suit specific requirements. Following are some of the free software packages and their benefits; but there are hundreds of others out there.

3D Tracking: We can observe our movements in detail with the help of Google Maps (www.maps.google.com) and Google Earth (www.earth.google.com) if we download 3D tracking software (www.free.3dtracking.net). It is navigation and tracking software that allows us to access our tracking system whilst on the go, ensuring we have complete control of our fleet and clients, at our fingertips. This software is usable on PDA or mobile phone with a GPS receiver. We can record the data with the help of the web server and can view the recorded data again at any time.

Cetus GPS: Scientists, explorers, GIS surveyors, and GPS enthusiasts can use *Cetus GPS* (www.cetusgps.dk) for field data collection and GPS tracking. *Cetus GPS* was originally used by the Swiss Army to carry out its operations. We can extend the features of our standard GPS equipment with this free downloadable software.

EasyGPS: With *EasyGPS* (www.easygps.com) we can edit, create, and transfer routes and waypoints between our Magellan, Garmin, or Lowrance GPS receiver and our computer. *EasyGPS* allows us to connect with the best information and mapping sites on the web and also gives us access to topographical maps and aerial photos with a single click. We can also read weather forecasts and exchange data with expert GPS users all over the globe.

Garmap Win: *Garmap Win* (www.harukaze.sakura.ne.jp/garmap/e_garma pwin.html) is somewhat similar to *EasyGPS* and can be used with *Garmin* receivers to create and edit routes and track logs and waypoints. We can download waypoints, tracklogs, and routes and can find the current location on a map with *Garmap Win*.

GPSActionReplay: *GPSActionReplay* (http://gpsactionreplay.free.fr) software allows us to replay and analyse GPS data which includes real-time animation, auto-focusing, and auto speed calculation. This is a Java-based application which runs on Windows, Mac, and Linux and also other web browsers within an applet.

While downloading the free GPS software, it is necessary to make sure that we recognise what we are receiving as these software come with provisions and restrictions for use. Sometimes the software comes with limited functionality and the entire product is available for purchase. In some cases some features are disabled after a particular period as they are provided as evaluation version. However, it is important to make sure the components of the software match our needs in order to enhance our receiver's performance.

EXERCISES

DESCRIPTIVE QUESTIONS

1. Describe, with diagram, the basic architecture of a GNSS receiver.
2. Explain the working principle of the RF section.
3. What do you know about the receiver antenna and preamplifier?
4. Explain the necessity and working principle of a radio modem.
5. Explain Doppler shift and time shift recognition by a receiver.
6. How does a receiver solve integer ambiguity? Explain briefly.
7. How can receivers be classified? Explain briefly.
8. What do you understand by RINEX? Why is it important?
9. What do you understand by 'sequential receiver' and 'continuous receiver'? Explain their working principles with relative advantages and disadvantages.
10. What are the different ways of classifying GNSS receivers? Provide examples.
11. Explain mapping and surveying receivers.

12. Which essential questions should we consider while purchasing a GNSS receiver? Elaborate your answer with explanations.
13. Can we use our mobile phone as a low-accuracy survey instrument? Explain the software and hardware concepts associated with it.

SHORT NOTES/DEFINITIONS
Write short notes on the following topics

1. Receiver antenna
2. Preamplifier
3. Beat frequency
4. Channel
5. Role of microprocessor
6. Control and display unit
7. Radio modem receiver
8. Waypoint
9. Track point
10. Track-log
11. OEM receiver
12. Mapping receiver
13. RINEX
14. SW Maps
15. Collector for ArcGIS

9 Geodesy

9.1 INTRODUCTION

In the previous chapters, we discussed satellite-based navigation and positioning techniques; the basic concepts, the hurdles, and how to overcome those hurdles. Position information is interesting but makes no sense without a common representation of the world. For example, a mariner may know where the vessel is and have the coordinates of the destination, but in order to navigate the route safely, the location of potential dangers must also be known, as well as of ports of haven, designated shipping lanes, restricted waters, etc. Therefore, we need to represent the positioning information on maps; and there is a need for a global coordinate referential in order to allow a position representation. This brings us to the discussion of *geodesy*.

The word geodesy comes from Greek, literally meaning 'dividing the earth'. The first practical objective of geodesy was the provision of an accurate framework for the control of national topographical surveys, and hence it is the foundation of a nation's maps (Smith 1997). To prepare the map, geodesy must define the basic geometrical and physical properties of the figure of the earth. The scientific objective of geodesy has therefore always been to determine the *size, shape,* and *gravitational field* of the earth (Clarke 2001; DMA 1984; Smith 1997; Seeber 1993; Hofmann-Wellenhof and Moritz 2005). Geodesy can be defined as the science which deals with the methods of precise measurements of elements of the surface of the earth and their treatment for the determination of the geographic positions on the surface of the earth including the gravity field of the earth in a three-dimensional time varying space. It also deals with the theory of size and shape of the earth. Satellite geodesy is the measurement of the form and dimensions of the earth, the location of objects on its surface and the figure of the earth's gravity field by means of satellite technique (Seeber 1993). Traditional astronomical geodesy, that includes astronomical positioning, is not commonly considered a part of satellite geodesy. This chapter will discuss the essentials of geodesy to understand GNSS positioning.

Positioning is the determination of the coordinates of a point on land, at sea, or in space with respect to a coordinate system. Positioning is solved by computation from measurements linking the known positions of terrestrial or extraterrestrial points with the unknown terrestrial position. This may involve transformations between or among astronomical and terrestrial coordinate systems. To represent these point positions in a spatial representation (e.g., maps) we often need to handle projected systems as well.

9.2   COORDINATE SYSTEM

In mathematics and applications, a *coordinate* (or *referential*) *system* is a system for assigning a tuple (ordered list) of numbers to each point in an *n*-dimensional space. Coordinate systems are used to identify locations on a graph or grid (Gaposchkin and Kołaczek 1981). For example, the system of assigning longitude and latitude to geographical locations is a coordinate system. There are various coordinate systems available to represent the location of any point. However, all of them fall into two broad categories—*curvilinear* and *rectangular*. Curvilinear system uses *angular* measurements from the origin to describe one's position, whereas rectangular coordinate system uses *distance* measurements from the origin. One example of a curvilinear system is the geographic coordinate system that uses angular latitude/longitude measurements, and one example of rectangular coordinate system is the Cartesian co-ordinate system that uses linear measurements.

9.2.1   CELESTIAL EQUATORIAL COORDINATE SYSTEM

In astronomy, a celestial coordinate system is a curvilinear coordinate system for mapping positions in the sky. There are different celestial coordinate systems in use. The *celestial equatorial* coordinate system is probably the most widely used celestial coordinate system (Cotter and Lahiry 1992; Kaler 2002). It is most closely related to the geographic coordinate system. Let us try to understand this coordinate system.

The orbit of the earth is an ellipse of low eccentricity ($e = 0.01673$), the sun being at one focal point. The earth rotates from west to east in 24 h on an ecliptic plane. Ecliptic plane is the geometric plane containing the mean orbit of the earth around the sun. The rotation axis of the earth deviates from the perpendicular of the ecliptic plane by an angle equal to 23°27'; in other words, there is an angle of 23°27' between the ecliptic plane and equatorial plane. The earth is also rotating around the sun in the same direction.

The *vernal point* (also called *vernal equinox*) is defined as being the point where the sun crosses the Equator when going from the southern hemisphere to the northern hemisphere. This apparent location of the sun is positioned directly over the earth's Equator. The intersecting line of the equatorial plane and the ecliptic plane indicates the direction of the vernal point (Figure 9.1). Considering the vernal point direction, a reference in the astronomical domain, a terrestrial location can be defined using both *right ascension* and *declination*, defined as follows (Figure 9.2a).

The right ascension (α) is the angle between the meridian that crosses the location and the vernal point from the centre of the earth. It is calculated in the equatorial plane counterclockwise from 0° to 360°. The declination (δ) is the angle of the direction of the location with the equatorial plane, evaluated on the local meridian, from 0° to 90° north or south. This is called the celestial equatorial coordinate system.

9.2.2   EARTH-CANTERED INERTIAL COORDINATE SYSTEM

From the celestial equatorial coordinate system, it is possible to describe earth cantered Cartesian (rectangular) coordinate system. It is called the *earth-cantered inertial*

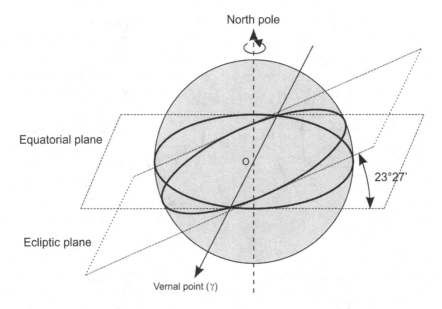

FIGURE 9.1   Vernal point.

(ECI) coordinate system, commonly used in astronomical calculations (Kaplan 1996; Parkinson and Spilker 1996; Samama 2008). In this system (Figure 9.2b), the centre of the earth is considered as the origin (O), the z-axis is the polar axis, x-axis is through the vernal point, the y-axis is chosen to form a right-handed system and lies in the equatorial plane at 90° counter clockwise from the x-axis.

However, neither the celestial equatorial nor the ECI system is really practical for positioning. Because the vernal point is a fixed point, it does not rotate with the rotation of earth about its axis; thus, a given location on earth does not keep the same coordinate values, as the earth is rotating on itself. These systems are mainly used to define star locations.

9.2.3   GEOGRAPHICAL COORDINATE SYSTEM

For positioning purposes, it is better to choose a referential that rotates with the earth, allowing a single coordinate value for a given location all the time. The *geographical coordinate system* (GCS) was designed for this purpose (Figure 9.2c). A GCS expresses every location on earth by two of the three coordinates of a curvilinear coordinate system, which is aligned with the spin axis of earth. A GCS uses a 3D spherical surface to define locations on the earth (Figure 9.3).

The location of any point on the earth's surface can be defined by a reference using *latitude* and *longitude*. Longitude and latitude are angles measured from the earth's centre to a point on the earth's surface. Thus, the origin of this coordinate system is the centre of the earth. The angle between the line joining the point to the centre of the earth and the equatorial plane is called *latitude*. The counterclockwise

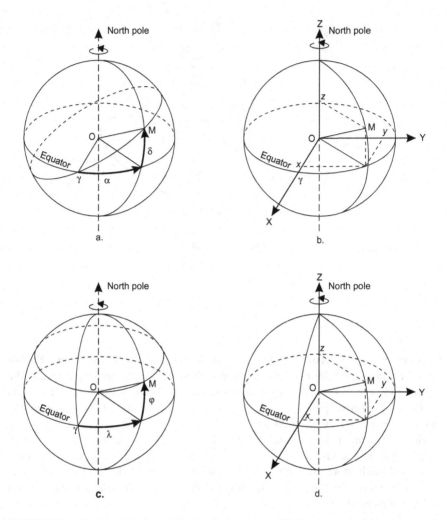

FIGURE 9.2 (a) Celestial equatorial coordinate system; (b) Earth-cantered inertial coordinate system; (c) Geographic coordinate system; and (d) Earth-cantered earth-fixed coordinate system.

angular distance between the prime meridian plane and the meridian plane of the point measured in equatorial plane is called *longitude*. In this system, 'horizontal lines', or east–west lines, are lines of latitude, or parallels. 'Vertical lines', or north–south lines, are lines of longitude, or meridians. These lines encompass the globe and form a gridded network called a *graticule*.

The arbitrary choice for a central line of longitude is that which runs through the Royal Observatory in Greenwich, England, and is hence known as the Greenwich meridian or the prime meridian. Lines of longitude are widest apart at the Equator and closest together at the poles. Lines of latitude lie at right angles to lines of

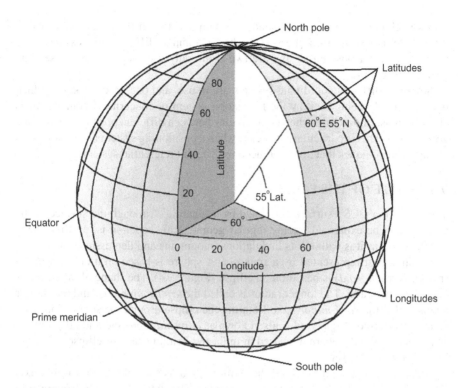

FIGURE 9.3 Geographic coordinate system.

longitude and run parallel to one another. Each line of latitude represents a circle running round the globe. Each circle has a different circumference and area depending on where it lies relative to the two poles. The circle with the greatest circumference is known as the Equator and lies equidistant from the two poles.

The origin of the graticule (0,0) is defined by where the Equator and prime meridian intersect. The globe is then divided into four geographical quadrants. North and south are above and below the Equator, and west and east are to the left and right of the prime meridian. Latitude and longitude values are traditionally measured either in decimal degrees or in degrees, minutes, and seconds. Remember, they are angular measurements.

9.2.4 EARTH-CANTERED EARTH-FIXED COORDINATE SYSTEM

From the GCS, it is also possible to define an equivalent Cartesian system, called the earth-cantered earth-fixed (ECEF) coordinate system (Kaplan 1996; Gopi 2005; Chatfield 2007). The main parameters are as follows (Figure 9.2d): The origin O is the centre of the earth. The z-axis is the polar axis, oriented towards north. The x-axis is on the equatorial plane and passes through the Greenwich meridian. The y-axis is chosen to form a right-handed system and lies in the equatorial plane at 90° counter clockwise from the x-axis. The GCS and ECEF coordinate systems are very practical for positioning and are two commonly used referentials for satellite navigation systems.

So, many different coordinate systems are in use, and one needs to know which one is currently being used by the navigation system of concern. Of course, this is also of utmost importance when comparisons between different positioning systems are required. This is particularly the case when a positioning system is an integration of various techniques that are referenced to different referential systems.

9.3   SHAPE OF THE EARTH

The shape of a GCS's surface is defined by a spheroid. Although the surface of the earth cannot be expressed by any regular geometrical shape—it is 'earth-shaped', i.e., the earth as it is with all its undulations (mountains and depressions). However, the earth is best represented by a *spheroid*. A sphere is based on a circle, while a spheroid (or *ellipsoid*) is based on an ellipse (Figure 9.4). The shape of an ellipse is defined by two radii. The longer radius is called the *semi-major axis*, and the shorter one is called the *semi-minor axis*. Rotating the ellipse around the semi-minor axis creates a spheroid. A spheroid is also known as *ellipsoid*. Spheroid and ellipsoid are synonymous. Word 'spheroid' is used in India and Britain whereas 'ellipsoid' is used in America and Russia.

A spheroid is defined by either the semi-major axis (a) and the semi-minor axis (b), or by semi-major axis and the flattening (f). The flattening is the difference in length between the two axes. The flattening is

$$f = (a-b)/a \tag{9.1}$$

Eccentricity (e) is another quantity which can also describe the shape of the earth as

$$e^2 = (a^2 - b^2)/a^2 \tag{9.2}$$

The earth is roughly an oblate ellipsoid, with semi axes of nearly 6378 and 6357 km; but it may depart from the ellipsoidal shape by some kilometres. Because

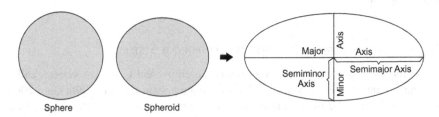

FIGURE 9.4   Sphere and spheroid.

of gravitational and surface feature variations, the earth is not a perfect spheroid. The earth has been surveyed many times to aid us in a better understanding of its surface features and their peculiar irregularities. The surveys have resulted in many spheroids that represent the earth. Generally, a spheroid is chosen to fit one country or a particular area. A spheroid that best fits one region is not necessarily the same one that fits another region.

There are many spheroid references that exist, each of which is a local *geoid* approximation. Geoid, in simple words, is the hypothetical surface of the earth that coincides with sea level everywhere (Bowditch 1995). It is nearly ellipsoidal but a complex surface. The geoid is almost the same as mean sea level, i.e., it may be described as surface coinciding *mean sea level* in the oceans and lying under the land at the level to which the sea would reach if admitted by small frictionless channels. The geoid on an average coincides with the mean sea level in open oceans. Ambiguity is due to mean sea level not being exactly an equipotential surface or due to periodic changes in the form of the geoid due to earth tides, but these will not be more than a metre. The mean sea level or geoid is the datum for measurement of heights above it. The geoid may depart from the spheroid by varying amounts (Figure 9.5), as much as 200 m or even more. The geoid unfortunately has rather disagreeable mathematical properties. It is a complicated surface with discontinuities of curvature, hence not suitable as a surface on which to perform mathematical computations. Therefore, spheroid is the only option for mathematical calculations.

In 1866, Sir Alexander Ross Clarke had taken measurements in Europe, Russia, India, South Africa, and Peru to form a spheroid for the mapping of the United States (Clarke 1880). An estimate of the spheroid allows calculation of the elevation

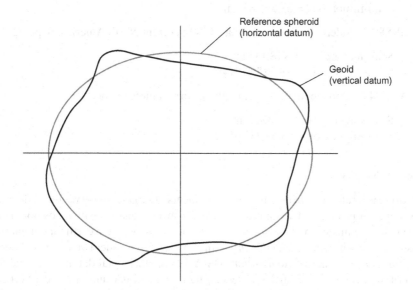

FIGURE 9.5 Spheroid and geoid.

of every point on earth, including sea level. In 1983, a new datum was adopted, based on measurements taken in 1980 and internationally accepted as the *Geodetic Reference System 1980* (GRS80). In 1984, the United States military refined the values of GRS80 ellipsoid and created the *World Geodetic System 1984* (WGS84 or WGS 1984) (Hofmann-Wellenhof and Moritz 2005).

Everest Spheroid is the reference surface for India and adjacent countries. It was named after Sir George Everest who was the surveyor general of India from 1830 to 1843. He was responsible for meridian arc measurements to estimate the size of the spheroid. It was originally defined in feet in 1830. Values in feet are given below:

Semi major axis, a = 20,922,931.80 feet
Semi minor axis, b = 20,853.374.58 feet
Flattening, f = 1/300.8017

The spheroids that are encountered most commonly are as follows:

Everest (Sir George) 1830: One of the earliest spheroids; appropriate for India

Semi-major axis = 63,77,276 m
Semi-minor axis = 63,56,075 m

Clarke 1886 for North America basis for United States Geological Survey (USGS)

Semi-major axis = 63,78,206.4 m
Semi-minor axis = 63,56,583.8 m

GRS80 (Geodetic Reference System, 1980); current North America mapping

Semi-major axis =63,78,137 m
Semi-minor axis = 63,56,752.31414 m

WGS84 (World Geodetic System, 1984), current global choice

Semi-major axis = 63,78,137 m
Semi-minor axis = 63,56,752.31 m

9.4 DATUM

A *datum* is a reference from which measurements are made. In surveying and geodesy, a datum is a reference point on the earth's surface against which position measurements are made, and an associated model of the shape of the earth for computing positions. This is called *geodetic datum*. In geodesy we consider two types of datum; *horizontal datum* and *vertical datum* (DMA 1984). Horizontal datums are used for describing a point on the earth's surface, generally in latitude and longitude. Vertical datums are used to measure elevations or underwater depths.

NOTE

Sea level is monitored at tidal observatories by hourly or continuous measurement of tides. Mean of high and low tides of measurements over a metonic cycle of 19 years is taken as mean sea level. A 19-year metonic cycle is used in order to include all possibly significant cycles through the 18.67 years period for the regression of the moon's nodes while still terminating on a complete yearly cycle. As there are irregular apparent secular trends in sea-level, averaging of tidal observations over a specific 19-year cycle is necessary to have a common reference of vertical datum (Bowditch 1995).

While a spheroid approximates the shape of the earth, a datum defines the position of the spheroid relative to the centre of the earth. A datum provides a frame of reference for measuring locations on the surface of the earth. It defines the origin and orientation of latitude and longitude lines. A GCS is often incorrectly called a datum, but a datum is only one part of GCS.

An earth-centred, or *geocentric datum*, uses the earth's centre of mass as the origin of the spheroid. The most recently developed and widely used geocentric datum is WGS 1984. It serves as the framework for measurement of locations worldwide. A local *geodetic datum* aligns its spheroid to closely fit the earth's surface in a particular area. A point on the surface of the spheroid is matched to a particular position on the surface of the earth (generally at mean sea level). This point is known as the origin point of the datum. The coordinates of the origin point are fixed, and all other points are calculated from it. The spheroid's origin of a local datum is not at the centre of the earth. The centre of the spheroid of a local datum is deviated from the earth's centre. Because a local datum aligns its spheroid so closely to a particular point on the earth's surface, it is not suitable for use outside the area for which it was designed. A geodetic datum is defined by a set of at least five parameters; semi-major axis, flattening or semi-minor axis, and coordinates of the origin of the spheroid adopted as reference surface. To realise the datum, an initial point is chosen on the surface of the earth. Coordinates (latitude and longitude) of this point are estimated by astronomical observations of stars; height is obtained by *spirit levelling* above mean sea level from the value of a known benchmark. Azimuth of one line is also obtained by astronomical observations. Coordinates of this point obtained by astronomical observations and spirit levelling are estimates of natural coordinates of the initial point. It is assumed that these coordinates are the same as geodetic coordinates of the initial point. Control networks then provide control points all over the country by adopting various methods. In the case of India, Kalyanpur in central India was chosen as the initial point. Potsdam is initial point for Europe, Meades Ranch for North America, and so on.

We should choose an appropriate spheroid for a specific datum. A few such examples are given here

Datums	Spheroids
Everest–India and Nepal	Everest Definition 1975
Kalyanpur 1880	Everest 1830
Kalyanpur 1937	Everest (Adjustment 1937)
Kalyanpur 1962	Everest Definition 1962
Kalyanpur 1975	Everest Definition 1975
Kandawala	Everest (Adjustment 1937)
Indian 1954	Everest (Adjustment 1937)
Indian 1960	Everest (Adjustment 1960)
Indian 1975	Everest (Adjustment 1975)
WGS 1972	WGS 1972
WGS 1984	WGS 1984
PZ-90	PZ-90
GTRF	GRS80
Beijing 1954	Krassowsky 1940
CGCS2000	CGCS2000

9.4.1 WGS 1984 Datum

Unlike local geodetic datums, which are essentially defined by parameters associated with a single 'origin' terrestrial station, datums used in satellite navigation system are defined by a combination of: (a) *physical models* such as the adopted model of the earth's gravity field, gravitational constant of the earth, the rotation rate of the earth, the velocity of light, etc.; and (b) *geometric models*, such as the adopted coordinates of the satellite tracking stations used in the orbit determination procedure, and the models for precession, nutation, polar motion, and earth rotation, that relate the celestial reference system (in which the satellite's ephemeris is computed) to the earth-fixed reference system (in which the tracking station coordinates are expressed).

Such a datum has the following characteristics:

- It is geocentric, because the geocentre is the physical point about which the satellite orbits.
- It is generally defined as a Cartesian system (although a reference ellipsoid is often also defined), with axes oriented close to the principal axes of rotation (z-axis) and the intersection of the Greenwich meridian plane and the equatorial plane (x-axis), with the y-axis forming a right-handed system.
- There are a number of different satellite datums, each associated with different satellite tracking technology (e.g., Satellite Laser Ranging, Transit Doppler, etc.) and different combinations of gravity field model, earth orientation model, and tracking station coordinates used for orbit computation.

The WGS84 is one such satellite datum, defined and maintained by the US National Imagery and Mapping Agency as a global geodetic datum. It is the datum to which all NAVSTAR GPS positioning information is referred by virtue of being the reference system of the broadcast ephemeris (Seeber 1993). WGS84 is an ECEF system fixed to the surface of the earth (see Figure 9.2d).

The defining parameters of the WGS84 reference ellipsoid are:

Semi-major axis: 6378137 m.
Ellipsoid flattening: $1/298.257223563^3$.
Angular velocity of the earth: 7292115×10^{-11} rad/sec.
The earth's gravitational constant (atmosphere included): 3986005×10^{-8} m^3/sec^2.

Reference systems are required to be periodically redefined for a number of reasons, such as when the primary tracking technology improves (e.g., when the Transit Doppler system was superseded by GPS) or the configuration of ground stations alters radically enough to justify a re-computation of the global datum coordinates. The result is generally a small refinement in the datum definition, and a change in the numerical values of the coordinates.

9.4.2 INDIAN GEODETIC DATUM

Indian geodetic datum is a local datum and based on Everest spheroid as reference surface was defined piecemeal at various times. Astronomical observations were carried out at least twice. More precise observations carried out at a later date were accepted. Hence meridional and prime vertical deflection of vertical, were defined at Kalyanpur. Parameters of the datum are given below (Agrawal 2004):

Initial point (origin)	Kalyanpur
Latitude of origin	24° 07' 11.26"
Longitude of origin	77° 39' 1 7.57"
Meridional deflection of vertical	−0.29"
Prime vertical deflection of vertical	+2.89"
Geoidal undulation	0 m
Semi major axis	6,377,301.243 m
Flattening	1/300.8017
Azimuth to Surantal (Madhya Pradesh)	190° 27' 06.39"

9.4.3 INTERNATIONAL TERRESTRIAL REFERENCE SYSTEM

The stations on the earth's surface with known coordinates are sometimes known as the *terrestrial reference frame* (TRF). It is important to note that there is a difference between a datum and TRF. A datum is a set of constants with which a coordinate system can be abstractly defined, not the coordinated network of monumented reference stations themselves that embody the realisation of the datum (Sickle 2004). However, instead of speaking of TRFs as separate and distinct from the datums on which they rely, the word 'datum' is often used to describe both the framework (the datum) and the coordinated points themselves (the TRF).

In 1991, the International Association of Geodesy decided to establish the International GNSS Service (IGS) to promote and support activities such as the maintenance of a permanent network of GNSS tracking stations, and the continuous computation of the satellite orbits and ground station coordinates (Dixon 1995). Both of these were preconditions to the definition and maintenance of a new satellite datum independent of WGS84. After a test campaign in 1992, routine activities commenced at the beginning of 1994. The network is an international collaborative activity consisting of about 50 core tracking stations located around the world supplemented by more than 200 other stations (some continuously operating, others only tracking on an intermittent basis).

The definition of the reference system in which the coordinates of the tracking stations are expressed, and periodically re-determined, is the responsibility of the International Earth Rotation Service. The reference system realisation is known as the International Terrestrial Reference Frame (ITRF) (Burkholder 2008), and its definition and maintenance is dependent on a suitable combination of *satellite laser ranging*, *very long baseline interferometry* and *GNSS coordinate results* (however, increasingly it is the GPS system that is providing most of the data). Each year a new combination of precise tracking results is performed, and the resulting datum is referred to as ITRFxx, where "xx" is the year (http://itrf.ensg.ign.fr). A further characteristic that sets the ITRF series of datums apart from the WGS, is the definition not only of the station coordinates but also their velocities due to continental and regional tectonic motion. Hence, it is possible to determine station coordinates within the datum, say ITRF98, at some 'epoch' such as 1 January 1999, by applying the velocity information and predicting the coordinates of the station at any time into the future (or the past).

Such ITRF datums, initially dedicated to geodynamical applications requiring the highest possible precision, have been used increasingly as the fundamental basis for the redefinition of many national geodetic datums. For example, the new Australian datum, known as the Geocentric Datum of Australia is a realisation of ITRF92 at epoch 1994.0 at a large number of control stations (Manning and Harvey 1994). Of course, other countries are free to choose any of the ITRF datums (it is usually the latest), and define any epoch for their national datum (the year of the GPS survey, or some date in the future, such as the year 2025). Only if both the ITRF datum and epoch are the same, can it be claimed that two countries have the same geodetic datum. However, differences in such datums can still be accommodated through similarity transformation models (Section 9.5).

9.5   ELLIPSOIDS AND DATUMS USED IN GNSS

As GNSS is not a local positioning system, rather a global system for positioning and navigation, every GNSS uses earth-centred geocentric datum, not local datum. However, the spheroids they use are different. The geocentric datums used by GPS and GLONASS are, respectively, WGS84 and PZ-90 (Parametry Zemli 1990, English translation: Parameters of the Earth 1990, also known as PE-90). The reference used by Galileo is Galileo Terrestrial Reference Frame (GTRF) aligned with the International Terrestrial Reference Frame (ITRF) and is covered by the ISO 19111 standard. Galileo GTRF is based on GRS80 ellipsoid. Chinese BeiDou uses China Geodetic Coordinate System 2000 (CGCS2000). However, earlier the BeiDou system used the Beijing 1954 datum that is based on Krassovsky 1940 ellipsoid. The IRNSS uses WGS84 datum and QZSS uses QZSS/PNT datum. The QZSS/PNT is very similar to ITRF; the difference between ITRF and QZSS/PNT is nominally within 20 mm.

In the WGS84 model, the plane sections parallel to the equatorial plane are circular. The equatorial radius equals 6378.137 km. The plane sections perpendicular to the equatorial plane are ellipsoidal. The section containing the z-axis has its semi-major axis equal to 6,378,137 m (the equatorial radius) and its semi-minor axis is equal to 6,356,752.31 m.

The PZ-90 model is similar to the WGS84, with a semi-major axis of the plane section of the equatorial plane being 6,378,136 m and a semi-minor axis of 6,356,751 m (GLONASS ICD 2002). The GTRF model is also similar (based on GRS80) with the semi-major axis of 6,378,137.31414 m and the semi-minor axis of 6,356,752 m (Zaharia 2009). The CGCS2000 has semi-major axis of 6,378,137 m and flattening of 1/298.257222101. The CGCS2000 is referred to ITRF97 at the epoch of 1 January 2000 (2000.0).

What a GNSS user is likely to have experienced is a wrong setup in their GNSS receiver relative to the map they use for navigation purposes. It is essential that the GNSS receiver be set up to conform to the map. In particular, the coordinate system and its datum need to be the same. For example, the GPS calculates the position based on WGS 84 spheroid and datum. However, users may encounter maps that use other datum and spheroids. The details for each specific map can generally be found in the margin information on the map, and must be entered correctly into the receiver if an accurate relationship between the receiver and the map is to be assured. For surveying purposes, it is wise to use the specific GNSS datum and spheroid (i.e., WGS84, or PZ-90, or GTRF, the applicable one), which can then be converted into any other local system with the help of software with no discernible loss of accuracy for virtually all situations (except for cases of very high accuracy requirements).

To convert the coordinates of any local geodetic system to any geocentric system (or vice versa) the following transformation parameters are required:

- dx, dy, and dz; the three coordinates of origin of local system with respect to origin of a geocentric system, known as translation parameters (Figure 9.6).
- ω, φ, and κ; the three directions of axes of local system with respect to directions of axes of a geocentric system, known as rotation parameters (Figure 9.6).
- da and db, the two differences in semi major axis and semi minor axes of the two systems, or simply ds the change in scale.

Some organisations prefer to ignore rotation parameters. A minimum of three points are required, where coordinates in both the systems are known, to calculate transformation parameters.

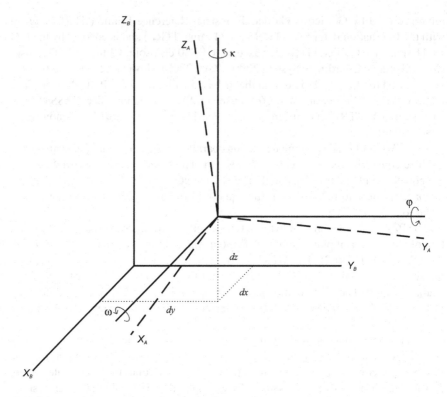

FIGURE 9.6   Coordinate transformation model.

Bursa-Wolf model (Vanicek 1995; Torge 1993) is normally being used for transformation by many geodesists as follows:

$$
\begin{pmatrix} X_B \\ Y_B \\ Z_B \end{pmatrix} = ds \times R \times \begin{pmatrix} X_A \\ Y_A \\ Z_A \end{pmatrix} + \begin{pmatrix} dx \\ dy \\ dz \end{pmatrix} \tag{9.3}
$$

where, X_A, Y_A, Z_A are the coordinates of a point in local system and X_B, Y_B, Z_B are the coordinates of the same point in geocentric system (Figure 9.6); R is a 3×3 orthogonal rotation matrix as follows:

$$
R = \begin{pmatrix} \cos\kappa\cos\phi & \cos\kappa\sin\phi\sin\omega + \sin\kappa\cos\omega & \sin\kappa\sin\omega - \cos\kappa\sin\phi\cos\omega \\ -\sin\kappa\cos\phi & \cos\kappa\cos\omega - \sin\kappa\sin\phi\sin\omega & \sin\kappa\sin\phi\cos\omega + \cos\kappa\sin\omega \\ \sin\phi & -\cos\phi\sin\omega & \cos\phi\cos\omega \end{pmatrix} \tag{9.4}
$$

For small rotation, this matrix may be approximated by:

$$R \approx \begin{pmatrix} 1 & \kappa & -\phi \\ -\kappa & 1 & \omega \\ \phi & -\omega & 1 \end{pmatrix} \qquad (9.5)$$

NOTE

Each GNSS has unambiguously defined geocentric geodetic datum. Coordinates obtained through GNSS receivers therefore pertain to WGS84, PZ-90, GTRF, or CGCS2000 datum. All national control and maps are, however, based on local geodetic datum. Coordinates obtained from the GNSS may therefore differ from the national system coordinates (local datum) up to 100 m or even more. The coordinates can, however, be transformed from a specific GNSS datum to any other local datum if reliable transformation parameters are available. So far, very accurate transformation parameters have not been found out in respect of Indian datum (Agrawal 2004). Keeping this problem in mind, the users should be well aware of it and recognise the utility of the GNSS receiver for their purposes.

9.5.1 GNSS AND HEIGHT MEASUREMENT

A point on the earth's surface is not completely defined by its latitude and longitude. As noted earlier, there is a third element—height. Surveyors have traditionally referred to this component of position as *elevation*. A level oriented at a point on the earth's surface defines a line parallel to the geoid at that point; this level is actually the height or elevation. Therefore, the elevations determined by level are elevation relative to geoid, called *orthometric elevation* (the distance from the geoid to a point, measured along a line normal to the geoid). In contrast, GNSS measures the relative elevations of points above the ellipsoid, called *ellipsoidal height,* which is different than the orthometric height. Therefore, the height values associated with GNSS coordinates may indicate that water flows uphill. Orthometric elevations are not directly available from the geocentric position derived from GNSS measurements. However, it is not difficult to convert ellipsoidal latitude, longitude, and height to orthometric measurements because the reference ellipsoid is mathematically defined and clearly oriented to the earth (Sickle 2008; Smith 1997; Zilkoski and Hothem 1989). Modern high performance receivers can process this mathematics internally and give the orthometric height as output (if we ask the receiver to do so).

The ellipsoidal (h) and orthometric (H) heights are closely related by the geoid height (N) (Figure 9.7). *Geoidal height* is the separation between ellipsoid and geoid—height of geoid above the ellipsoid (it may have a negative value if the geoid is under the ellipsoid). A network of benchmark observations, gravity observations,

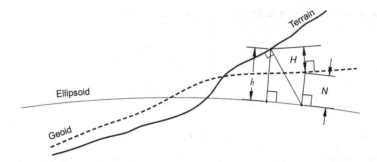

FIGURE 9.7   Relation between orthometric and ellipsoidal height.

and elevation models are used to develop a geoid model from which geoidal heights can be estimated (RMITU 2006; Zilkoski *et al.* 2005). The accuracy of these geoid heights is dependent upon the accuracies of the various measurements used to construct the model. Surveyors must apply the geoidal height to GNSS derived height values to obtain orthometric heights (which are related to mean sea level). GNSS derived ellipsoidal height, when combined with geoidal height, can give usable orthometric height. The following height equation is used:

$$h = H + N \tag{9.6}$$

Note that this equation is only an approximation, as orthometric height is measured along a curved plumb line normal to the geoid surface, while the ellipsoidal and geoid heights are measured along straight lines normal to the ellipsoid surface. The accuracy of geoid models used to estimate the geoidal height is generally of the order of a few centimetres. Further, users must be aware that geoid is not exactly matched with the mean sea level and may vary up to 10 m. Therefore, the heights derived by GNSS measurements are often with some error.

Many of the commercially available GNSS software packages and data collectors (controllers) enable the required conversion of ellipsoidal height into orthometric height. The surveyor must provide estimates of the geoid undulation at three points which are well spaced throughout the survey area. The geoid undulation is determined by occupying points with known orthometric height during the survey. The software then uses this information to estimate the necessary parameters and applies the transformation to all remaining points.

9.6   PROJECTION

The earth is divided into various sectors by the lines of latitudes and longitudes. This network is called a *graticule*. A map projection denotes the preparation of the graticule on a flat surface. Theoretically, map projection might be defined as 'a systematic drawing of parallels of latitude and meridians of longitude on a plane surface for the whole earth or a part of it on a certain scale so that any point on the earth surface may correspond to that on the drawing'. Whether we treat the earth as a sphere or a spheroid, we must transform its 3D surface to create a flat map sheet. This mathematical transformation is commonly referred to as a map projection (Yang *et al.* 2000; Bugayevskiy and Snyder 1995).

The earth can ideally be represented by a globe. However, it is not possible to make a globe on a very large scale. For instance, if anyone wants to make a globe on a scale of one inch to a mile, the radius of the globe will be 330 ft. It is difficult to make and handle such a globe and uncomfortable to carry it in the field for reference. Therefore, a globe is least useful or helpful in the field for practical purposes. Moreover, it is neither easy to compare different regions over the globe in detail, nor convenient to measure distances over it. Therefore, projection system has evolved to transform the three-dimensional earth onto a two-dimensional plane. However, instead of a single projection system, various projections are being used by different countries. The reason for this is that each country would like to be represented in correct shape and size, and a single projection is not capable of projecting every area of the earth with the same accuracy due to the irregular shape of the earth. In practice, a plane representation of the earth is used, in order to:

- Provide us with a plane surface representing a part of the ellipsoid.
- Transform angular metrics into a more practical metric system.
- Allow metric measurements of distances.

9.6.1 SELECTION OF PROJECTION

It is not an easy task to flatten a spheroid that contains a variable surface onto a plane. It is like the stretching and tearing that we would have to do to a basketball to make its curved surface lie flat on the ground. Different projections cause different types of distortions (inaccuracies). There is no ideal map projection, but representation for a given purpose can be achieved. The selection of projection is made on the basis of the following issues (Yang *et al.* 2000; Bhatta 2020; ESRI 2004):

Conformality (shape): When the scale of a map at any point on the map is the same in any direction, the projection is conformal. Meridians (lines of longitude) and parallels (lines of latitude) intersect at right angles. Shape is preserved locally on conformal maps.

Distance: A map is equidistant when it portrays distances correctly from the centre of the projection to any other place on the map.

Direction (bearing): A map preserves direction when azimuths (angles from a point on a line to another point) are portrayed correctly in all directions.

Area: When a map portrays areas over the entire map so that all mapped areas have the same proportional relationship to the areas on the earth that they represent, the map is an equal-area map.

Some projections minimise distortions in some of these properties at the expense of maximising errors in others. Some projections are designed to distort all of these properties moderately.

TABLE 9.1

Classification of Projections Depending on Different Bases

Basis	Classes
Method of construction	1. Perspective
	2. Non-perspective
Preserved qualities	1. Homolographic/equal area
	2. Orthomorphic/conformal/true in shape
	3. Azimuthal (true bearing, i.e., direction and angle)
Developable surface area	1. Cylindrical
	2. Conical
	3. Azimuthal/zenithal/planar
	4. Conventional
Position of tangent surface	1. Polar
	2. Equatorial or normal
	3. Oblique
Position of viewpoint or position of light	1. Gnomonic
	2. Stereographic
	3. Orthographic

9.6.2 CLASSIFICATION OF PROJECTIONS

A great number of map projections have been developed. The characteristics of these projections are so complex that they often possess one or more common properties. There is no projection which can be grouped in a single class. Moreover, it is rather difficult to obtain a rational classification of a map projection. There can be as many classifications as there are bases (Table 9.1).

Although there are various classifications of map projection considering various factors, they can be classified into the following four general classes:

1. Cylindrical
2. Conical
3. Azimuthal/Zenithal/Planar
4. Miscellaneous

9.6.2.1 Cylindrical Projections

Cylindrical projections result from projecting a spherical surface onto a cylinder (Figure 9.8). When the graticule is prepared on the surface of a hollow cylinder, it is called cylindrical projection. In this projection, meridians are geometrically projected onto the cylindrical surface, and parallels are mathematically projected. This produces graticular angles of 90°. The cylinder is 'cut' along any meridian to produce the final cylindrical projection. The meridian opposite the cut-line becomes the *central meridian*. The meridians are equally spaced, while the spacing between parallel lines of latitude increases towards the poles. This projection is conformal and displays true direction along straight lines.

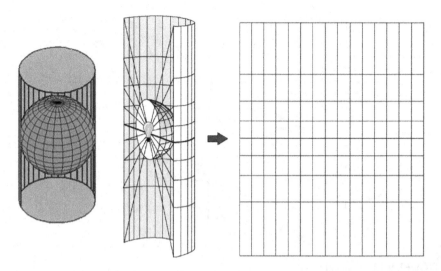

FIGURE 9.8 Cylindrical projection.

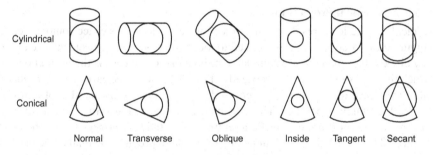

FIGURE 9.9 Various cases of cylindrical and conical projections.

For more complex cylindrical projections, the cylinder is rotated and/or made tangent or secant (Figure 9.9).

Tangent case: When the cylinder is tangent to the sphere, contact is along a great circle (the circle formed on the surface of the earth by a plane passing through the centre of the earth).

Secant case: In the secant case, the cylinder touches the sphere along two lines, both small circles (a circle formed on the surface of the earth by a plane not passing through the centre of the earth).

Normal case: When the axis of the cylinder passes through the poles of the earth, the case is said to be 'normal.'

Transverse case: When the cylinder upon which the sphere is projected is at right angles to the poles, the cylinder and resulting projection are transverse.

Oblique case: When the cylinder is at some other, non-orthogonal, angle with respect to the poles, the cylinder and resulting projection is oblique.

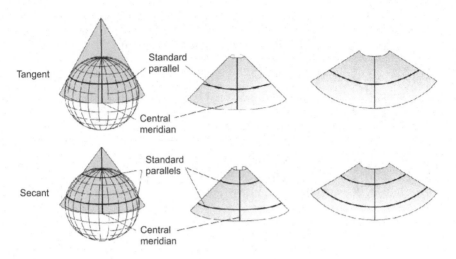

FIGURE 9.10   Conical projection.

9.6.2.2   Conical Projections

Conical projections result from projecting a spherical surface onto a cone (Figure 9.10). A cone may be imagined to touch the globe of a convenient size along any circle (other than the great circle), but the most useful case is the normal one in which the apex of the cone lies vertically above the pole on the earth's axis produced and the surface of the cone is tangent to the sphere along some parallel of latitude. It is called 'standard parallel'. The meridians are projected onto the conical surface, meeting at the apex of the cone. Parallel lines of latitude are projected onto the cone as rings. The cone is then 'cut' along any meridian to produce the final conic projection which has straight converging lines for meridians, and concentric circular arcs for parallels. The meridian opposite the cut line becomes the *central meridian.*

More complex conic projections contact the global surface at two locations (Figure 9.10). These projections are called secant projections and are defined by two standard parallels. It is also possible to define a secant projection by one standard parallel and a scale factor. The distortion pattern for secant projections is different between the standard parallels than beyond them. Generally, a secant projection has less overall distortion than a tangent projection. On still more complex conic projections, the axis of the cone does not line up with the polar axis of the globe. These types of projections are called *oblique.*

NOTE

The concept of two parallels for secant case is applicable for cylindrical projections as well in conical projections, but, unlike conical projections, they have the same radius.

9.6.2.3 Azimuthal Projections

Azimuthal projections result from projecting a spherical surface onto a plane (Figure 9.11). This projection is also known as planar projection or zenithal projection. In this projection, a flat paper is supposed to touch the globe at one point and project the lines of latitude and longitude on the plane. This type of projection is usually tangent to the globe at one point, but may be secant as well. The point of contact may be the North Pole, the South Pole, a point on the equator, or any point in between. This point specifies the aspect and is the focus of the projection. The focus is identified by a central longitude and a central latitude.

In respect of the plane's position touching the globe, zenithal projection is of three main classes.

- Normal or equatorial zenithal (where the plane touches the globe at the Equator).
- Polar zenithal (where the plane touches the globe at a pole).
- Oblique zenithal (where the plane touches the globe at any other point).

According to the location of the view point zenithal projection is of three types:

- Gnomonic/central (view point lies at the centre of the globe).
- Stereographic (view point lies at the opposite pole).
- Orthographic (view point lies at the infinity).

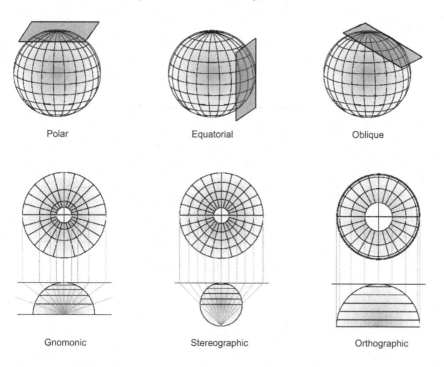

FIGURE 9.11 Azimuthal projection.

9.6.2.4 Miscellaneous Projections

The projections discussed earlier are conceptually created by projecting from one geometric shape (a sphere or a spheroid) onto another (a cone, cylinder, or plane). Many projections are not related as easily to a cone, cylinder, or plane. Miscellaneous projections include unprojected ones, such as rectangular latitude and longitude grids; and other examples of that do not fall into the cylindrical, conic, or azimuthal categories.

9.6.3 PROJECTION PARAMETERS

A projection is not enough to define a projected coordinate system. There are some other issues to be answered, like, where is the centre of the projection, was a scale factor used, etc. (ESRI 2004). Without knowing the exact values for the projection parameters, the earth cannot be projected.

Each projection has a set of parameters that we must define. The parameters specify the origin and customise a projection for our area of interest. Angular parameters use the GCS units, while linear parameters use the projected coordinate system units.

Projections require a point of reference on the earth's surface. Most often this is the centre, or origin, of the projected coordinate system (Figure 9.12). As a result, the west side of y-axis and the south side of x-axis are negative. Some of the parameters are used to avoid these negative values.

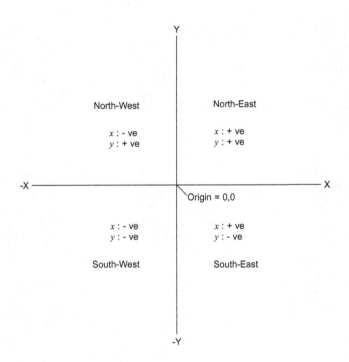

FIGURE 9.12 Projected coordinate system.

9.6.3.1   Linear Parameters

False easting: Linear distance applied to the origin of the x-coordinates.

False northing: Linear distance applied to the origin of the y-coordinates.

Usually, the origin of a projected coordinate system is considered the centre point of the map to reduce the overall distortion. This implies that our map has negative coordinate values for the west side of y-axis and the south side of x-axis. False easting and northing values are usually applied to ensure that all x or y values become positive. For example, we have considered false easting as 50,000 m and false northing as 60,000 m. In this case, one false origin will be considered at 50,000 m left of y-axis and 60,000 m below the x-axis; all measurements will be carried out from this false origin instead of the true origin. Thus, we avoid negative values for x- and y-coordinates, which are within 50,000 m west of y-axis and 60,000 m south of the x-axis. We can also use the false easting and northing parameters to reduce the range of the x- or y-coordinate values. For example, if we know all y values are greater than five million meters, we could apply a false northing of –5,000,000.

Scale factor: A unit-less value applied to the centre point or line of a map projection. The scale factor is usually slightly less than 1.0. The Universal Transverse Mercator (UTM) coordinate system, which uses the transverse Mercator projection, has a scale factor of 0.9996. Rather than 1.0, the scale along the central meridian of the projection is 0.9996. This creates two, almost parallel, lines approximately 180 km away, where the scale is 1.0. The scale factor reduces the overall distortion of the projection in the area of interest.

9.6.3.2   Angular Parameters

Azimuth: Defines the rotation of the centre line of a projection. The rotation angle measures in compass direction from north.

Central meridian: Defines the origin of the x-coordinates, the meridian opposite of the cutline.

Longitude of origin: Defines the origin of the x-coordinates. The central meridian and longitude of the origin parameters are synonymous.

Central parallel: Defines the origin of the y-coordinates.

Latitude of origin: Defines the origin of the y-coordinates. This parameter may not be located at the centre of the projection. In particular, conic projections use this parameter to set the origin of the y-coordinates below the area of the interest. In that instance, it is not necessary to set a false northing parameter to ensure that all y-coordinates are positive.

Longitude of centre: Defines the origin of the x-coordinates. It is usually synonymous with the longitude of origin and central meridian parameters.

Latitude of centre: Defines the origin of the y-coordinates. It is almost always the centre of the projection.

Standard parallel 1 and standard parallel 2: Used with conic projections to define the latitude lines where the scale is 1.0 (no distortion). When defining a conic projection (tangent case) with one standard parallel, the standard parallel defines the origin of the y-coordinates. For other conic cases (secant case), the y-coordinate origin is defined by the latitude of origin parameter.

9.6.4 COMMON PROJECTIONS

It has been mentioned earlier that there are numerous projection systems adopted to project the earth surface on to a 2D plane to create maps. Different projection systems are appropriate for different applications (ESRI 2004). It is beyond the scope of this book to describe all of them. However, a few very common projection systems have been explained in the following sections to provide a better understanding of the concepts.

9.6.4.1 Polyconic Projection

The name of this projection translates into 'many cones'. This refers to the projection methodology. This system is more complex than the regular conic projections, but still a simple construction. This projection is created by lining up an infinite number of cones along the central meridian. The central meridian is straight. Other meridians are complex curves. The parallels are non-concentric circles. The scale is true along each parallel and along the central meridian. This affects the shape of the meridians. Unlike other conic projections, the meridians are curved rather than linear (Figure 9.13).

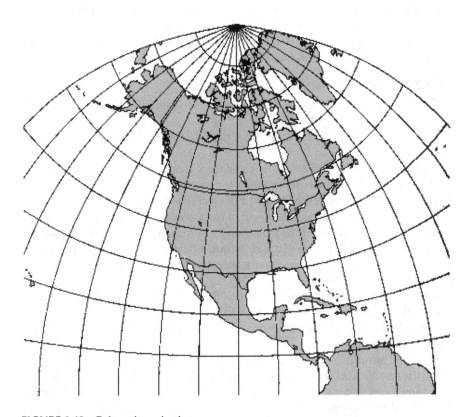

FIGURE 9.13 Polyconic projection.

Polyconic projection preserves the area, shape, distance, and azimuth for small areas; best for north–south extents; the scale increases away from the central meridian and it is applied for topographic maps. It is generally considered that the scale distortion is acceptable only up to 9° away from the Central Meridian and is not recommended for larger areas because of distortion.

9.6.4.2 Lambert's Azimuthal Equal-Area Projection

This projection was first presented by Johann Heinrich Lambert (1728–77) of Alsace in 1772. This map projection is one of the most popular projections used in atlases today to map large areas such as an entire country, polar area, oceanic mapping, etc. This projection can accommodate all aspects of azimuthal projection, i.e., equatorial, polar, and oblique.

This projection distorts shape minimally. Area is equal; scale is true only at the centre in all directions that decreases with distance from the centre along the radii and increases from the centre perpendicularly to the radii. Only the centre is free from distortion. Distortion is moderate for one hemisphere but becomes extreme for a map of the entire earth.

Meridians are equally spaced lines intersecting at the poles (Figure 9.14). Parallels are unequally spaced circles, centred at the pole (Figure 9.14a). Spacing of the circles gradually decreases away from the pole. Central meridian is a straight line. Other meridians are complex curves, unequally spaced along the Equator and intersecting at each pole.

9.6.4.3 UTM Projection

The UTM (Universal Transverse Mercator) system is a specialised application of the Transverse Mercator projection. The choice of the transverse Mercator is probably now used more than any other projection for accurate mapping.

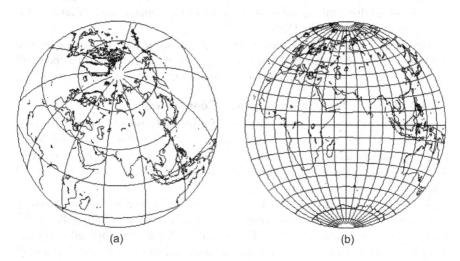

(a)   (b)

FIGURE 9.14 Lambert azimuthal equal-area projection.

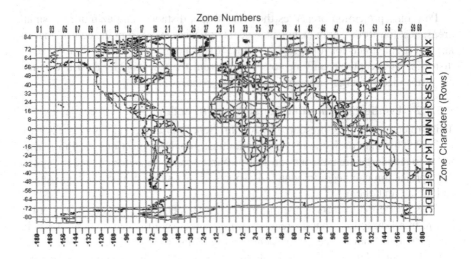

FIGURE 9.15 The UTM projection system.

The UTM coordinates define 2D, horizontal positions. UTM zone numbers designate 6° longitudinal strips extending from 80° South latitude to 84° North latitude (Figure 9.15). The UTM zone characters (or rows) designate 8° zones extending north and south from the Equator. There are special UTM zones between 0° and 36° longitude at row-X and a special zone (zone-32) between 56° and 64° north latitude.

The origin for each zone is its central meridian and the Equator. Each zone has a central meridian. Zone 44, for instance, has a central meridian of 81° east longitude. The zone extends from 78° to 84° east longitude.

Eastings are measured from the central meridian (with a 500,000 m false easting to ensure positive coordinates). Northings are measured from the Equator (with a 10,000,000 m false northing for positions south of the Equator and 0 m false northing for positions north of the Equator).

The UTM is conformal, best for north–south extents; scale is true along the two meridians halfway between the Central Meridian and the edge of the zone (too small between these lines and too large outside of these lines).

9.6.4.4 Latitude/Longitude Geographic Coordinates

The latitude/longitude GCS (also known as unprojected latitude/longitude system) is not a map projection. Unprojected maps include those that are formed by considering longitude and latitude as a simple rectangular coordinate system (Figure 9.16). Scale, distance, area, and shape are all distorted with the distortion increasing towards the poles.

The earth is modelled as a sphere or spheroid. The sphere is divided into equal parts usually called degrees; some countries use grads. A circle is 360° or 400 grads. Each degree is subdivided into 60 minutes, with each minute composed of 60 seconds. Precision below degree is expressed as minutes and seconds, and decimals of seconds, in one of two formats: either plus or minus DDMMSS.XX, where DD are

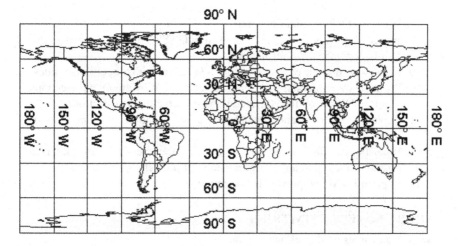

FIGURE 9.16 Latitude/longitude GCS.

degrees, MM are minutes, and SS.XX are decimal seconds; or alternatively, as DD.XXXX or decimal degrees.

The GCS consists of latitude and longitude lines. Each line of longitude runs north–south and measures the number of degrees east or west of the prime meridian. Values range from −180° to +180°. Lines of latitude run east–west and measure the number of degrees north or south of the Equator. Values range from +90° at the North pole to 90° at the South pole. The standard origin is where the Greenwich prime meridian meets the Equator. All points north of the Equator or east of the prime meridian are positive.

EXERCISES

DESCRIPTIVE QUESTIONS

1. What is geodesy? Why is it important in satellite navigation and positioning?
2. Explain celestial equatorial and earth-cantered inertial coordinate systems with appropriate sketches.
3. Explain geographical coordinate system and earth-cantered earth-fixed coordinate system with appropriate sketches.
4. What do you understand by 'geoid' and 'ellipsoid'? What are the differences between them? What is a datum? How is a vertical datum determined?
5. Explain WGS84 and Indian geodetic datum? How can you convert coordinate values of Indian geodetic datum to WGS84?
6. What do you understand by 'spheroid' and 'datum'? Mention the difference(s) between geocentric and local datum. Briefly describe the different spheroids and datums used in different GNSS.

7. What is ITRF? Explain in detail.
8. Which ellipsoids and datums are used in GNSS? What is the mathematics of coordinate transformation?
9. How do we determine height in GNSS? Explain with proper illustrations.
10. What do you understand by 'map projection'? Why are map projections required? What are the considerations for selecting a projection?
11. Explain cylindrical and conical projection. Mention different cases of them.
12. What do you understand by 'projection parameters'? Briefly describe essential linear and angular parameters of projection.
13. Explain UTM and polyconic projection.

SHORT NOTES/DEFINITIONS

Write short notes on the following topics

1. Geodesy
2. Referential system
3. Vernal point
4. Graticule
5. Flattening
6. Eccentricity
7. PZ-90
8. False easting
9. Standard parallel
10. Central meridian
11. Geocentric datum
12. Horizontal and vertical datum
13. Mean sea level
14. Geoid
15. Orthometric height
16. Geoidal height

10 Applications of GNSS

10.1 INTRODUCTION

GNSS has a variety of applications on land, at sea, and in the air. Basically, GNSS allows us to record coordinates of locations and helps us to navigate from one place to another. GNSS can be used everywhere except where it is impossible to receive the signal (such as 'normally' inside buildings, in mines and caves, parking garages and other subterranean locations, and under water). A vital role of GNSS is to help in-orbit navigation and positioning of remote sensing satellites.

The GNSS technology has matured into a resource that goes far beyond its original design goals. It has exceeded all expectations; it is an amazing and proud achievement. Nowadays, scientists, sportsmen, farmers, soldiers, pilots, surveyors, hikers, delivery drivers, sailors, dispatchers, lumberjacks, fire-fighters, and people from many other walks of life are using GNSS in ways that make their work more productive, safer, and sometimes even easier. Today, most cellular phones, smart-watches, and car stereos have receivers for GNSS satellite signals.

While this chapter outlines some of the varied uses of GNSS and focuses on the major achievements of GNSS, it does not describe how GNSS can be implemented in or detail the requirements for specific applications.

10.2 CLASSIFICATION OF GNSS APPLICATIONS

Satellite navigation is being used in many different application areas, from purely commercial to highly scientific. There are many professional domains that have found a great interest in using GNSS, mainly in order to reduce the time and complexity issues of previously used systems. This is notably the case in the civil engineering and construction industries, where high accuracy receivers have been in use for years now.

The classification of applications can be carried out in many ways: mass market versus professional, commercial versus scientific, or by main application domains, such as surveying, environment, and agriculture. Some possible classifications are provided in Table 10.1. It is important to note that some of these classifications are overlapping. If we look at Table 10.1, we shall understand the vastness of GNSS applications. However, all of these applications can be grouped into five broad categories as follows (Bhatta 2020).

> *Location*—*determining a basic position.* The first and most obvious application of GNSS is the simple determination of a 'position' or location. GNSS is the first positioning system to offer highly precise location data for any point on the planet, in any weather, at any time. This capability alone would be sufficient to qualify it as a major utility.

TABLE 10.1
Applications of GNSS

Application Areas	Some Examples
Surveying and mapping	• Geodetic control survey. • GIS mapping. • Structural deformation survey. • Construction stakeout and grading. • Coastal engineering surveys. • Photogrammetric mapping control. • Remote sensing applications control survey. • Geophysics, geology, and archaeological survey.
Navigation	• Automobile. • Aircraft (including drone). • Maritime. • Space flight. • Heavy equipment. • Office/home delivery. • Cyclists, hikers, climbers, and pedestrians.
Tracking	• Delivery systems. • Precision farming. • Personal emergencies. • Fleet management. • Tracking spacecrafts. • Tracking people. • Tracking animals.
Geodynamics	• Global and continental plate movements and deformation analysis. • Regional crustal movements and deformation analysis. • Local crustal movements, subsidence/rise, and monitoring of crustal deformation, landslides, ground deformation, plate tectonic studies, and local geo-tectonics. • Earth rotation monitoring. • Polar motion studies.
Engineering	• Determination of control points for cartography, GIS, photogrammetry, remote sensing (ground control points), geophysical surveys, inertial surveys, hydrographical surveys, expeditions of all kinds, archaeological mapping. • Monitoring movements by repeated or continuous measurements for ground subsidence or rise, dam deformation, subsidence of offshore structures, settlement of buildings. • Control points for engineering projects, for example, tunnel surveys, bridge construction, road surveys, pipe lines, power and communication lines, waterways, canal, and drain surveys. • Real-time guidance and control of vehicles, such as taxi fleet management, transport vehicles and cars, construction vehicles, dumpers, and excavators in open pit mining.

(Continued)

TABLE 10.1 (CONTINUED)
Applications of GNSS

Application Areas	Some Examples
Mining	• Surveying for preparation of mine plans. • Control surveys for surveying and mapping. • Connection of mine plans to national datum and grid. • Surveys for acquiring land for mines and payment of compensation. • Ore body delineation. • Tracking and monitoring of heavy equipment. • Precise positioning of drill equipment. • Precisely navigate electric drills to designed blast hole patterns and exact target depths. • Tracking of materials and vehicles. • Calculate surface to surface volumes. • Navigation of vehicles. • Design and construction of roads. • Determination of profile and volumes in open pit mines. • Land reclamation at surface mines. • Proximity warning system.
Photogrammetry, remote sensing, and GIS	• Ground control points (GCP). • Ground truthing. • Navigation of airplanes/drones. • Sensor platform coordinates and orientation. • GIS Mapping survey.
Science and research	• Tropospheric studies of precipitable water vapour. • Ionospheric monitoring. • Investigations of tide gauge benchmark motions for long-term sea level studies. • Orbit determination. • Earthquake prediction. • Weather forecasting. • Animal surveillance.
Space applications	• Positioning and navigation of other space vehicles. • Radar altimetry. • Mapping from space. • Gravimetry and gravity gradiometry from space.
Location-based services (LBS)	• Emergency response. • Delivery. • Medical. • Advertising and marketing.
Military applications	• En route navigation. • Low-level navigation. • Target acquisition (tracking). • Reconnaissance.

(*Continued*)

TABLE 10.1 (CONTINUED)
Applications of GNSS

Application Areas	Some Examples
	• Remotely operated vehicles.
	• Updating inertial navigation systems.
	• Sensor emplacement.
	• Missile guidance.
	• Command and control.
	• Monitoring nuclear detonations.
	• Precise bombing.
	• Search and rescue.
	• Nuclear detonation detection.
Time transfer	• Time signal transmission by laboratories.
	• Astronomy.
	• Wristwatches.
	• Telecommunication.
	• Banking.
Special applications	• Railroad corridor surveying.
	• Pesticide spraying and seed control in agriculture.
	• Forest management.
	• Glacial geodesy.
	• Attitude control of aircrafts and ships.
Other applications	• Games.
	• Global Maritime Distress and Safety System (GMDSS) is interfaced with GPS. The system is mandatory since 1 February 1999.
	• Road usage pricing.
	• Skydiving.
	• Marketing.
	• Social networking.
	• In modern warfare, GNSS is essential. During Kosovo and Iraq wars, GNSS was used for several purposes.
	• Study of animal behaviour.
	• GNSS has been used to map malaria-infested areas in Africa.

Navigation—*getting from one location to another.* GNSS helps us to determine exactly where we are, but sometimes it is important to know how to get elsewhere. It was originally developed for navigation—in the air, on the water, and on the land. GNSS is rapidly becoming commonplace in automobiles. Sophisticated systems can show the vehicle's position on an electronic map display, allowing drivers to keep track of where they are and look up street addresses, restaurants, hotels, and other destinations. Some systems can even automatically create a route and give turn-by-turn directions to a designated location.

Tracking—monitoring the movement of people and things. GNSS used in conjunction with communication links and computers can be used for tracking any moving vehicle (on the water, on the land, and in the air). So, it is no surprise that police, ambulance, and fire departments have adopted systems like GNSS-based automatic vehicle location managers to pinpoint both the location of the emergency and the location of the nearest response vehicle. Transport and traffic departments are also rapidly adopting GNSS-based tracking.

Mapping—creating maps of the world or parts of it. It is a big world out there, and using GNSS to survey and map precisely, we can save time and money in these applications. Today, GNSS makes it possible for a single surveyor to accomplish in a day what used to take weeks with an entire team. They can even do their work with a higher level of accuracy than ever before.

Timing—bringing precise timing to the world. Although GNSS is well-known for positioning, navigation, tracking, and mapping, it is also widely used to disseminate precise time, time intervals, and frequency. Time is a powerful commodity, and exact time is rather more powerful. There are three fundamental ways in which time is useful to us. As a universal marker, time informs us when things happened or when they will. As a way to synchronise people, events, and other types of signals, time helps keep the world on schedule. As a way to inform how long things last, time provides an accurate, unambiguous sense of duration. The GNSS satellites carry highly accurate atomic clocks. In order for the system to work, our GNSS receivers here on the ground synchronise themselves to these clocks. This implies that every GNSS receiver is, in essence, an atomic accuracy clock and can provide time information in nanoseconds.

Applications of GNSS range from earth sciences to social science, space to ground, land to air and water, entertainment to serious business, travel, transport, surveying and mapping, navigation, geodynamics, research and exploration, tracking, environment, disaster management, treasure-hunting and spying, modern warfare to peace initiatives, to enhancement of quality of life and what not—even terrorist activities (Kaplan 1996; Clarke 1996; Czerniak and Reilly 1998; Agrawal 2004; Samama 2008). It is almost impossible to enumerate all possible applications of GNSS. GNSS being an all-weather real-time continuously available, the most economic precise positioning and time determination technique contributes to its unlimited use. Although GNSS has countless end uses, in the following sections we shall try to indicate the important uses of it starting from surveying and mapping. Most of these applications are applicable for defence and as well as civilian endeavours; however, some of the applications are defence-specific, e.g., weapon guidance.

10.3   SURVEYING AND MAPPING

Surveying and mapping applications are equally important to the military and civilians. These applications may have several implementations or dimensions based on the requirement of the project (Kaplan 1996; Sickle 2008; Owings 2005; Leick 2004; RMITU 2006; Trimple 2001; Zilkoski and Hothem 1989). The following sub-sections are intended to describe some applications of GNSS in surveying and mapping. However, in surveying and mapping, there are several issues involved and necessary to be addressed from a practical point of view. These issues will be addressed in Chapters 11 and 12.

10.3.1 GEODETIC CONTROL SURVEY

A control survey is an exercise that is designed to mark points; these marks will form the framework for subsequent surveys. A geodetic survey can be considered a high-level control survey with special accuracy requirements. Geodetic surveys are considered distinct from control surveys in that the marks being coordinated are assumed to be stable, well-built structures.

Establishing or densifying survey control is one of the major uses of GNSS technology (Czerniak and Reilly 1998; Sickle 2008; RMITU 2006; USACE 2007). GNSS is often more cost-effective, faster, accurate, and reliable than conventional (terrestrial) survey methods. The quality control statistics and large number of redundant measurements in GNSS networks help to ensure reliable results. Primary horizontal and vertical control monuments are usually set using static GNSS survey methods for high accuracy (Figure 10.1), although some post-processed kinematic methods may also be employed. These primary monuments are typically connected

FIGURE 10.1   Static GNSS control survey.

to national horizontal and vertical reference datums. From these primary monuments, supplemental site plan mapping or vessel/aircraft positioning is performed
using real-time kinematic (RTK) techniques. Field operations to perform a GNSS
static control survey are relatively efficient and can generally be performed by one
person per receiver. GNSS is particularly effective for establishing primary control networks as compared with conventional surveys because inter-visibility is not
required between adjacent stations.

10.3.2 GIS Mapping

Real-time and post-processed techniques can be used to perform topographic mapping surveys and Geographic Information System (GIS) base mapping (Kaplan 1996;
Sickle 2008; Owings 2005; USACE 2007). Depending on the accuracy requirement,
either code or carrier phase techniques may be employed. In general, most topographic mapping is performed using RTK methods. Real-time topographic or GIS
feature data is usually collected using backpack antenna mounts (Figure 10.2a) or
pole mounted antenna (Figure 10.2b). Data are logged on standard data collectors
similar to those used for terrestrial total stations. Data collector software is designed
to assign topographic and GIS mapping features and attributes. Code differential
techniques may be used for GIS mapping features requiring meter-level accuracy.
If only approximate mapping accuracy is needed (10–30 m), handheld standalone
code-based GNSS receivers with absolute positioning may be used (Figure 10.2c).
Refer to Chapters 11 and 12 for detailed discussion.

10.3.3 Structural Deformation Survey

A deformation survey is performed during excavation operations in order to
check that the construction region is not moving due to earthwork operations
(USACE 2002a). Another example of a deformation survey occurs in the routine

  (a)   (b)   (c)

FIGURE 10.2   (a) RTK survey with backpack mounted antenna; (b) RTK survey with pole
mounted antenna; (c) Surveying with handheld standalone code-based receiver.

monitoring of dam walls. In these types of surveys, marks in the region of inter-est are regularly surveyed and related to other marks which are not in the same vicinity and are considered stable. If movement in the marks in the survey region is detected, then action can be taken to maintain safety (Cruz *et al.* 2006). In general, the accuracy associated with monitoring surveys is extremely high (in the millimetres).

GNSS survey techniques can be used to monitor the motion of points on a civil structure relative to stable reference monuments (Sickle 2008; USACE 2003; USACE 2002a). This can be done with an array of antennae positioned at selected points on the structure (Figure 10.3) and on the reference monuments. Baselines are formulated between the points on the structure and the monuments to monitor relative movements. Measurements can be made on a continuous basis. A GNSS structural deformation system can operate unattended and is relatively easily installed and maintained. Alternatively, periodic monitoring observations are taken using RTK or post-processed kinematic techniques. Prior to performing structural monitoring surveys, the stable reference network (reference monuments) must be accurately positioned. Long-term static GNSS observations are typically used to perform this task.

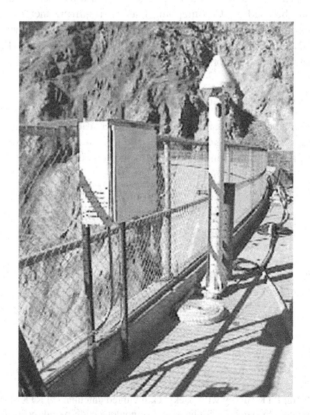

FIGURE 10.3   GNSS antenna positioned on a structure to monitor structural deformation.

FIGURE 10.4   RTK stakeout for construction (Courtesy: Trimble).

10.3.4   Construction Stakeout and Grading

Surveyors are frequently asked to undertake construction surveying work, which primarily includes the setting out of features such as buildings and roads as well as infrastructure, such as pipelines. Historically, this work has been undertaken using more conventional survey equipment, e.g., the total station. However, the advantages of static and RTK have been widely recognised for this work. The rapid nature of RTK in particular makes this work efficient and cost reductive (Figures 10.4 and 10.5). Automatic systems for bulldozers/graders use the cut/fill information to drive the hydraulic controls of the machine to automatically move the machine's blade to grade (Figure 10.4). Use of 3D machine control dramatically reduces the number of survey stakes required on a job site, reducing time and cost.

Survey-grade GNSS receivers are now designed to perform all traditional construction stakeouts (Figure 10.5), e.g., lots, roads, curves, grades, etc. Typical applications include staking out baselines, boring rig placement, topographic survey and facility or utility construction alignment. GNSS can also be used to control and monitor earthmoving operations, such as grading levees or beach construction (Trimble 2001).

NOTE

Stakeout (or setting out) survey is performed in order to establish location of structures, like buildings and roads or infrastructure, like drainage or property

FIGURE 10.5 Construction survey with GNSS—stakeout of plot/lot and road (a), baselines (b), topography (c), and alignment (d) (Courtesy: Leica Geosystems).

corners based on proposed plans. It is the movement of features from the construction plan to the actual site by transferring dimensions from the layout plan to the ground. It is probably the most critical step in the entire construction process.

10.3.5 COASTAL ENGINEERING SURVEYS

Differential GNSS positioning and elevation measurement techniques have almost replaced conventional survey methods in performing beach surveys and studies (USACE 2003; USACE 2002b). Depth measurement sensors (physical or acoustical) are typically positioned with RTK methods. Vessels and other platforms usually merge RTK observations with inertial measurement units in order to reduce surf heave (Figure 10.6). Land sections of beach profile surveys are usually controlled using RTK topographic methods.

10.3.6 PHOTOGRAMMETRIC MAPPING CONTROL

The use of an airborne GNSS receiver, combined with specialised inertial navigation, LIDAR, and photogrammetric data processing procedures, can significantly

FIGURE 10.6   Survey vessel equipped with RTK GNSS and other sensors (Courtesy: Titan Environmental Surveys Limited).

reduce the amount of ground control for typical photogrammetric projects. In effect, each camera image or LIDAR scan is accurately positioned and oriented relative to a base reference station on the ground. In the past, the position and orientation of the camera was back-computed from ground control points (Bhatta 2020). Traditionally, these mapping projects require a significant amount of manpower and monetary resources for the establishment of the ground control points. Therefore, the use of airborne GNSS technology significantly lessens the production costs associated with wide-area mapping projects. Tests have shown that ground control coordinates can be developed from an airborne platform using GNSS kinematic techniques to centimetre-level precision in all three axes if system-related errors are minimised and care is taken in conducting the airborne GNSS and photogrammetric portions of the procedures (USACE 2002c).

10.3.7   REMOTE SENSING APPLICATIONS CONTROL SURVEY

Survey of ground control points (GCPs) for geometric correction is a very common application of GNSS in remote sensing (Figure 10.7a). In the remote sensing application

(a)

(b)

FIGURE 10.7   (a) Survey of ground control point; (b) Ground truthing with GNSS.

called *ground truthing* (Bhatta 2020), it is important to know the exact location of field-observed reference data (Figure 10.7b). In order to achieve accurate geometric correction as well as ground truthing, GCPs with known coordinates are required.

The requirements of GCPs are that the points should be identical and recognisable, both on the remote sensing image as well as on the ground, and should be measurable (Bossler *et al.* 2002). In such cases, earlier traditional control surveys were performed. However, today, GNSS can provide these geographic coordinates in a very short time. Carrier-based classic static survey or the code/carrier-based differential/relative approach is ideal for GCP survey; however, standalone code-based techniques may also be adopted for coarser resolution imageries.

10.3.8   GEOPHYSICS, GEOLOGY, AND ARCHAEOLOGICAL SURVEY

High precision measurements of crustal strain can be made with differential GNSS by finding the relative displacement between GNSS receivers. Multiple stations situated around an actively deforming area (such as a volcano or fault zone) can be used to find strain and ground movement. These measurements can then be used to interpret the cause of the deformation, such as a dike or sill beneath the surface of an active volcano. GNSS data have been used to address a wide range of unresolved questions in geophysics, including several questions related to earthquake prediction (refer to Section 10.10.2).

Mapping of soil and hydrological sampling locations is an important task for the geologist and is used for several purposes, such as interpolation of hydrologic data, soil contamination, soil erosion, water contamination mapping. Getting the coordinates of these sampling locations was never as easy as with the GNSS.

GNSS has direct contribution in surface mines and indirect contribution to underground mines (Table 10.1). Static and RTK mode is used in mining. Point positioning, relative positioning, and Differential GNSS (DGNSS) are used for different types of applications. DGNSS information is used to efficiently manage the mining of an ore body and the movement of waste material by installing GNSS receivers in the trucks. RTK GNSS position information is used by blast-hole drills and to control the depth of each hole that is drilled. Automated drills are used in surface mines to increase safety and productivity. A single operator, located in the safety of the control room, can operate and monitor multiple automated drills. GNSS can help in digging with a uniform height over a whole site, allowing geological real-time analysis of the shape of the mine. In some cases, when the digging is deep, a problem of satellite visibility can occur; that is, not enough satellites are in view to provide good positioning, either because the minimum number of satellites is not reached or the dilution of precision (DOP) values are too high. In such cases, the use of pseudolites has been reported (Samama 2008).

When archaeologists excavate a site, they generally make a three-dimensional map of the site, detailing where each artefact is found. GNSS plays a vital role to record these site locations precisely (Garrison 2003).

10.4 NAVIGATION

Navigation is perhaps the most well-known of all the applications of GNSS. Once we know our location, we can, of course, find out where we are on a map. Using the GNSS coordinates, appropriate software can perform all manner of tasks, from locating the unit, to finding a route from our present location to a destination or dynamically selecting the best route in real-time.

These systems need to work with map data, which does not form part of the GNSS system, but is one of the associated technologies. The availability of high-powered computers in small, portable packages has led to a variety of solutions which combines maps with location information to enable the user to navigate. One of the first such applications was the car navigation system that allowed drivers to receive navigation instructions without taking their eyes off the road, via voice commands. Then there are handheld GNSS units, which are commonly used by those involved in outdoor pursuits; they provide limited information, such as the location, and possibly store waypoints. More advanced versions include aviation systems, that offer specific features for those flying aircraft, and marine systems, that offer information pertaining to marine channels, tide times, etc. The last two require maps and mapping software that differ vastly from on-road GNSS navigation solutions and can often be augmented with other packages designed to allow the user to import paper maps or charts.

There are even GNSS solutions for several games. Golf GNSS systems help the player calculate the distance from the tee to the pin or determine exactly where they are in relation to features, such as hidden bunkers, water hazards, or greens. Again, specific maps are needed for such applications.

10.4.1 AUTOMOBILE NAVIGATION

Automobiles can be equipped with GNSS receivers (Figure 10.8) at the factory or as aftermarket equipment. Units often display moving maps and information about location, speed, direction, and nearby streets and points of interest. Once only rich people used GNSS receivers in their high-end cars; but these days, GNSS systems are offered even with entry level cars. Car GNSS systems come with street maps, turn-by-turn voice prompts, touch-screen operation, video games, radio, DVD or CD players, cell phone connectivity and computer links to update software.

There are several reasons that make GNSS attractive for automobile navigation (Krakiwsky 1991; French 1995; Czerniak and Reilly 1998). First, the coverage is global and available 24 hours a day. This allows for the same system to be mass-produced and installed on any car in the world. Another is that non-government consumer interest in GNSS has driven down the cost of receivers. The absolute nature of GNSS positioning allows for no degradation of position accuracy with distance. GNSS can be integrated with other navigation techniques, such as *map-matching*, to provide the user with excellent location and navigation information. Map-matching technique is used in combination with other methods to correlate the position of the vehicle with a map (Richard and Mathis 1993).

There are a few problems associated with GNSS automobile navigation. First, the errors associated with the GNSS signal do not allow sufficient accuracy. Another is the 'urban canyon' phenomenon. When in a downtown city environment, the buildings often block almost all of the satellites from view. This seriously affects the ability to provide accurate location information in cities. Each of the above-mentioned problems has a solution. To get around the errors associated with GNSS, differential GNSS (DGNSS) can be employed. Using a combination of navigation techniques along with GNSS can also help to overcome the problems associated with GNSS automobile navigation. Many existing systems use GNSS to periodically correct *dead-reckoning* (Chapter 6, Section 6.7.2). Some of the newer systems use GNSS as the primary navigation tool and use dead-reckoning (or some other form of navigation) to provide coverage in urban areas (Richard and Mathis 1993).

FIGURE 10.8   Cars equipped with GNSS receivers.

10.4.2 Aircraft/UAV Navigation

Aircraft navigation systems usually display a 'moving map' and are often connected to the autopilot for en route navigation (Figure 10.9). Cockpit-mounted GNSS receivers use ground-based augmentation systems for increased accuracy (Clarke 1996; Kaplan 1996; CASA 2006; Chatfield 2007). Many of these systems may be used for flight navigation, and some can also be used for final approach and landing. A GNSS flight recorder also logs GNSS data for verifying their departure, route, and arrival.

All unmanned aerial vehicles (UAVs—commonly known as drones) are automatically guided by a GNSS-based application. The autonomous navigation and landing of a UAV is challenging as the navigation system has to deal with the movement of the UAV caused by the wind. The GNSS+ inertial navigation system (INS) is used in UAVs. A GNSS antenna is mounted somewhere on the vehicle that receives location and time data from GNSS satellites. This data is then usually fed into the avionics, autopilot, or navigation systems of the vehicle, and can also be used to determine the velocity. In addition to navigation, unmanned vehicles may use GNSS to georeference/geotag the gathered data, avoid collisions, provide a tracking facility, 'return to home', follow a series of preset waypoints, or deliver goods to a predetermined location. However, UAV image geotagging faces an additional challenge. Usually, autopilot triggers the camera and records the coordinate it has at that moment. When the UAV is flying, for example, at 20 m/s the autopilot will have position readings only at 4 m intervals (because normally GNSS works at 5 Hz). This is not suitable for precise georeferencing/geotagging. In addition, there is always a delay between the trigger and the actual moment the photo is taken. That means the camera clicking

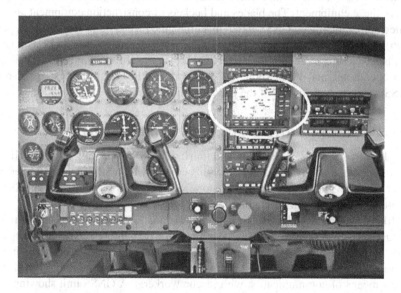

FIGURE 10.9 GNSS navigation system installed in aircraft (marked with white ellipse).

interval and/or the speed of UAV must be sacrificed. Therefore, the GNSS receiver must be connected to the camera shutter and an RTK receiver should be used.

10.4.3   MARITIME NAVIGATION

Boats and ships can use GNSS to navigate all of the world's lakes, seas and oceans (Bowditch 1995; Sweet 2003). In maritime navigation, GNSS is being used to accurately determine the position of ships when they are on the open sea and also when they are manoeuvring in congested ports. Maritime GNSS units include functions useful on water, such as 'man overboard' functions that allow for instantly marking the location where a person has fallen overboard, which simplifies rescue efforts. GNSS may be connected to the ship's self-steering gear and chartplotters. GNSS can also improve the security of shipping traffic. By integrating accurate GNSS position information into distress beacon signals, GNSS is also revolutionising search and rescue operations. It can also be used to improve the accuracy of fishing activities. GNSS, in combination with SBAS, can be used to monitor and protect environmentally vulnerable areas, such as marine parks, or to monitor and prevent illegal fishing.

10.4.4   MACHINE CONTROL AND NAVIGATION

There are many types of unmanned vehicles, including ground vehicles, aerial vehicles, underwater vehicles, and surface vehicles. Unmanned heavy equipment vehicles on the ground can use GNSS in construction, mining, and precision agriculture. GNSS technology is being integrated into equipment such as bulldozers, excavators, graders, pavers, and farm machinery to enhance productivity in the real-time operation of these equipment. The blades and buckets of construction equipment are controlled automatically by GNSS-based machine guidance systems (Ryan and Popescu 2006). Agricultural equipment may use GNSS to steer automatically or as a visual aid displayed on a screen for the driver. This is very useful for controlled traffic, row crop operations, and when spraying. Harvesters with yield monitors can also use GNSS to create a yield map of the paddock being harvested.

10.4.5   NAVIGATION FOR BICYCLERS, HIKERS, CLIMBERS, AND PEDESTRIANS

Bicycles often use GNSS in racing and touring. GNSS navigation allows cyclists to plot their course in advance and follow this course, which may include quieter, narrower streets without frequent signal stops. Some GNSS receivers are specifically adapted for cycling with special mounts and housings.

Hikers, climbers, and even ordinary pedestrians in urban or rural environments can use GNSS to determine their position, with or without reference to separate maps. In isolated areas, the ability of GNSS to provide a precise position can greatly enhance the chances of rescue when climbers or hikers are disabled or lost (if they have a means of communication with rescue workers). A GNSS unit showing basic waypoint and tracking information is typically required for outdoor sport and recreational navigation. GNSS equipment for the visually impaired is also available.

10.4.6 SPACE FLIGHT AND SATELLITE NAVIGATION

The NASA Johnson Space Center initiated GPS navigation projects for the space shuttle, International Space Station and the X-38 (prototype Crew Return Vehicle) in the 1990s. While application of GNSS technology to the shuttle and the ISS was successful, far more technical difficulties were encountered than were originally anticipated, and many lessons were learned. In space flight navigation, GNSS receivers should be treated as computers rather than just 'plug and play' devices. With the advent of GNSS technology came the potential for more accurate, more responsive, and autonomous spacecraft navigation in orbit comparing the early methods.

Low earth orbit satellites (e.g., remote sensing satellites) can use GNSS system for their orbit control and correction. NASA JPL is investigating the development of an autonomous space navigation system called the *Deep Space Positioning System* (similar to earth-bound GNSS but for interplanetary space). This system will facilitate interplanetary onboard navigation capability beyond the low earth orbit. NASA is also planning for a GPS-like system for lunar navigation.

NOTE

Tourist Information Systems. The combination of a system that has the ability to drive us from one location to another and potentially huge databases leads quite naturally to tourist information systems. The main advantage is that tourists are usually moving in unknown environments, where navigation exhibits highly satisfactory levels, and they typically want to access geographical sites efficiently. They do not have the same time constraints as workers, but it is more relaxing to know this kind of help is available. Such a guidance system could also propose different features such as a quick one-day visit of an optimised number of sites or a thorough 'inspection' of a single site. The availability of a car park or hotel, for example, plus detailed information, provides the user with a wide range of configurations. The rapid development of memory capacity does not really limit any needs imaginable today; the real difficulties lie rather in the collection of information than in the storage. Google Maps now allows its users to add places to the app along with pictures and reviews, thereby growing its database rapidly.

10.5 TRACKING

GNSS tracking systems meet both civil and military user requirements, though initially there were originally two very different levels of service that separated such interests. Until 2000, there was distinct GNSS tracking service for civilians and separate real-time GNSS (GPS) tracking system for the US Government, which had drastic differences in accuracy and range. With the discontinuation of SA in May of 2000, GNSS tracking systems are being used for a wide range of activity employing its strengths to make life safer and easier for everyone from dispatchers to worried pet owners.

GNSS tracking systems can be used to improve crop yield. Farming soil sensors and other monitors mounted on a UAV can help to pinpoint locations where changes in watering, fertilisation, or weed control are necessary. Some other common uses of GNSS tracking are: personal emergencies, roadside assistance, landscaping, architecture, endangered animal preservation, animal migration tracking, stolen vehicle location, fleet management, illegal parking, and business logistics. These are all practical uses for GNSS tracking systems that are relatively simple and extremely effective. Satellite-based or ground-based augmentation system or an inverted DGNSS can be very useful for tracking applications (French 1995; Czerniak and Reilly 1998).

10.5.1 FLEET MANAGEMENT

An important application of GNSS tracking is fleet management systems (Ryan and Popescu 2006; Broida 2004; Czerniak and Reilly 1998). The idea is to integrate a GNSS receiver and a telecommunication system; the satellite navigation part allows individual locations of the GNSS-equipped trucks or vehicles, and the telecommunication part is designed to get all the positioning data from the mobile receivers back to a central control and management station (also called the control station). Thus, the central controller has real-time visibility of the status and positioning reports of the fleet (Figure 10.10).

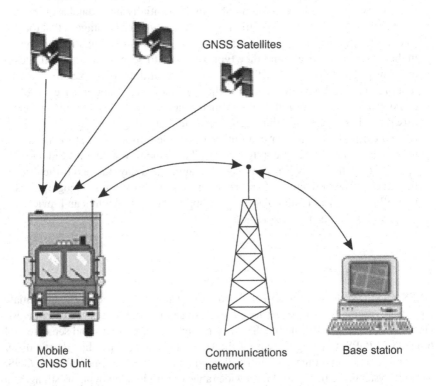

Mobile
GNSS Unit

Communications
network

Base station

FIGURE 10.10 GNSS tracking in fleet management.

This system was first implemented for truck fleets and was soon followed by taxi companies and bus fleets. In the case of trucks, the main objective was to follow the goods and to check the route for both surveillance purposes and retrieval after theft. In the case of a taxi fleet, the goal was to optimise the service to customers (improving time to pickup and ensuring correct dropoff). Knowing the location of all the taxis in a fleet is bound to help in getting one to the customer's place as quickly as possible. Once again, the fundamental features of current in-car navigation systems were incorporated in these deployments. Of course, many other additional pieces of data are required. What is the taxi fare? Are the relevant roads jammed? Among all the data of interest, one can state that traffic information in real-time is certainly one of the most important things.

In the case of bus fleets, the purpose is different. The difficulty with buses, compared to the metro or train, is the lack of centralised information for users and, conversely, the difficulty in planning a route using several different buses. Also important is the fact that when travelling on dedicated lanes (which is frequent nowadays), it is quite easy to give an accurate time for all the stops, but not when travelling on 'open' lanes. Then, when a user arrives at the bus stop, he can never be sure if the bus has already left. Thus, an interesting approach is to give live information to the user, such as when the next bus is going to arrive or how to reach a destination, together with a time estimation of the journey and live locations of the buses (Samama 2008).

10.5.2 PARKING AUTOMATION

GNSS is used for parking automation. When a customer pays for parking, they enter the license plate of their vehicle, a code that identifies the parking area and the amount of parking time required. This information is sent to a database. As the monitoring vehicle drives along the street, the vehicle cameras capture the license plates of the parked cars. The license plate number, along with the time and position provided by the GNSS receiver, is compared to the database of paid parking. If a vehicle is not found in the database, the photograph can be sent to the Parking Authority. However, the monitoring vehicle may face several difficulties: (1) poor DOP because of building obstructions, (2) no satellite signal in case of basement parking, and (3) positional inaccuracy. By switching from a GNSS-only system to a GNSS+INS system, the monitoring vehicle can overcome these challenges. In the Canadian city of Calgary, a similar system was successfully implemented.

10.5.3 TRACKING OF SPACECRAFT

Spacecraft are now beginning to use GNSS. The addition of a GNSS receiver to a spacecraft (Figure 10.11) allows precise orbit determination without ground telemetry-tracking. This, in turn, enables autonomous spacecraft navigation, formation flying, and autonomous orbit correction. In India, satellite IRS-1D was launched on 29 September 1997, the first satellite with a GPS receiver. Using the GPS receiver in the low earth orbit satellite made it possible to determine its position in orbit. Today many spacecraft are equipped with a GNSS tracking system for orbit determination.

FIGURE 10.11   GNSS technology used in satellites.

10.5.4 TRACKING OF PEOPLE

GNSS tracking can be used to track prisoners (Samama 2008). Some prisoners, at the end of their prison term or under parole, are freed on condition that they will report regularly to their parole officer. In the past, an ex-prisoner had to be present at a given location, at the prison, police station, or even at home, at a pre-determined time. In a later, more advanced system, the police could check the presence by way of an electronic device that carried out proximity detection. If, at a given time, the presence of the device was not detected, an alert was sent to the police. There were disadvantages to this method; in particular, that once the alert had been given, the police were obligated to then find the ex-prisoner.

A new method is currently being implemented, using both GNSS for location and GSM for data transmission. The complete system is included in an electronic bracelet and allows restricted location-based displacement of the prisoner. It can also be used the other way round: to eliminate suspicion in the case of theft or aggression. Several such commercial systems do exist and are available to keep track of elderly people.

This bracelet can also be used in other applications where there is a real need to be able to follow, to find, or locate people. One such device, in the form of a bracelet, is intended to provide peace for those dealing with people suffering from Alzheimer's disease (relatives and doctors). Such a patient can 'stray' and forget where they are; and in such cases, even if they are very near the hospital or house, it may be impossible for them to find their way back (possibly causing legal issues for care facilities). The tracking bracelet will increase safety for the patient and decrease headaches for caretakers (Samama 2008).

Today, companies keep track of delivery personnel for the safely of their employees and goods by using GNSS technology. This type of system uses a combination of navigation and tracking.

10.6   TIME-RELATED APPLICATIONS

As we know, GNSS satellites are equipped with atomic clocks, accurate to nanoseconds. As part of the position determining process, the local time of GNSS receivers becomes synchronised with the very accurate satellite time. This time information, by itself, has many applications, including the synchronisation of communication systems, electrical power grids, and financial networks. GNSS can be used as a reference clock for time code generators or Network Time Protocol (NTP) time servers. Sensors for seismology or other monitoring applications can use GNSS as a precise time source so events may be timed accurately. Time division multiple access (TDMA) communications networks often rely on this precise timing to synchronise radiofrequency generating equipment, network equipment, and multiplexers.

Some very demanding fields, in term of synchronisation, exist in different domains: seismology, communication networks, such as the Internet and wireless telephones, banking and finance. Experts believe that this tight relationship between time and positioning is the essence of the future revolution of our daily lives, not only because of the need for precise time in order to achieve accurate positioning but also due to the main goal of our modern lives: that is, optimising our time.

For telecommunication systems for instance, the transmission of the transmitter time to the receiver is currently achieved through the use of so-called 'synchronisation bits' or 'sequence'. This approach requires a given amount of resources to be devoted to this task and also needs a sufficient bandwidth to be available. In the case of microsecond synchronisation, this is considerable, but not really in the nanosecond domain. Let us now imagine that every mobile phone is equipped with a GNSS receiver for navigation purposes. The synchronisation function will then be available in nanoseconds without the involvement of additional resources and wider bandwidth, ever with better performance.

10.7   GEODESY

The use of GNSS techniques in geodesy has revolutionised the way geodetic measurements are made. The ability of GNSS geodesy to estimate 3D positions with millimetre-level precision with respect to a global terrestrial reference frame has contributed to significant advances in geophysics, seismology, atmospheric science, hydrology, and natural hazard science. An increasing number of national governments and regional organisations are using GNSS measurements as the basis for their geodetic networks. Once the network is established, GNSS receivers can offer a simple way to follow the evolution of an earthquake (explained in Section 10.10.2). It is also an interesting way for geodesy cartography or for topographical measurements (surveying) to be carried out (USACE 2007). Very accurate receivers are then required, such as dual-frequency and carrier phase measurements (in order to achieve centimetre accuracy, or even better, as required in these cases).

10.8   CIVIL ENGINEERING

Civil engineering is another community that saw a great advantage in using high-accuracy GNSS positioning receivers. Nothing was impossible in civil engineering before GNSS equipment, but GNSS receivers have highly simplified certain phases. High-accuracy GNSS receivers (down to the centimetre level) are now used for many different tasks from absolute positioning of a road as well as the heights of the various layers (gravel, bitumen, and so on). Engineering surveying and mapping have benefited greatly from GNSS as discussed in Section 10.3.

10.9   LOCATION-BASED SERVICES

A location-based service (LBS) is an information and entertainment service, accessible with mobile devices through the mobile phone network and utilising the ability to make use of the geographical position of the mobile device (Gutierrez 2008). LBS services include services to identify a location of a person or object, such as discovering the nearest banking cash machine or the whereabouts of a friend or employee. LBS services include parcel tracking and vehicle tracking services. They include weather and geophysical services (Lee 2008), and even location-based games. They are an example of telecommunication convergence with GNSS.

It is forecast that the two major domains of applications of the GNSS will be transport and LBS, which will represent more than 70% of the total market (Samama 2008). The term LBS is used to designate all services that use the location of the user as input. Usually, the mobile terminal is a communication device, such as a mobile phone. Thus, the interesting part of LBS lies in the fact that the user could access services and providers could sell services, thus generating revenues, potentially on a wide scale. Large companies have decided to push the LBS sector forward, although it appears that take up is slower than expected.

The first real difficulty is to cope with the problem of personal navigation only by using existing technologies of positioning. We have already seen that indoor positioning is not yet provided by the GNSS constellations, or by any other means currently in place. In these cases, positioning would be possible within 1 m both outdoors and indoors with the integration of telecommunication and additional instrumentation. Thus, until now, LBS is mainly directed at the professional sector that wants to find industrial solutions with guarantees of services, and also recreational activities where the aforementioned limitations are acceptable. Of course, this constitutes a real brake on the development of LBS. The main applications are then mainly due to the integration of telecommunication services with the positioning capabilities of a mobile device. Thus, when available, the terminal's location is used to reduce the search domain for weather information, points of interest selection, or to help to find a person.

The accuracy requirements of some LBS are very high. For example, as a person walks from store to store in a mall, a path is created. In this era of location, critical decisions, such as a piece of geospatial information is vital for managers of the mall. Technology now provides new opportunities to capture, store, and process geospatial data. It addresses the growing demands of emerging markets, especially concerning location enabled devices. Geospatial information is no longer confined to conventional cartographic uses and temporal aspects of data are becoming increasingly important for preferences we choose in daily life.

The entertainment value of locational information is also a related opportunity. Gaming software in association with position technology has shown new avenues for next-generation gaming and interactive entertainment technology. In such contexts, open development platforms such as *Android* (www.code.google.com/android) and *Open Handset Alliance* (www.openhandsetalliance.com) are steadily gaining focus.

There is another application of LBS that is certain to provide the positioning industry, in general, with an enabling feature: the emergency call. Such calls, in many countries, are specially reserved numbers anyone can dial in the case of an emergency. Local relays are then activated and specific operators will route the call as quickly as possible to the appropriate service, e.g., police, fire brigade, medical centre, and so on (Figure 10.12).

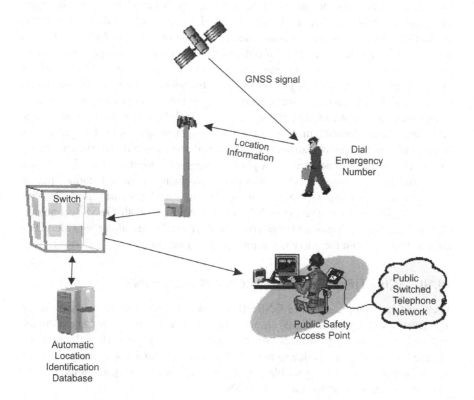

FIGURE 10.12 Emergency call service using GNSS and telecommunication.

FIGURE 10.13   GNSS receiver with phone in jacket for emergency call.

The basic idea of an emergency call is the possibility for an emergency call centre to locate the caller precisely. Indeed, the first difficulties arose a long time ago, with the first analogue telephone centres. When a user called the fire brigade, they often forgot to tell the operator their location because they were too busy coping with the fire. There was then no possibility for the operator, once the caller had hung up, to find the location where he should send the brigade. With the advent of electronic centres, the problem disappeared as it was very easy, by searching the database, to go back to the address for the telephone number, which was visible to the operator. The problem arose once again with the advent of mobile phones, where the location of the user was once again unknown. Because of the obvious need for this kind of service, many countries decided to oblige telecom companies to provide a user's location in the case of an emergency call. GNSS receivers are now widely available in mobile phones and even in jackets and undergarments (Figure 10.13). Receivers in jackets and undergarments are specially designed to support emergency services. They come with a panic button; and, upon pressing this button, it becomes active and periodically sends location information to the emergency call centre. The geographic location of a mobile phone can also be determined and transmitted to the emergency call centre in case of emergencies. A phone's geographic location may also be used to provide location-based services including advertising or other location-specific information. LBS using GNSS is now helping in social networking. All mobile phone companies are marketing cellular phones equipped with GNSS technology, offering the ability to pinpoint friends on custom created maps, along with alerts that inform the user when the party is within a programmed range.

10.10   SCIENTIFIC AND RESEARCH APPLICATIONS

There are many applications of GNSS that exist in scientific domains because the high accuracy (a few millimetres) allows numerous interests in seismology, climatology, and geophysics, for instance. As a matter of fact, a good part of the progress achieved in recent years is due to new or improved data analysis techniques jointly with a growing variety of available measurements. The following sections describe three major scientific applications of GNSS.

10.10.1 ATMOSPHERIC STUDY

GNSS signals are affected while travelling through the atmosphere because the refraction index varies in relation to a vacuum. It is thus possible, through GNSS measurements, to carry out an analysis of the various levels of the atmosphere—troposphere and stratosphere for the lower layers and ionosphere for the upper one. Of course, regarding the importance of ionosphere propagation-induced error in GNSS positioning, such analyses are also likely to improve accuracy in mass-market receivers.

An analysis of the troposphere can be carried out with permanent GNSS stations. From continuous phase measurements of a static GNSS dual-frequency receiver, it is possible to estimate the correction value for the wet part of the troposphere. This is acceptable at least at the vertical to the station. The accuracy of such measurements can be as small as 1–2 mm. The dry fraction is estimated by local pressure measurements. If the number of stations is sufficient, and if simultaneous observations are carried out from many different satellites, it is also possible to estimate the dissymmetry of the atmosphere, notably when clouds are passing in the sky. The collected data from the high number of stations throughout the world allow the precise estimation of water vapour content and its evolution (Samama 2008).

The stratosphere lies above the troposphere (Chapter 5, Figure 5.1). Data concerning this layer (temperature, for instance) are used in meteorology and also long-term climate studies. The main idea is to use satellites located in low earth orbits (typically 750 km) in order to analyse occultation phases, which occur when the GPS satellite moves around to the other side of the earth. It is then possible, through Doppler measurements, to determine the pressure and temperature gradients, in three dimensions, of the stratospheric layer.

The ionosphere is the upper layer of the atmosphere (Chapter 5, Figure 5.1). It is an ionised layer and thus leads to delay due to collisions between the GNSS signal and the ionised particles. This is a real disturbance as the principle of GNSS positioning relies on time measurements. If the velocity of the signal is estimated incorrectly, the positioning will exhibit corresponding errors. A correction of the propagation model is possible through the use of different frequencies, which also allows determination of the total electron content (TEC), an important parameter that could be used for modelling the ionosphere. These measurements can be carried out either by static ground receivers or by low earth orbiting satellites, once again using the occultation method. Then, ionosphere maps can be drawn on local or even global scales. The main uses for these are GNSS ionosphere propagation modelling for positioning purposes, but also for satellite telecommunication systems.

10.10.2 TECTONICS AND SEISMOLOGY

The earth's lithosphere is broken up into several *tectonic plates*—there are seven major and many minor plates (Figure 10.14). These plates move in relation to one another at a very slow rate. The use of very accurate positioning (a few millimetres)

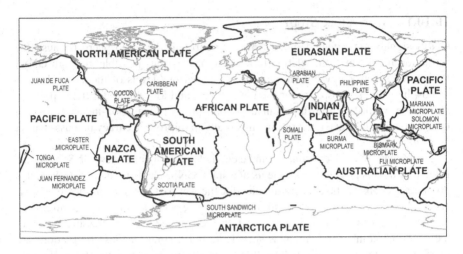

FIGURE 10.14   Earth's tectonic plates (Courtesy: USGS).

allows observation of slow-motion movements of tectonic plates over long periods. Tectonic plates are moving at a few centimetres per year; and a long-term analysis of static receivers allows the movement of plates to be precisely determined. This information can help in earthquake prediction (Aydan 2006; Kato *et al.* 1998; Jiang *et al.* 2007). The same applies to studies concerning, for example, the impact of tides on the earth's surface.

The analysis of earthquakes and, more generally, seismology means dating the various observations in order to allow for correlations. Precision dating is possible with GNSS due to the fine time management required for these systems. Thus it helps in determining whether two distant phenomena are linked to each other or not, using an underground wave propagation model to estimate the time bias that would have occurred in the case of related events.

The ionosphere observation can also provide prediction of earthquake and post-disaster terrestrial events. It has been shown that earthquakes or tsunamis (before and after) induce changes in the constitution of the ionosphere that can be observed through GNSS readings (Samama 2008). Therefore, studying ionosphere can also help in earthquake prediction. The augmentation of the number of available satellites with the combination of GPS and Galileo constellations should increase observation capabilities.

10.11   ANIMAL SURVEILLANCE AND WILDLIFE APPLICATIONS

Positioning systems could enable animal management in different types of applications. At first sight, it could help in defining migration movements of wild animals. This has already been achieved through the installation of miniaturised GNSS receivers coupled to transmitting devices; which allows animals to be followed continuously in real-time. It is also helpful in the case of protected species. By permanent monitoring, any harm done to the animal can be precisely dated and located,

FIGURE 10.15 GNSS for tracking animals.

allowing optimised pursuits. This can also help human populations located near habitats of dangerous wild animals, in detecting their presence and coping with sharing the same environment. This locating feature can be used to study very specific wildlife behaviour, such as the sense of orientation developed by travelling pigeons. Equipped with miniaturised recording receivers, it has been possible to know the route followed by pigeons. Of course, even with this information, the mystery has not yet been solved, but this is a valuable tool to study. GNSS is also used for tracking of domestic animals and pets (Figure 10.15).

10.12 MILITARY APPLICATIONS

The first applications of GPS or GLONASS were planned clearly for the military, more precisely, concerned with the ability to provide a good departure location for intercontinental missiles launched at sea (Kaplan 1996). The military applications of GNSS span many purposes, some of which are similar to the civilian applications. Following are some of the important military applications of GNSS.

Navigation. GNSS allows soldiers to find objects in the dark or in unfamiliar territory and to coordinate the movement of troops and supplies. The GNSS-receivers used by the commanders and soldiers are called the *Commanders Digital Assistant* and the *Soldier Digital Assistant*, respectively.

Target tracking. Various military weapons systems use GNSS to track potential ground and air targets before they are flagged as hostile. These weapon systems pass geographic co-ordinates of targets to precision-guided munitions to allow them to engage the targets accurately. Military aircraft, particularly those used in air-to-ground roles, use GNSS to find targets (for example, gun camera video from AH-1 Cobras in Iraq show geographic coordinates that can be looked up in Google Earth).

Missile and projectile guidance. GNSS allows accurate targeting of various military weapons including intercontinental ballistic missiles, cruise missiles, and precision-guided munitions. GNSS navigation and guidance is providing effective, low-cost means for precision targeting. This targeting option is used primarily against fixed or relocatable targets where the location of the target is expected to remain fixed for the duration of planning and execution of the attack. GNSS-guided weapons are provided with an integral multi-channel GNSS receiver and inertial measurement unit (IMP) that monitors the weapon's locations and attitude to adjust its flight path to accurately impact the target. In low-cost un-powered weapons, the guidance system adjusts the weapon's free fall to hit a pre-selected point fed into the weapon prior to take-off. During free fall, the bomb's GNSS receiver determines location data that are corrected with the help of INS data. The guidance computer uses the GNSS/INS data to adjust the movable tail fins and correct the bomb's trajectory toward the designated target's coordinates (Figure 10.16). However, GNSS weapons are not designed for engagement of moving targets.

Search and rescue. In case of any mishap, the military personnel can be tracked and rescued, e.g., downed pilots can be located faster if they have a GNSS receiver. To reach the rescue location, troops can benefit from GNSS navigation.

Reconnaissance and mapping. The military uses GNSS extensively to aid mapping and reconnaissance.

Drone piloting. The drone autopilot is dependent on GNSS guided navigation and return. Drones are used extensively in military applications, including reconnaissance, logistics, target and decoy, mine detection, search and rescue, research and development, and missions in unsecured or contaminated areas.

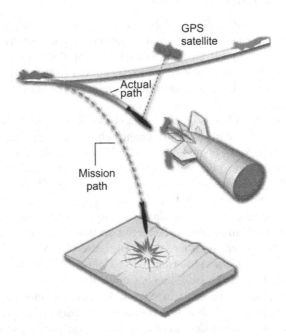

FIGURE 10.16 Precision-guided munitions.

Nuclear detonation detection. The GPS satellites also carry a set of nuclear detonation detectors consisting of an optical sensor (Y-sensor), an X-ray sensor, a dosimeter, and an electromagnetic pulse sensor (W-sensor) which form a major portion of the United States Nuclear Detonation Detection System.

10.13   PRECISION AGRICULTURE

Precision farming or precision agriculture is an agricultural concept relying on the existence of in-field variability. It is about doing the right thing, in the right place, in the right way, at the right time. Precision agriculture evolved from a concept to an emerging technology recently. Precision agriculture is often described as the next great evolution in agriculture and is considered a concept, management strategy, and even a philosophy. GNSS provides the agriculturist with a new capability of gathering information for implementing decision-based precision agriculture (NRC 1997; Srinivasan 2006).

GNSS can aid in soil sampling, mapping and preparing a land information system (LIS), and mobile mapping (Shanwad *et al.* 2002). Mobile mapping is the ability to collect field data, with unique geospatial location, time tags, and attributes, for integrating into or updating a GIS or LIS. Mobile mapping provides the freedom to collect data anytime, anywhere, in any manner. Mobile mapping is essentially useless without the GNSS component. The GNSS component not only provides the location for all data collected but also provides the time in which it was collected. GNSS also enables the user to navigate back to any particular location anytime thereafter. Once the field data has been collected using mobile mapping, the data can be downloaded into a desktop GIS. The GIS then provides the producer the ability to consider all the options for production. The producer can then use the positional data and the decisions that were made with the GIS to carry out the mechanised part of precision agriculture.

GNSS is also very useful for navigating and tracking heavy equipment used in agriculture. Agricultural equipment may use GNSS to steer automatically or as a visual aid displayed on a screen for the driver. This is very useful for controlled traffic, row crop operations, and when spraying. Harvesters with yield monitors can also use GNSS to create a yield map of the paddock being harvested. The main uses of GNSS in agriculture consist of having an analysis tool in order to optimise the spraying of fertilisers and other herbicides and insecticides, and the management of set-aside lands. The installation is achieved in the best possible conditions: tractors and other agricultural machines move slowly, have enough electric power to supply the receiver, and the typical accuracy needed is 1 m. Specific software allows one to have a graphical representation of the farm work together with automatic time alerts for cultivation purposes. With the increased importance of ecological and environmental matters, this approach can also be used in order to demonstrate and enhance the changing agricultural practices in this field. This is certainly a good motivation to develop this market.

10.14   OTHER APPLICATIONS

Besides the fact that specific games can be developed in the light of positioning capabilities, where the players are localised and the evolution of the game depends on these locations, another activity is *Geocaching* (www.geocaching.com). This is based on a user 'hiding' an object in a given location. The idea is then to allow other players to find this object (Cameron 2004; Peters 2004; Broida 2004). This popular activity often includes walking or hiking to natural locations. Treasure-hunting then becomes the logical extension of Geocaching. Some other games are Geodashing, GeoPoker, GPS Golfing, etc. (Broida 2004). The website www.gpsgames.org provides several GNSS-based games.

The GNSS Road Pricing system can charge road users based on the data from GNSS sensors inside vehicles (Aigong 2006). Advocates argue that road pricing using GNSS permits a number of policies such as tolling by distance on urban roads and can be used for many other applications in parking, insurance, and vehicle emissions. However, critics argue that GNSS could lead to an invasion of people's privacy.

Another GNSS application can be found in skydiving. Most commercial drop zones use a GNSS to aid the pilot to 'spot' the plane to the correct position relative to the drop zone that will allow all skydivers on the load to be able to fly their canopies back to the landing area. The 'spot' takes into account the number of groups exiting the plane and the upper winds. In areas where skydiving through cloud is permitted, the GNSS can be the sole visual indicator when spotting in overcast conditions. Altitude can also be determined by GNSS before diving.

GNSS has its application in marketing also (Keohane 2007). Some market research companies have combined GIS systems and survey-based research to help companies to decide where to open new branches and target their advertising according to the usage patterns of roads and the socio-demographic attributes of residential zones.

GNSS also found its way into the management and control of epidemic/pandemic. It was widely used worldwide during the 2020 Covid-19 pandemic. Several mobile apps were developed during this time. These apps were used to track infected or recovered people. Once a user installed the app, their movements were tracked using GNSS technology and the information fetched was compared with the government data on the whereabouts of those who have been diagnosed. Some apps used GNSS locations to detect users when they were in proximity to each other and alert them anonymously if they were in contact with someone who had tested positive. The GNSS positioning was also used for queue/crowd management, response management, supply of foods/medicines, and drone-based crowd identification applications.

Other than the above, there are several applications where GNSS has been or is being implemented to benefit from several aspects. This brief journey of GNSS applications should help us to understand its range of application and how this positioning and navigation technology is helping us. In conclusion, GNSS systems are now reaching their maturity phase, although there is still room for improvement.

EXERCISES

DESCRIPTIVE QUESTIONS

1. How can you classify the applications of GNSS? Explain in brief.
2. Briefly discuss the surveying and mapping applications of GNSS.
3. How is GNSS used in automobile navigation?
4. Discuss the tracking applications of GNSS.
5. How can you use GNSS for structural deformation survey and construction stakeout?
6. Discuss applications of GNSS in geophysics, geology, and archaeology.
7. Discuss aircraft and UAV navigation applications.
8. Discuss the space applications of GNSS.
9. What are the time-related applications of GNSS? How can GNSS help in geodesy?
10. Elaborate on LBS in the light of GNSS.
11. How can GNSS help in civil engineering and mining? Explain in brief.
12. Describe some scientific applications of GNSS.
13. How can GNSS be used to boost tourism? 'GNSS can aid in animal surveillance'. Explain.
14. Briefly outline the military applications of GNSS.
15. How can GNSS be applied in agriculture and entertainment?
16. Discuss 'safety of life' using GNSS.

SHORT NOTES/DEFINITIONS

Write short notes on the following topics

1. Navigation
2. Tracking
3. Surveying
4. Mapping
5. Geodetic control survey
6. Fleet management
7. LBS
8. Precision farming
9. Geocaching
10. Drone navigation

11 Surveying with GNSS

11.1 INTRODUCTION

This chapter discusses the integration of GNSS techniques into surveying operations, describing recommended methods and procedures needed to attain a desired survey accuracy standard. The guidelines in this chapter are based upon several sources of literature (Jones 1984; Wells 1986; Hoffmann-Wellenhof *et al.* 1994; Kaplan 1996; Trimble 2001; RMITU 2006; Sickle 2008; Ghilani and Wolf 2008) as well as practical experience.

Use of GNSS technology in surveying requires specialised equipment, data collection techniques, and data processing algorithms. Here we provide a theoretical and practical foundation for surveyors as they begin to embrace GNSS technology and integrate GNSS equipment in their daily survey operations. GNSS-based survey is an evolving technology, as GNSS hardware, data collection techniques, and processing software are improved and new guidelines replace the existing ones.

11.2 SURVEYING TECHNIQUES

Surveyors generally use GNSS for *control surveys*, *topographic surveys* (mapping surveys), and *stakeouts*. Control surveys provide horizontal and/or vertical position data for the support or control of subordinate surveys and establish control points in a region of interest. Control points are established on previously built survey monuments (also called survey marks/markers, survey benchmarks, or geodetic marks) as shown in Figure 11.1. These control points are used as references in the subsequent surveys. Therefore, they need to be very accurate and reliable. Baselines are measured using careful observation techniques (a baseline in static survey is a line between two control points on the earth's surface). These baselines form tightly braced networks (called *control networks*), and precise coordinates result from the rigorous adjustment of the networks (called *network adjustment*, refer to Section 11.5.10). *Static* (also called *classic static*) and *fast-static* (also called *rapid static*) observation techniques, combined with a network adjustment, are best suited to control survey.

Static or classic static GNSS surveys provide the highest possible accuracy, which occupies a point for longer periods of time. Static systems may take one hour to several days to collect a single point if very high accuracy is required. A static survey may even run continuously if extremely high accuracy is needed (e.g., monitoring plate tectonic movements); this is called continuously operating static survey. The equipment setup varies significantly, depending on how long the receiver will be operational. Continuously operating permanent stations require many of the same techniques as classic static systems, but require a high level of precision and detail in

FIGURE 11.1 Different types of control marks.

their installation. Semi-permanent and permanent surveys involve extended deployment of a station beyond 48 hours and potentially for many years. The advantage of a permanent survey is continuous data collection, which enables high-precision (millimetre level) positioning. These installations require significant knowledge in geodesy and processing techniques, which are beyond the scope of this book. However, field techniques for site selection, logistics, installation, and execution are similar and can be applied for the whole range of applications. In general, the additional requirements of permanent installations include high-precision mounting devices, uninterrupted power supply, high battery backup (in case of power failure), extremely rugged high-precision receivers, excessive memory capacity, and constant obstacle-free sky.

A static survey involves a single receiver that individually records satellite observations that are post-processed using a variety of techniques to receive a position (multiple receivers may collect multiple points simultaneously but independently). Remember, this post-processing is not a kind of post-processing we use in differential and kinematic GNSS survey, where rover data are corrected with respect to the base data. This post-processing involves network adjustment, least squares adjustment, loop closure, and several other statistical methods. DGNSS and kinematic techniques require a minimum of two receivers—base and rover; however, static survey uses only one independent receiver to collect a point.

Rapid static surveys are very similar to static survey; however, they use shorter occupation times to collect a greater number of points compared to static survey. They do this by occupying a given point for a longer period of time than most kinematic surveys but much less than the static surveys. A rapid static solution may take 8–30 min (or more) depending on the receiver type, baseline length, and the number and geometry of the satellite. Rapid static technique also requires post-processing similar to static surveys.

A static or rapid static survey is advantageous because of the higher accuracy and simplicity of the setup, using a single antenna/receiver combination that has half as many components as used in kinematic survey. Static/rapid static survey is required for longer baselines when radio communications or baseline distances for a survey exceed the capability of a kinematic technique. However, static surveys typically have reduced accuracy compared to a kinematic survey of the same occupation time; static/rapid static can provide accuracies higher than kinematic survey if we allow a very long observation time.

Kinematic GNSS surveys are used to rapidly collect large numbers of high-precision survey positions. Kinematic GNSS survey requires the simultaneous observation of the same four (or more) satellites by at least two receivers—one (or more) base receiver and one (or more) rover. The base receiver is located over a known control point for the duration of the survey. The rover is moved to the points that are to be surveyed or staked out. When the data from these two receivers is combined, the result is a 3D vector between the base antenna phase centre and the rover antenna phase centre. This 3D vector is commonly referred to as *baseline*; however, it is different than the static survey baseline. In the case of static survey, baseline means vector between two control points; whereas, in kinematic survey, baseline defines the vector between base and rover. Establishing a base receiver is not required if we can get the base data from an agency (either government or private) that has a control network of continuously operating base stations. Kinematic survey works for shorter baselines (within 10 km) and is suitable for high-accuracy local survey with reference to a base receiver. This type of survey may also be performed over an extended area, but it requires establishing multiple base stations.

The basis of the kinematic system is a mobile rover, that takes initial positions, and a base station that allows for corrections of the rover's position. A rover is an antenna and receiver combination, which can be mounted to a range pole or backpack and is carried to each site for measurements. The rover is carried to each observation point and kept there for few seconds to complete the survey. The rover's position is processed against the static base receiver's position.

This correction may take place either in real-time via communication link or later by means of post-processing. RTK techniques use a communication link to transmit base observations to the rover for the duration of the survey. Post-processed kinematic (PPK) techniques require data to be stored and resolved sometime after the survey has been completed; communication link is not required in this case. From the preceding discussions, it may appear that whether processed after or during the survey, these methods produce the same results; however, this is not true. PPK may achieve higher accuracies compared to RTK (explained in Section 11.6).

Topographic surveys determine the coordinates of significant points in a region of interest. They are usually used to capture physical and cultural features over the earth surface to produce maps. Kinematic techniques (real-time or post-processed) are best suited to topographic surveys with higher accuracy requirements because it takes shorter occupation time for each point. Code-based differential technique may also be used in topographic survey for lower accuracy requirements. *Stakeout* is the procedure whereby predefined points are located and marked in the field. To stake out a point, we need results in real-time. Real-time kinematic (RTK) is the only technique that provides centimetre-level, real-time solutions. Differential code solution in stakeout can provide accuracies up to 1 m. DGNSS is rather simple and similar to RTK in concept of operation; however, it is based on code correlation. Code-based single point positioning is also used by some people to capture geographic features.

TABLE 11.1
Details of Widely Used Survey Techniques

Parameter	Differential GNSS	Real-Time Kinematic (RTK)	Post-Processed Kinematic (PPK)	Static/Fast-Static
Real-time solution	Yes	Yes	No	No
Post-processed	Yes	No	Yes	Yes
Measurement	Code-based	Carrier-based	Carrier-based	Carrier-based
Initialisation required?	No	Yes	Yes	No
Long observation required?	No	No	No	Yes
Typical observation time	Few seconds	Few seconds	Few seconds	45 min or longer for static high precision 8–30 min for fast-static (depends on the receiver type, baseline length, and the number and geometry of the satellites)
Accuracy	Around 1 m	Centimetre	Centimetre	Few millimetres

However, theorists do not consider this technique a survey technique, rather it is a technique used in navigation (navigation solution). We have fully detailed DGNSS and navigation solution techniques in Chapter 6. In this chapter we will focus on carrier-based solutions. Table 11.1 summarises details of frequently used basic survey techniques (they may have different implementations). Some other techniques also do exist but not widely used because of their insignificant relative advantages. Table 11.2 summarises relative advantages and limitations of different survey techniques. Generally, the technique we choose depends on factors such as the receiver configuration, the accuracy required, time constraints, and whether or not we need real-time results. Figure 11.2 shows a flowchart that will help to choose the appropriate survey technique. Observation time directly influences the achievable accuracies and thus the survey techniques. Figure 11.3 shows a graph to provide a rough idea of relations among observation time, accuracy, survey technique, and associated applications.

11.3   EQUIPMENT

In the past, in high-precision survey grade receivers, the antenna (with preamplifier) and the receiver (with or without the radio modem) were separate units. The controller was another unit or it was integrated with the receiver unit. Separate power units were also used. These units were connected through coaxial cable. Figure 11.4 shows one such setup in assembled and disassembled condition. This king of antenna-receiver setup was messy and difficult to handle. Modern antennas integrate the receiver in a single unit, commonly called a *receiver* or *smart antenna*;

TABLE 11.2

Merits and Demerits of Different GNSS Survey Techniques

Survey Technique	Merits	Demerits
Code-based single point positioning	Instantaneous solution and used in navigation and tracking.	Very low accuracy is the main drawback. It was not designed for survey.
DGNSS	Real-time DGNSS provides corrected positions in the field and can be used in navigation, tracking, and survey. Post-processed DGNSS requires processing after data collection and used in survey. Very fast capture of points with low accuracy.	Requires two receivers. In case of real-time DGNSS, additional instrumentation for data transmission from base to rover is required. Post-processed DGNSS cannot provide corrected positions in the field, and post-processing facility is required.
RTK	Real-time corrected positions are available in the field. Can be used in navigation as well as non-engineering survey. Very fast capture of points with high accuracy.	Necessary to set up base stations at benchmarks. Significantly increased equipment cost and logistics. Must have radio connection between base and rover. Must set up base stations on a known mark.
PPK	Reduced logistics, cost, and complications. Sufficient for most non-engineering surveys. Very fast capture of points with high accuracy. May provide better accuracy than RTK.	Corrected data is not available until processed. Necessary to set up base stations at benchmarks.
Rapid static	Reduced equipment expense and complication compared to PPK. Local base stations are not required. Higher accuracy than PPK.	Requires much longer occupation times than RTK/PPK, with fewer potential measurements. Lower accuracy than classic static.
Static (classic static)	Higher precision than rapid static, less equipment than RTK or PPK. Local base stations are not required.	Requires very long occupation times to reach similar accuracy to RTK or PPK. Requires more precise mounting and additional data than RTK, PPK, and rapid static.
Continuously operating static	Highest possible precision and accuracy. Local base stations are not required.	Requires complex infrastructure, precision mounting, and extremely long occupation times.

the control/display is housed in another unit (called a *controller*). The antenna/receiver unit also houses a radio modem. Both of these units have separate microprocessor, storage, and power units. The data can be logged either in the receiver or in the controller. Receiver and controller are connected via Bluetooth. Figure 11.5 shows a tripod mounted base (or static) receiver and a range pole mounted rover with controller.

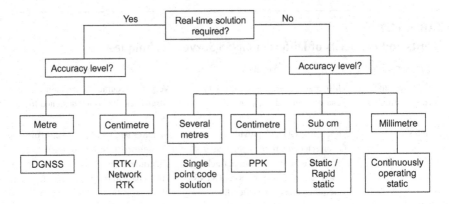

FIGURE 11.2 Flowchart for selection of appropriate survey technique.

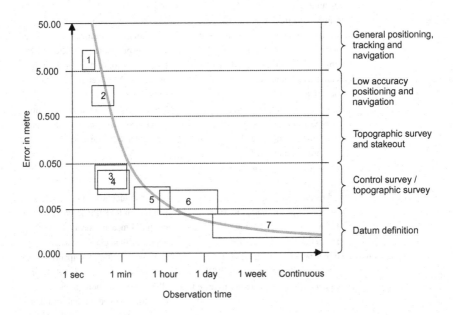

FIGURE 11.3 Relations among occupation time, survey techniques, accuracies, and associated applications.

When performing specific type of GNSS surveys (e.g., carrier-based static, fast-static, kinematic, or code-based differential), receivers and software must be suitable for that survey, as specified by the manufacturer. Whenever feasible, all antennas (or integrated antenna/receiver) used for a project should be identical. For vertical control surveys, identical antennas must be used unless software is available to accommodate the use of different antennas. If different antennas are used, they should be from the same manufacturer. For primary GNSS control surveys, tower-mounted antennas with a ground plane attachment must be used, otherwise antennas should

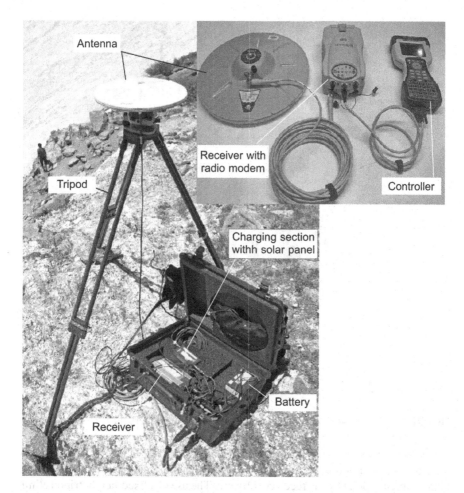

Antenna

Receiver with
radio modem

Tripod

Controller

Charging section
withh solar panel

Battery

Receiver

FIGURE 11.4   External tripod mounted antenna, separate receiver, and controller.

be mounted on an adjustable tripod (Figure11.6a). A ground plane is a flat surface of metal that a magnet mount or body mount antenna uses as an integrated part of it (Figure 11.7a). When tripods or towers are used, optical plummets (Figure 11.7b) are required to ensure accurate centring over marks. The use of range poles and/ or stakeout poles (Figure 11.6b) to support antennas should only be employed for kinematic surveys.

The antenna must be placed directly above the point on the ground with the same attention to detail as in any other survey. A setup error translates directly into a position error. One of the most common errors in GNSS surveys is the incorrect reading or recording of the height of the antenna. Antenna height error affects all three position parameters (x, y, and z), but is more critical for elevation surveys. The height of the antenna should be measured for every setup at least twice, once before the first observation and immediately after the last observation. If the height of the antenna

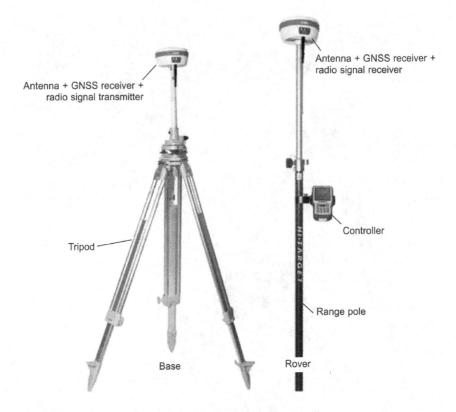

Antenna + GNSS receiver +
radio signal transmitter

Antenna + GNSS receiver +
radio signal receiver

Controller

Tripod

Range pole

Base

Rover

FIGURE 11.5   Integrated antenna/receiver and separate controller setup.

is measured manually, then it must be measured with two independent measuring systems (Metric/English) to eliminate blunders. This measurement is then reduced to the reference point by the receiver software. The use of a fixed-height tripod eliminates the antenna measurement and the possibility of an incorrectly recorded height; however, it may suffer from levelling difficulties.

A GNSS receiver is a device that new surveyors should familiarise themselves with before undertaking the survey operation. This can save time, resolve problems in operation, and prevent accidental data loss during the survey. Reading the instrument manual is essential: knowing the specifications of the GNSS receiver model, set-up, functionalities, how to use, and the limitations of the gadget are very important. One should take the time to practice using the receiver before conducting an actual survey; this is important for set-up procedures, data collection, and data deletion. Observation of signal strength and level of accuracy for different weather conditions and locations is also required. Battery life of the receiver is another concern. This is important in planning the survey especially in cases where there are no available spare batteries for the unit. Average running time for batteries depends on manufacturer design, but typically they last 4–10 hours for internal setups; larger batteries (or solar-charged batteries) are used with external battery setups.

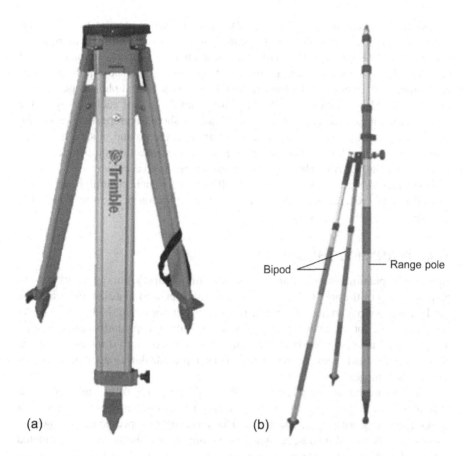

FIGURE 11.6 (a) Adjustable tripod, (b) Bipod attached with range pole.

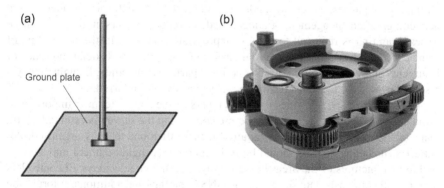

FIGURE 11.7 (a) Ground plane with tower, (b) Optical plummet.

Although manufacturers test and verify quoted accuracy specifications for their products, it is the responsibility of the professional surveyor to validate all new and old equipment (RMITU 2006). All equipment must be properly maintained and regularly checked for accuracy. Errors due to poorly maintained equipment must be eliminated to ensure valid survey results. Level vials, optical plummets, and collimators should be calibrated at the beginning and the end of each survey. If the survey duration exceeds a week, these calibrations should be repeated weekly for the duration of the survey. If any instrument drops to the ground, it should be calibrated before using. Successful validation also demonstrates the competence of the surveyor in using GNSS technology to achieve the required accuracy.

GNSS receivers are designed to provide ellipsoidal height in general. If we require geoidal height (general demand), an appropriate definition of the geoid model is essential in the controller software.

11.4 PLANNING THE SURVEY

Appropriate planning is essential before conducting a survey. Survey teams need to prepare a map of the area to be surveyed and have it printed in a size and scale that can be easy to write on and read while in a moving car or out in the field (A3 or A4 size is suitable for better handling). Nowadays, online maps and satellite images (e.g., Google Earth/Google Map/Bing Map) are also widely used if internet access is available in the field. Some of the controllers even provide downloaded offline maps and satellite images.

If no one in the survey team is familiar with the places to be surveyed, local resources must be engaged. Planning of survey routes on the maps is essential. If working in teams, we assign a specific area for surveying to a specific team. However, alternate areas should also be assigned to a team in case the survey is completed early or if problems arise for primary areas.

Hazards and risks that may be encountered in the survey areas must be considered (flash flood areas, road conditions, local criminal gangs, insurgents, bulwarks of rival political parties, etc.) to avoid delays and prevent any untoward incidents from happening. Often engagement of a local guide or influential person may solve these issues. Planning is required to include appropriate surveyors in the team, type of vehicle to use in certain areas to avoid physical (rough road) and social (presence of insurgents or areas controlled by rival political parties) constraints. Bringing along a resource person for the areas during the survey is always preferred.

Assignment of specific roles to the members of the survey team is important to avoid chaos. Preparing all materials before conducting the survey is essential. Extra maps of the route plan would be required if more than one team is organised. All batteries (including spares) should be fully charged. A digital camera may be carried to take pictures of the areas being surveyed (although nowadays all controllers have inbuilt cameras). Data sheets for the GNSS readings and additional information (*observation log*) must be prepared. This serves as the backup for the data and is a much easier way to write down notes. A GNSS log sheet may contain the following items: project ID, job ID, date, location, operator's name, mask angle, dilution of precision (DOP) values, accuracy reading, receiver/antenna model, antenna height, ellipsoid, geoid, unit, point ID, point name, other attributes, and so on. Today, all GNSS controllers support digital databases that can be linked with each point. This has dramatically reduced the need for an observation log.

Setting up the instrument before going out to the field is important. Here are some of the settings that should be applied: datum (ellipsoid and geoid), units, bearing, latitude/longitude units (*decimal-degree* is preferred in GIS), projection. DOP values need to be evaluated in advance. Free online software from Trimble for predicting DOP for any location and many constellations is available at www.gnssplanning.com. Based on the DOP estimates, one may need to adjust their survey time. In regard to the estimated DOP values, remember that the planning software does not consider physical obstructions (e.g., tall buildings) in the field. Once the receiver and the team are ready to survey, it is necessary to take a few test readings. These readings should be taken in an open space with a good view of the sky for better accuracy. The achieved accuracy should be noted; this will serve as an estimate for the next readings.

11.5 GENERAL FACTORS FOR GNSS SURVEYING

We have now learned about the various GNSS surveying methods that can be adopted based on the application and accuracy requirements. However, there are other factors that impact the precision of GNSS irrespective of the surveying method(s) adopted. As with most surveying tasks, GNSS surveys are more likely to be successful if properly planned and designed. There are a number of issues that need to be considered before and during a GNSS survey. This section presents a number of issues that should be considered before surveys are attempted and during the survey as well.

11.5.1 ACCURACY

Perhaps the first question that needs to be answered pertains to which GNSS techniques are capable of achieving the accuracy required by the project. Manufacturer specifications indicate that carrier-phase surveys are capable of achieving centimetre to sub-centimetre accuracy, plus one or two parts per million (ppm) of the baseline length. Surveyors must be aware that these specifications usually correspond to the standard deviation of computed baselines. Doubling (and sometimes greater) the specified value often provides a more realistic assessment of the capabilities of a receiver. Users should also be aware that errors due to factors such as multipath are not considered in these accuracy values. Another factor to consider is the vertical component; it is generally not as accurate as the horizontal component. A rough rule of thumb relates the height error to the horizontal error by a factor of 1.5 to 3.0 depending on satellite geometry (DOP).

11.5.2 OBSTRUCTIONS

In order to apply GNSS technology to surveying applications, a clear view of the satellites is required. This precludes the use of GNSS technology in tunnels, under bridges, and in areas with many tall buildings. GNSS technique may not be suitable for surveying of features in well-established areas. In these situations, portions of the survey, such as the control work, can be performed using GNSS technology. The remainder of the survey can be completed by using a total station. Another approach to overcome the problems due to obstruction is *offsetting* (refer to Section 11.5.12). Professional experience will decide whether the site is suitable or not. It is important to note that the points of interest must be free of overhead obstructions, not simply the area.

11.5.3 Length of Baselines

The distance between the two receivers, or *baseline length*, is another consideration in GNSS surveying, as the survey accuracy degrades as the separation between the receivers increases. This is due to spatial correlation of errors at both sites not being as high as if the receivers are adjacent to each other. This fact is reflected by the ppm component of most accuracy specifications. In addition, the time required to successfully resolve the integer ambiguity generally increases as the baseline length increases. This results in surveys which are less accurate and not as efficient to perform.

For static surveys, the occupation (or observation) time is generally quite long in order to ensure ambiguity resolution as well as to average random measurement and multipath effects. As a result, baseline length is not a critical a factor. Users must be aware that a limitation exists if single frequency receivers are used as the ionospheric error will cause problems over baselines greater than 10–15 km. Baseline lengths should be kept below this length, especially, in the peak of the sunspot cycle activity. For rapid static surveys, surveyors must use dual frequency receivers. The aim of such surveys is to resolve the ambiguities as quickly as possible. The most efficient rapid static surveys are performed when baseline lengths are less than 5 km. Baselines longer than this can be observed, however, the time required to resolve the ambiguity may result in a static survey campaign being more efficient.

Kinematic surveys (including RTK) are the most sensitive to baseline length, as the resolution of the integer ambiguities in an efficient manner is what enables the use of short occupation times. Unsuccessful ambiguity resolution results in surveys which do not meet required accuracy levels. Most manufacturers recommend that RTK surveys be performed over baseline lengths of less than 10 km. Although longer distances are observed, best results are achieved when the base and rover are separated by less than 5 km. Remember, that the distance also depends on the capacity of radio transmitter and obstructions, such as buildings.

These guidelines may appear to be restrictive if misinterpreted. To be clear, the survey may extend beyond these specified baseline lengths; it is the base–rover separation that should stay within these limits. If multiple base stations are used, surveys can be performed successfully over extremely large areas.

11.5.4 Occupation Time

For a static survey, the occupation time per point surveyed is selected to provide sufficient measurements to enable the integer ambiguities to be resolved. Users must be aware that a change in satellite geometry during the occupation period is required to enable the ambiguities to be solved. This is partly due to the ambiguity being a value which is extremely close to the distance between the satellite and receiver at the beginning of the survey. As the range to the satellites changes, the ranges and ambiguities start to separate. This enables statistical methods to identify the correct number of integer cycles more easily. Therefore, 100 measurement epochs collected at a 1 sec rate are most likely insufficient to resolve the ambiguities, whereas 100 epochs at a 15 sec rate (i.e., 25 min) are more likely to be sufficient. This highlights that occupation time, rather than number of measurement epochs, is the key factor in performing static surveys.

The occupation time required for static surveying is a function of a number of elements including the baseline length, number of satellites, satellite geometry, atmospheric conditions and multipath conditions. In general, by using modern surveying receivers, 20–30 min of dual frequency measurements are usually sufficient to resolve the ambiguities over baseline lengths of less than 10 km. An additional 10 min may be sufficient to extend the baselines to 10–20 km. Both these estimates presume continuous tracking of at least five satellites. It should be noted that, in the presence of obstructions, it may be necessary to increase the occupation time in order to achieve clean measurements. Single frequency users are advised to acquire measurements for twice as long, i.e., 40–60 min over baseline lengths of less than 10 km. If there are six, seven, or even eight satellites being observed, experienced users who are familiar with the performance of their equipment may wish to observe for shorter periods than this, say 10 min, particularly if performing rapid static surveys over shorter baselines. For baseline lengths longer than 25–30 km, dual frequency receivers should be used, and observation times should not be shorter than one hour to ensure successful ambiguity resolution.

Kinematic surveys utilise a short initialisation technique to resolve the integer ambiguities. This initialisation procedure may take several forms; however, the end result is successful constraint of the integer terms required during processing. Once the survey is initialised, each point of interest only needs to be occupied for few epochs. In order to acquire sufficient measurements to detect if a bad epoch has been recorded, surveyors are recommended to acquire at least ten epochs while at rover sites. The recording rate for kinematic surveys is usually higher than that of static surveys, thus, ten epochs of measurement may correspond to 30 sec or less. To be conservative, occupation periods of 30–60 sec should be used for kinematic occupations.

11.5.5 Recording Rate

The recording rate represents the rate at which positioning measurements are stored. This rate is often termed the *data rate* or *epoch rate*. For static surveys, there is little advantage in storing measurements at a low rate. Typically, a recording rate of 10 or 15 sec is used for static occupation periods of 20 min or more. For longer sessions which may involve several hours of measurement, a rate of 30 sec or even 60 sec is suitable. For static surveying, it is important to assess the amount of work to be performed and weigh the amount of data against the available volume of storage space. For example, it may be feasible to perform four observation sessions of 45 min duration in one day. In this instance, the user should verify that the rate chosen is such that 180 min of data can be stored. It should also be noted that the amount of data storage required will depend on the number of satellites observed and the manufacturer's ability to compress the acquired measurements into efficient data structures. For rapid static surveys, similar considerations apply. The primary difference between a rapid static survey and a static survey is the shortened occupation period. In order to provide the processing algorithm with sufficient measurements to perform statistical operations, a higher data rate is generally used for rapid static surveys. For example, a 10 min occupation should be performed at a data rate of 5 or 10 sec rather than 30 or 60 sec.

The data requirements of a kinematic survey are quite different from a static or rapid static survey. Kinematic surveys are designed to be more efficient than static surveys by employing shorter occupation times. In general, a minimum of ten epochs at each rover site is recommended. If the epoch rate is set to 60 sec, the performance of the kinematic survey is no different from a rapid static survey. A recording rate of 3 or 5 sec is, therefore, usually adopted for stop-and-go kinematic surveys. This enables the rover to occupy marks for less than 1 min, while still providing sufficient epochs to enable gross measurement errors to be identified. By purchasing additional memory chips or data cards or using a computer with a large hard disk, higher sampling rates can be used; in all cases, an increased cost is incurred. The alternative is to transfer the acquired measurements to an appropriately powered computer during the survey in the field. If these options are not feasible, a compromise can be reached by recording at a slower rate, say 10 sec, and occupying rover sites for closer to 2 min. However, with the advancement of technology, nowadays controllers come with more than sufficient amount of internal memory.

The final survey types to be considered are the continuous kinematic survey (one type of PPK) and RTK survey in which the position of the receiver while it is in motion is the focus. The recording rate needs to be carefully selected to provide points at desired intervals. For example, a survey vehicle is travelling at a speed of 60 km/h. The selection of the data rate must, therefore, be computed based on a desired point spacing and the estimated speed of the host platform. Users may find that surveys need to operate at rates of 1 sec to be effective for the chosen application. If so, it may be necessary to invest in additional data storage to enable surveys of a practical observation period to be performed.

One vital point that must be remembered is that the base receiver must record measurements at least at the same rate as the rover. The base receiver may record faster than the rover, as long as the rate is evenly divisible by the rover rate. For example, a rover rate of 10 sec and a base rate of 2 or 5 sec is satisfactory. A base rate of 3 sec and rover rate of 2 sec will mean that two of every three rover epochs are ignored. The measurement epochs in receivers are determined by dividing the GNSS time by the recording rate. A remainder of zero indicates that the measurement should be stored. This means that surveyors do not need to synchronise receivers as such, as the receiver clocks perform this function automatically.

11.5.6 MEASUREMENT REDUNDANCY

Professional surveyors prefer redundancy in their survey procedures. Control work is performed by traversing, computing, and distributing measurement errors. Radiations are often checked using a right angle offset from a traverse line. Even a detail survey has a check of sorts as anomalous terrain variations and large bends in fence lines may indicate measurement errors. Each of these techniques provides the surveyor with the ability to detect gross measurement errors. Surveying with GNSS is also similar; the only difference is that the observation procedure is less prone to user error as almost everything else is done automatically by the receiver. The most likely source of human error is coordinating incorrect marks or naming marks incorrectly, improper definition of ellipsoid/geoid, and erroneous entry of antenna height details. The fact is, any errors due to the receiver measurement procedure are often difficult to detect.

GNSS surveys can be designed to contain sufficient measurement redundancy to enable gross errors to be detected. Surveyors should be aware of any requirements that demand redundant measurements in GNSS survey. For example, static control surveys may require the occupation of each mark on at least two separate occasions. For kinematic surveys (including RTK), each point may need to be coordinated (synonymous to surveyed or occupied or collected) from two reference stations. It may also be remembered that the kinematic occupations are to be independent, in other words, the two reference receivers cannot operate at the same time to enable each point to be occupied once. Surveyors can increase the integrity of their results by planning redundant measurements in their survey procedures. The use of loop closures and least squares adjustments can then be used to isolate problematic measurements (refer to Sections 11.5.9 and 11.5.10).

One additional method by which checks can be performed is to occupy as many previously coordinated marks as possible. A minimum number of control points must be occupied and integrated into surveys to enable coordinates to be computed relative to the appropriate coordinates. Integration of additional coordinates serves two purposes—it assists in identifying erroneous GNSS baselines as well as integrating the survey into the control coordinate system. This also verifies the accuracy of control coordinates.

11.5.7 SATELLITE GEOMETRY

Satellites which are well spaced will tend to provide better results than constellations which are poorly spaced. The indicator used to describe the instantaneous satellite geometry is termed the dilution of precision (DOP). A high DOP value indicates poor satellite geometry; a low value indicates strong geometry. The DOP value is calculated from the inverse of the normal matrix in a point positioning least squares adjustment (refer to Chapter 5, Section 5.9.2).

The DOP is highly dependent on the number of visible satellites. A minimum of five satellites should be observed simultaneously. During the observing session, the geometric dilution of precision (GDOP) should never be greater than 8. The position dilution of precision (PDOP) should not be greater than 5. Satellite signals shall be observed from a minimum of two quadrants that are diagonally opposite each other. The position of the satellites above an observer's horizon can be obtained from sky plot and critical consideration in planning a GNSS survey (refer to Chapter 5, Section 5.9.2).

Obstructions that are 20 degrees or more above the horizon should be noted on an obstruction diagram. The effect of obstructions should be minimised by proper reconnaissance prior to observations. Satellites below an elevation mask of 10 degrees should not be used in baseline measurements. Signals from these satellites will have higher multipath effect. In the urban environment, the mask angle should be kept to around 15 degrees.

11.5.8 CONTROL REQUIREMENTS

Almost all surveys require the computed coordinates to be related to an existing set of coordinates. Even a straightforward re-establishment survey will also require the survey to be rotated onto the datum used by a previous survey. This may be performed by setting up the survey instrument on a mark occupied in the previous survey and sighting along a direction determined from the previous survey. This provides the bearing datum, or in effect, determines the necessary rotation parameter to apply to determined coordinates. If the total station being used has been calibrated, the distances can be considered correct and the survey can proceed.

GNSS surveys are a little different from total station surveys. The coordinates generated from measurements are referenced to the WGS84 datum for GPS (refer to Chapter 9, Section 9.5 for datums used by other GNSS systems) and are presented in terms of Cartesian coordinate differences between the reference and rover. If the desired coordinate system is different from this, a transformation needs to be applied (refer to Chapter 9, Section 9.5). For surveying applications where the coordinates of the local control points need to be considered when integrating new points, a global or regional set of pre-determined transformation parameters is often inadequate for application to new points, as the parameters are not sensitive to errors in the local coordinates. Therefore, surveyors must occupy points with known coordinates in order to integrate new points in local coordinate systems.

The number of control points required depends on the application. If horizontal coordinates are required, then a minimum of two points are required with known easting and northing coordinates in the desired coordinate system. This enables a scale factor, rotation and two translation components to be computed (refer to Chapter 9, Section 9.5). Note that any error in the local coordinates will be difficult to detect, as there is no redundancy in the transformation parameter estimation process. It is beneficial to observe a third control point in such circumstances. If the height of points is also required, sufficient information must be available to compute the geoid-spheroid separation. If the survey extends for less than 10 km, geometric geoid modelling techniques can usually be applied to good effect. Geometric techniques require the survey to be connected to three points with known horizontal and height coordinates. An additional point with known height only is sufficient to check the success of the geoid modelling technique.

11.5.9 LOOP CLOSURES AND BASELINE DIFFERENCES

The GNSS surveying techniques are capable of generating centimetre accuracy results if the carrier-phase ambiguities are correctly identified and constrained during data processing. The results are generally presented as Cartesian coordinate differences. These coordinate differences, or vectors, represent the three-dimensional coordinate difference between the base and rover. In addition to Cartesian coordinates, the vectors can be presented in terms of east, north, and height differences. This is commonly performed using a local projection. Regardless of the manner in which the vectors are presented, closures of connecting baselines can aid in the detection of erroneous measurements. In the same manner in which a traverse misclose is computed, the three-dimensional mis-close of GNSS vectors can also be determined. GNSS surveys are not performed to generate traverse measurement equivalents; therefore, surveyors use manually selected baselines to form loops of baselines. The closures can be performed using a calculator; however, some GNSS surveying systems provide loop closure utilities with the data processing software. Intelligent use of loop closures can enable erroneous baselines to be identified.

In order for a loop closure to be performed, baselines are required from more than one observation session. If only one session is used, the baselines are correlated and loop closures will tend to always indicate excellent results. This is due to the 'correlation' between the baselines rather than the 'quality' of the baselines. When multiple sessions are observed, a number of strategies for detecting poor quality vectors can be adopted. Consider the example shown in Figure 11.8 where several redundant baselines have been observed. One strategy which may be adopted is to check each triangle while trying to isolate any triangle which reveals poor results. If each triangle is closed, it is likely that a bad baseline will affect more than one triangle.

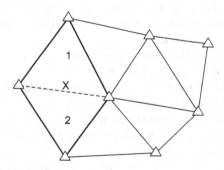

FIGURE 11.8   Redundant baseline observation.

This technique enables checking of correlated baselines from the same session. It is also likely, however, that a session which was too short to enable the ambiguities to be correctly resolved will highlight a minimum of two low-quality baselines. Comparing all triangles will enable such instances to be detected if sufficient baselines are observed.

In Figure 11.8, if baseline X is erroneous, it can be anticipated that triangles 1 and 2 will highlight a poor closure. By performing a closure around the four-sided perimeter of triangles 1 and 2, the poor baseline can be highlighted. In addition, several of the points should be occupied on more than one occasion. Performing loop closures will aid in detecting whether antenna height errors are present in the data set. In order for the processing software to be able to check for poor baselines it is important that the baselines are measured more than once (i.e., over multiple sessions) to obtain independent baselines. For example, in the case described earlier, the baseline X could be measured in the first and second sessions (by holding this baseline fixed), to enable the processing software to do a comparison.

Loop closures and baseline differences in repeat of baseline measurements are to be assessed to check for blunders and to obtain initial estimates for internal network consistencies. Software used for processing the raw data must be capable of producing results that meet the accuracy standards specified for the survey. The software must be able to produce, from the raw data, relative position coordinates and corresponding variance covariance statistics, which, in turn, can be used as input to three-dimensional network adjustment programs.

In the case of loop closure analysis, the error of closure is the ratio of the length of the line representing the equivalent of the resultant errors in the GNSS baseline vector components to the length of the perimeter of the figure to be analysed in the survey loop. A loop is defined as a series of at least three independent connecting baselines, which start and end at the same station. Each loop shall have at least one baseline in common with another loop. Each loop shall contain baselines collected from a minimum of two independent sessions. The following minimal criteria for baselines loops shall be used for static and rapid static procedures for first-order classification (1:100000) network adjustment:

- Baselines in the loop shall be from a minimum of two independent observation sessions. Loop closures incorporating only baselines determined for one common observation session are not valid for analysing the internal consistency of the network.
- Baselines in the loop shall not be more than 10 in total.
- Loop length shall not exceed 100 km.
- Percentage of baselines not meeting criteria for inclusion in any loop shall be less than 20% of all independent baselines.
- In any component (x, y, z) maximum misclosure shall not exceed 25 cm.
- In any component (x, y, z) maximum misclosure in terms of loop length shall not exceed 12.5 ppm.
- In any component (x, y, z) the average misclosure in terms of loop length shall not exceed 8 ppm.

For repeat baseline difference:

- Baseline lengths shall not exceed 250 km.
- In any component (x, y, z) maximum difference shall not exceed 20 ppm.

NOTE

When a survey connects back to itself, it is called a *loop*. If a loop is surveyed perfectly, the survey should come back to the exact point at which it started. If the loop doesn't come back to the exact starting point, it is called a *closure error*. Closure errors indicate that some or all of the survey measurements within a loop have errors. The kind of error found in a loop has a big effect on how the loops can be closed. There are three types of error in survey data: *random errors*, *systematic errors*, and *blunders*.

Random errors are generally small errors that occur during the process of surveying. They result from the fact that it is impossible to get absolutely perfect measurements each time. Systematic errors occur when something causes a constant and consistent error throughout the survey. The key to systematic errors is that they are constant and consistent. If we understand what has caused the systematic error, we can remove it from each measurement with simple mathematics. Blunders are fundamental errors in surveying process. Blunders are usually caused by human error. Blunders are mistakes in the process of taking reading, transcribing, or recording survey data.

Least-square is a mathematical approach that works by solving a large equation that take into account all the redundant data and minimises the errors. Least-square assumes that the errors in the survey are random. If the errors are not random, the data violates the mathematical model of least-squares and the process will not work properly. As a result, if we are going to use least-squares, special steps must be taken to detect any blunders and compensate for them.

11.5.10 NETWORK ADJUSTMENT

When establishing networks of GNSS baselines, a least squares adjustment of the generated baselines is often performed once processing is complete. These networks may comprise static and kinematic baselines, however, static baselines are generally observed. The network adjustment procedure has several functions in the GNSS surveying process. The adjustment provides a single set of coordinates based on all the measurements acquired as well as providing a mechanism by which baselines that have not been resolved to sufficient accuracy can be detected. A series of loop closures should be performed before the network adjustment procedure to limit the number of erroneous baselines entering in the adjustment process. A further feature of the network adjustment stage is that transformation parameters relating the GNSS vectors to a local coordinate system can be estimated as part of the adjustment. The adjustment process can be done in several ways. The following are the major elements of the adjustment process.

11.5.10.1 Minimally Constrained Adjustment

Once the processed Cartesian vectors have been loaded in the adjustment module, an adjustment should be performed where no coordinates are constrained. The adjustment should be performed using the datum of respective GNSS constellation (e.g., WGS84 for GPS). In actual fact, the processor does constrain one point internally to enable this adjustment to be solved. This solution provides a mechanism by which GNSS baselines which are not sufficiently accurate can be detected. Once the minimally constrained adjustment has been performed, the surveyor should analyse the baseline residuals and statistical outputs (which will differ between adjustment programs) and ascertain whether any baselines should be removed from subsequent adjustments. This process relies on the baseline network being observed in such a manner to ensure that redundant baselines exist. It is the redundant baselines that enable erroneous baselines to be detected.

11.5.10.2 Constrained Adjustment

Once the minimally constrained adjustment has been done and all unsatisfactory baseline solutions removed, a constrained adjustment can be performed. The constrained adjustment is performed to compute transformation parameters, if required, and yield coordinates of all unknown points in the desired coordinate system. The surveyor must ensure that sufficient points with known coordinates are occupied as part of the survey. The user should analyse the statistical output of the processor to ascertain the quality of the adjustment. Large residuals at this stage, after the minimally constrained adjustment has been performed, will indicate that the control points are non-homogeneous. It is, therefore, important that additional control points are occupied to ensure that such errors can be detected.

11.5.10.3 Error Ellipses

The standard deviations of the estimated coordinates are derived from the inverse of the normal matrix generated during formulation of the least squares process. Error ellipses for each point can be computed from the elements of this matrix. The ellipse presents a one standard deviation confidence region in which the most probable solution based on the measurements will fall. Surveyors should base the quality of the adjustment process on the magnitude of these ellipses. Many contracts will specify the magnitude of error ellipses for both the minimally constrained and fully constrained adjustments as a method of prescribing required accuracy levels. The product documentation for the adjustment program will further indicate the manner in which the ellipse values are generated.

11.5.10.4 Independent Baselines

For the least squares adjustment process to be successful, the surveyor must ensure that independent baselines have been observed. If more than one session is used to build the baseline network, then independent baselines will exist. In instances where one session is observed and all baselines are adjusted, the measurement residuals will be extremely small. This is due to the correlation that exists between the baseline solutions as they are derived from common data sets. This is not a problem as long as the surveyor is aware of the occurrence and does not assume that the baselines are of as high accuracy as implied from the network adjustment results. The inclusion of independent baselines is an important component of GNSS survey design and leads to a strong network configuration.

11.5.11 INDEPENDENT REOCCUPATION OF STATIONS

GNSS surveys require redundancy of observations which are used to detect blunders and to obtain statistically sound results. Redundancy is achieved by reoccupying some points in different sessions with different geometric combinations. The following criteria pertain to static, rapid static and reoccupation procedures for network adjustments:

- 10% of all stations should be occupied three times or more.
- All vertical control stations should be occupied twice or more.
- 25% of horizontal control stations should be occupied twice or more.
- All 'station pairs' for azimuth control should be occupied simultaneously twice or more.
- 100% of new stations should be occupied twice or more.
- For sessions where stations are occupied in succession, the antenna/tripod must be physically moved and reset between the sessions to be classified as an independent occupation.

11.5.12 POINT OR LINE OFFSET

One technique to avoid multipath and signal attenuation is *offsetting* (Figure 11.9). Offsetting can also be adopted where the survey-location is physically (or otherwise)

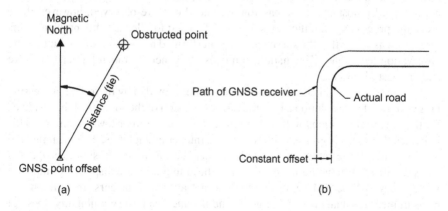

FIGURE 11.9 (a) Point offset, (b) line offset.

inaccessible. The offset point must be established far enough from the original position to avoid an obstructed signal, but close enough to prevent unacceptable positioning error. The length of time (distance between the original point and offset point) may be measured by an external laser, a laser cabled directly into the GNSS receiver, or even a tape and clinometers (for measuring angles). A point can be surveyed from a distance by using three different approaches: (i) by defining distance and bearing (angle with respect to north) (Figure 11.9a), (ii) by measuring distances from two different offset locations and defining the side on which the occupation point is located considering we are moving from first offset location to the second, and (iii) by measuring distances from three offset locations. The last two use trilateration technique to determine the desired point. However, the first method definitely reduces the time involvement, complexity, and chances of blunders.

A technique unique to RTK and differential GNSS, and used especially in mobile GNSS applications, is the creation of dynamic lines. The GNSS receiver typically moves along a route to be mapped logging positions at predetermined intervals of time or distance. Obstructions along a route present a clear difficulty for this procedure. Where it is impossible or unsafe to travel along the line to be collected in the field, it may be collected with a constant offset (Figure 11.9b). This technique is especially useful in the collection of roads and railroads where it is possible to estimate the offset with some certainty due to the constant width of the feature. It is also possible to collect routs with individual discrete points with short occupations where that approach recommends itself.

11.5.13   FLOAT SOLUTION

If insufficient data has been acquired to successfully resolve the carrier-phase ambiguities, a float solution is generated. In the float solution, the ambiguities are not constrained to integers rather they are left to 'float' as real numbers. Most commonly, the floating ambiguities will not be integers (or close to integers) in this instance. The most precise results are obtained when the ambiguities are constrained to be integers; therefore, a float solution generally implies that the required accuracy will not be met. In most cases, the baseline will need to be re-observed, however, there are some processing modifications that may be used to alleviate this problem. This discussion assumes that the survey has been performed using the static or rapid static observation technique. Kinematic techniques are generally much harder to resolve and are best re-observed.

The surveyor should closely analyse the output provided by their data processing program. In all float solution cases, the ratio of the sum of the squares of the residuals for the potential solutions will be a number that is close to or lower than one. This indicates that there is no clear solution to the integer ambiguities. The meaning and nature in which the ratio is computed will depend on the processing package and users should refer to their documentation. If the ambiguities are displayed as numeric values, they will be real numbers which do not approach integers. In addition, the measurement residuals may be large. In the instance that every ambiguity does not appear to be an integer, the baseline is probably best re-observed. If, however, there is one satellite which is not close to an integer, but the other values are quite close, the data should be re-processed after eliminating this satellite from processing. Similarly, analysing the residual graphs may reveal one or two noisy satellites that can be eliminated from processing for trying to generate a fixed ambiguity solution. Another modification worth trying is raising the elevation mask. If data has been observed at ten degrees and residual graphs reveal that measurements are particularly noisy when satellites are low to the horizon, the mask can be raised to, say, fifteen degrees and the processing software run again. This may solve the problem and generate fixed ambiguities.

In general, when the problems cannot be resolved using the above suggestions, the baseline must be re-observed. However, by analysing the output provided from the processing program and looking for satellite measurements which may be causing the problem, some baselines may be able to be processed to a satisfactory accuracy.

11.6 OBSERVATION METHODS

GNSS surveying generally uses carrier-phase observable in relative positioning approach for high accuracy measurements. The following sections describe these methods with their specifications. This discussion should help to choose an appropriate GNSS surveying method for a specific application. Surveyors must decide which technique is most suitable to the application of concern. In most cases, a combination of techniques is desirable. For example, static survey procedures may be used to connect the survey to control points. Kinematic techniques can then be used in the local survey region and a total station used to complete the obstructed portions of the survey.

There is one other critical aspect of the selected surveying technique that must be decided. The question pertains to whether the survey should be post-processed, or processed in real-time. If the coordinates of the points of interest are required in the field, then the survey must be processed in real-time. If the coordinates are not required in the field, a decision needs to be made based on a number of factors. If the survey is post-processed, the coordinates will be more reliable as the data can be manipulated several times and in a variety of ways (e.g., removal of noisy satellite, changing the mask angle). If long occupation times are chosen to enable errors to average, the measurements should be post-processed. The disadvantage of post-processed techniques is that significant amounts of data need to be stored and the success of the survey is not known until the survey and post-processing is completed. Real-time techniques do not suffer from this problem as the surveyor can see the computed coordinates in front of them while surveying. The disadvantage of real-time techniques is the need for the radio modem for communication link. The difficulties of the data link need to be weighed against the real-time coordinate update to decide whether post-processed or real-time surveying is the most applicable to a specific project.

Post-processing applications do not suffer from communication link restrictions. This translates to less equipment and, hence, less cost. However, perhaps the largest drawback of the post-processing approach is the difficulty in assessing the correct amount of data to collect to ensure that the integer ambiguities are resolved. Users are advised to adopt a more conservative approach when performing such surveys because insufficient data acquisition usually results in the survey being re-observed. Therefore, post-processed techniques suffer from a lack of productivity when compared with real-time techniques. Surveyors should use their best judgment while selecting an appropriate technique for a particular survey.

11.6.1 CLASSIC STATIC TECHNIQUE

Static (or classic static) GNSS survey procedures allow various systematic errors to be resolved when high accuracy positioning is required. This method is essentially used for baselines greater than 20 km. Static procedures are used to produce baselines between stationary GNSS units by recording data over an extended predetermined period of time during which the satellite geometry changes. This is the classical GNSS survey method for long baselines and higher accuracy, typically for 1:100000 scale or greater. In order to resolve integer ambiguities between satellite and receiver, points are occupied for long sessions (30 min to 6+ h), depending on the number of visible satellites, baseline length, intended accuracy, etc. Baselines are re-observed in different sessions according to a predetermined observation scheme.

Some of the specifications of this method are as follows:

- Static observations are required for all baselines greater than 20 km in length. Static observations may be required for baselines less than 20 km depending on particular project requirements. If the baseline length is within 10 km, the ionosphere-free fixed solution may be used (dual frequency receiver is required).
- A minimum of three receivers should be used simultaneously during all static GNSS sessions.
- A minimum of five satellites must be observed simultaneously for a minimum of 30 min, plus 1 min/km of baseline length per session. Remember, sessions that are a bit longer than this minimum will provide worthwhile redundancy that could make data processing more robust and improve project results.
- Data sampling should have an epoch rate of 30 sec or less.
- Typical achieved accuracy: sub-centimetre level (5 mm + 1 ppm).

The static observation procedure is the most commonly used GNSS observation technique due to its reliability and ease of data collection. The results generated from static observations are the most robust of the satellite positioning solutions due to the increased length in observation period. All control surveys over reference/rover separations of several kilometres are performed using static surveying techniques. The static procedure requires satellite measurements to be acquired simultaneously at multiple sites by stationary receivers.

Static surveys are performed by setting receivers on stable platforms, usually a tripod or survey pillar, and leaving them to record measurements at predetermined intervals for a period of time. Observations are usually collected at a rate of one epoch every 5, 10, 15, or 30 sec. The rate of collection is not of prime importance in static work, and measurements should be acquired at rates which are related to the amount of available data storage space. More importantly, data needs to be collected for a sufficient time period to enable the integer cycle ambiguities to be resolved. In addition, the effects of multipath and random measurement error can be reduced by observing for longer periods. There are no hard rules for determining how long the data should be collected. The time required to achieve a suitable accuracy is a function of the number of visible satellites, baseline length, multipath conditions, atmospheric conditions, and satellite geometry. Using past experience and knowledge, the surveyor may choose an observation period that they feel is sufficient to resolve ambiguities and obtain an accurate position (e.g., one hour for an engineering control network or 5 days continuous tracking for geodynamic surveys).

To make full use of the acquired measurements, static surveys should be post-processed. This requires the storage of the observables to be merged in the processing software at a later time. This implies both that results are not available in the field and that the time period required to obtain a desired accuracy is a 'calculated guess'. Experience under similar survey conditions generally defines the observation period; however, for accurate results, observation periods of less than 30 min should not be used for lines greater than 5 km in length. It should be noted that the longer the observation session, the more accurate the calculated position will be. If post-processing reveals that results are unsatisfactory, the baseline will need to be observed again. It is, therefore, wise to use caution when estimating observation periods; e.g., additional 5–10 min per point may be sufficient to prevent further observation.

The main drawback of static procedures is the lack of productivity. If, for example, observation periods of 45 min per point are adopted, it may only be possible to collect five or six points per day depending on the time required to move between marks (refer to Section 11.8). To increase the efficiency, multiple receivers can be used simultaneously. Many receiver manufacturers sell receivers in groups of three for this reason. Each receiver remains stationary at the same time to enable three points to be occupied. Each observation period is termed as a session. Once each session is completed, the receivers move and begin acquiring measurements simultaneously at three other stations. For each three-receiver session, three baselines can be generated. Technically, only two of the baselines are independent as the third uses measurements already used by the other two lines. However, it is recommended that all baselines are processed and adjusted using a least squares estimation procedure. The statistical output for each baseline will reflect the correlated nature of the third baseline. This use of sessions with multiple receivers enables surveys to generate networks of baselines. The greater the number of receivers, the greater the productivity.

The number of baselines per receiver combination can be computed by summing the number of receivers minus one, down to one. Three receivers yield 2+1=3 baselines, four receivers yield 3+2+1=6 baselines, five receivers yield 4+3+2+1=10 baselines, so on. It should be noted that increased receiver numbers generally require additional personnel. Figure 11.10 shows how three receivers can

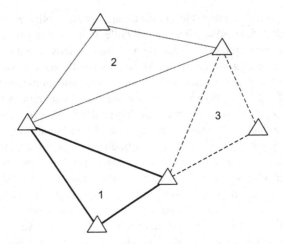

FIGURE 11.10   Three receivers can be used to collect six stations in three sessions.

be used to collect six stations using three sessions. These three sessions provide a network of nine connected baselines.

11.6.2   RAPID STATIC TECHNIQUE

Rapid static (also known as *fast-static*) GNSS surveys are similar to static surveys, but with a shorter observation period and usually with a faster epoch rate. This technique utilises shorter observation times for shorter baselines with accuracies usually less than 1:100000. It can be done with a classical network design or as radial surveys.

Specifications of this method are as follows:

- Rapid static procedures may be used on baselines up to 20 km in length.
- A minimum of three receivers shall be used simultaneously during all rapid static GNSS sessions.
- A minimum of 5 satellites should be observed simultaneously for a minimum of 5 min, plus 1 min/km of baseline length per session. Slightly longer session than this minimum could prove rewarding. Typical observation times range from 5–20 min.
- Data sampling shall have an epoch rate of 5 sec or less.
- Typical achieved accuracy: centimetre level (10 mm + 1 ppm).

The rapid static surveying technique was developed in an attempt to improve the efficiency of the static survey. Users should note that the observation procedures for rapid static surveying are the same as those for static surveying. The only difference is the length of the occupation period which is less than that required for static surveys. This reduced occupation length is facilitated by mathematical improvements

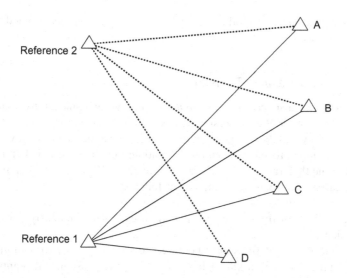

FIGURE 11.11 A series of radiation type vectors from two reference stations.

in processing software which enable the integer ambiguities to be determined using a smaller number of observations. In order to perform surveys with maximum efficiency, a dual frequency receiver which is capable of pseudorange and carrier-phase measurements on both carrier frequencies is required. The use of the pseudorange measurements (after they have been smoothed) also assists in the rapid determination of the ambiguities; therefore, receivers capable of code measurements are desirable.

As the occupation period of rapid static surveys is shorter than that of static surveys, it can often be extremely difficult to manage the movement of receivers in sessions (setting the tripod, placing all instruments and connecting them, antenna height measurement, and so on). Therefore, rovers on vehicles generally occupy points as efficiently as possible. Occupation times as short as 10 min are often sufficient to resolve the integer ambiguities over short baselines when at least six satellites are tracked. This results in a series of radiation type vectors from the reference station (Figure 11.11). To provide some redundancy, a second reference station may be used (Figure 11.11). This provides two vectors to each point. It should be noted that single reference method of data collection is unsuitable for detecting erroneous vectors if the rover station is the source of the error. An example of such an error is incorrect entry of the antenna height. A second occupation of each station, ideally using a different reference receiver, is preferable; however, this reduces survey efficiency. Surveyors must use their professional discretion in the manner in which rapid static surveys are conducted.

The rapid static survey procedure is most efficient when dual frequency receivers are used and baselines are kept below of 5 km. If the points of interest are free of overhead obstructions and six or more satellites are observed, surveys can be performed in a matter of minutes. In this environment, the time required to set up the antenna on a tripod makes significant contribution to the time spent at each point. To improve the efficiency, a tripod mounted on a vehicle or a handheld bipod arrangement may be used.

11.6.3 PSEUDOKINEMATIC TECHNIQUE

Pseudokinematic (also called *reoccupation*) surveys are similar to rapid static, except that the length of each session is short (1 min). Each point is re-visited after at least an hour, when the geometry of the satellites becomes significantly different. However, it requires revisit time and reoccupation; therefore, ultimately it may take even more time than rapid static. In practice, we generally avoid this technique.

Specifications of pseudokinematic are as follows:

- A minimum of three receivers should be used simultaneously during all sessions.
- Minimum of 5 satellites must be observed simultaneously for a minimum of 1 min plus 1 min/km of baseline length per session. All points surveyed should be re-occupied after at least one hour has elapsed to allow for a different alignment of the satellites. This method is not recommended unless the satellite configuration or site conditions do not permit fast-static techniques.
- Data sampling shall have an epoch rate of 5 sec or less.
- Typical achieved accuracy: 5–10 mm + 1 ppm.

11.6.4 STOP-AND-GO TECHNIQUE

The three techniques that we have explained in the preceding sections are actually different variations of static methods. The kinematic survey procedure was developed in the mid-1980s as an attempt to improve the productivity of surveying with GNSS receivers, especially for GIS applications. In applications where a large number of points are necessary to be captured within a short distance (say 100 points in 1 or 2 km), static procedures are inefficient and are, generally, not cost effective. The kinematic survey technique is ideally suited to such applications where points are closely spaced and easily accessed (e.g., lamp posts).

Once the carrier-phase integer cycle ambiguities have been determined, they do not change value if continuous tracking of the satellites is maintained, i.e., there are no cycle slips. The kinematic survey procedure is based on this characteristic of carrier-phase positioning. A short initialisation procedure is performed with the primary aim of determining the integer ambiguity values. Once this initialisation is completed, the roving receiver occupies points of interest for a short period, generally, less than 1 min (Figure 11.12). As the ambiguities do not change when satellites are continuously tracked, centimetre level accuracy can be obtained with a brief stationary occupation. Once a point has been occupied, the rover moves to the next point of interest where it acquires another minute of data. During the period in which the receiver is being transported between the two sites, satellite tracking must be

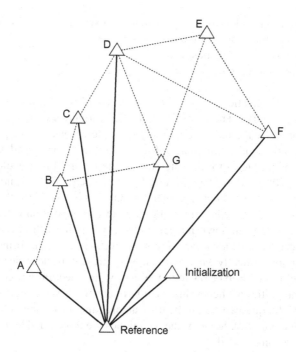

FIGURE 11.12   After a short initialisation, the roving receiver occupies points of interest.

maintained. This technique of continually moving, then stopping briefly, is termed as *stop-and-go kinematic* surveying (also called *semikinematic*).

Stop-and-go technique is applicable for baselines up to10 km. In this method, two receivers are used—one as a temporary reference (base) station and another as a rover. After initialisation, or resolution of the integer ambiguities, one receiver is left as stationary and another is allowed to roam. This results in the determination of the baseline vector between the stationary and roving receivers. We may use multiple rovers. The rover is initialised to establish an accurate relative position between the rover and the base. Following the initialisation, each new point is occupied for less than 1 min. Lock on the satellites must be maintained at all times or a new initialisation must be performed.

Specifications:

- A minimum of two receivers are necessary to be used simultaneously during all stop-and-go sessions. This procedure shall be limited to baselines of 10 km or less.
- A minimum of 5 satellites should be observed simultaneously for a minimum of 5 epochs.
- Initialisation of the rover can be accomplished by occupying a known point for a minimum of 5 epochs or making a rapid static observation of at least 5 min on the first point and then moving to other points to be surveyed.

- Data sampling shall have an epoch rate of 5 sec or less. A minimum of 5 epochs must be recorded for each point located.
- Collected data are to be post-processed.
- Typical achieved accuracy: 1–2 cm + 1 ppm.

The stop-and-go kinematic surveying procedure is performed to coordinate the position of stationary marks. The receiver must still continuously track satellites while in motion to preserve the integer ambiguity estimates. However, the position of the receiver while moving is not of interest. If cycle slips are experienced while moving, the integer ambiguity term for the interrupted satellite must be calculated again. If at least four satellites are still being tracked, this can be performed automatically by the processing software without user intervention. This is possible as the receiver can calculate its position using the remaining satellites. This position is then held fixed (constrained) and the unknown integer ambiguity for the fifth is estimated. This technique is termed the *known baseline initialisation* and is performed frequently by the receiver automatically when cycle slips are present. If the cycle slips cause the number of satellites to drop below four, the survey must be reinitialised to determine the integer ambiguities as their values will have changed. There are four ambiguity initialisation techniques which can be used at any time throughout a kinematic survey (refer to Section 11.7). Some of the techniques are more reliable than others and more strongly recommended.

11.6.5   Continuous Kinematic Technique

The continuous kinematic technique is similar to stop-and-go. The base and rovers are initialised as in the stop-and-go. After initialisation, the rover is constantly moving and measuring positions. Lock on the satellites must be maintained at all times or a new initialisation must be performed. This technique is now mostly used for detailed surveys over very large open area (such as beaches, paddocks, and open undulating country) where the rover is attached to vehicle or other continuously moving platform.

Specifications:

- A minimum of two receivers should be used simultaneously during all sessions.
- Initialisation of the rover can be accomplished by occupying a known point for a minimum of 5 epochs or making a rapid static observation on the first point and then moving to other points to be surveyed.
- Data sampling should have an epoch rate of 2 sec or less.
- Post-processed baseline solutions.
- Typical achieved accuracy: 1–2 cm + 1 ppm.

The continuous kinematic survey and the stop-and-go techniques are almost identical in field procedure, except the position of the receiver while it is in motion is now of interest. As long as the satellites are tracked without interruption, the position of the antenna can be estimated at each measurement moment. The epoch rate must be set carefully to ensure that position estimates are computed at a desirable frequency. An example of where a continuous kinematic survey may be practical is in the coordination of a train track. The antenna can be placed on the train and driven to *digitise* the track. Surveyors should build redundancy into kinematic surveys by occupying marks on multiple occasions or, in the continuous kinematic case, re-traverse the same route.

One feature of continuous kinematic surveys that must be considered is whether the height of the receiver is of interest while the receiver is in motion. If this is the case, then the height of the antenna above the ground must be kept uniform. This can be accomplished using a range of devices, many of which are best developed by the surveyor for a specific use.

Both stop-and-go and continuous kinematic techniques require a stationary base station receiver tracking continuously and also a roving receiver being used to determine necessary details and features. Data from both are downloaded at the end of the survey and then post-processed to obtain corrected data. The stop-and-go and continuous kinematic collectively are more commonly known as *post-processed kinematic* (PPK).

11.6.6 REAL-TIME KINEMATIC (RTK) TECHNIQUES

From earlier discussions, it is clear to us that the satellite carrier-phase measurements collected at both reference and roving receivers can be stored using a number of different techniques, then combined in a computer for post-processing. The restriction of this approach is that the results of the survey (positioning information) are not known until the survey is complete. RTK processing technique utilises a data link, usually in the form of a radio, to transfer raw measurements acquired at the reference receiver (located on a precisely surveyed mark) to the roving receiver (Chapter 6, Figure 6.6). The microprocessor in the roving receiver then combines the reference and rover information and computes the rover coordinates as the survey is being performed, i.e., in real-time.

RTK is rapidly becoming widespread and the most commonly preferred GNSS surveying technique. RTK is an advanced form of relative carrier-phase surveying in which the reference station transmits its raw measurement data to the rover(s), which then compute a vector baseline from the base receiver to the rover. Then it computes the coordinates (x, y, z) of the rover's position by using the known coordinates of the base receiver and measured vector baseline. The computation is done nearly instantaneously, with minimal delays between the time that the base receiver measures, and the time these data are used for baseline processing at the rover. RTK is only suitable for environments with reasonably good tracking conditions (limited multipath, obstructions, and RF noise), and with continuously reliable base to rover communication. To achieve the real-time positioning, a radio or other electronic data transmitting device (i.e., CDMA cell phones) link between the base receiver and the rovers must be established. RTK surveying techniques are an ever-changing, dynamic technology.

Conventional RTK surveys, by nature, result in a radial pattern of baselines (normally up to 10 km) from the base receiver to the rover(s). Baseline components (Δx, Δy, Δz) are produced on a specific datum. From these values, the coordinates of the occupied points by the rover are produced. This does not produce strong network geometry. Depending on the distance from the base station, the positions as determined may result in poor local accuracy if a direct connection is not observed between the two. If a second base station is observed, additional baselines should be formed to each RTK point. This will add redundancy and improve the positional reliability.

There are numerous and widespread applications for RTK-type surveying. Typical applications include, but are not limited to, hydrographic surveys, location surveys, real-time topographic surveys, construction stakeout surveys, and photogrammetric control surveys.

Specifications:

- The project area should contain and be enclosed with RTK control base stations.
- A minimum of two receivers must be used simultaneously during all RTK sessions. One base receiver shall occupy a reference point and one or more receivers shall be used as rovers.
- Initialisation of the rover(s) should be made on a known point to validate the initial vector solution. A check shot should be observed by the rover unit(s) before the base station is taken as reference.
- Each RTK point should have 2 different independent occupations based on a time offset. Each occupation consists of at least 10 epochs. The second occupation shall be made under a different satellite constellation, offset from the first occupation by at least 3 hours. However, in many mapping applications, the second occupation is ignored to increase the productivity.
- It is recommended that the second occupation be made from a different base station.
- To ensure good local accuracies between new RTK points and nearby existing stations, all previously established base stations, control points, and station pairs are to be RTK positioned, when feasible, for consistency.
- Data sampling should have an epoch rate of 2 sec or less (1 sec is recommended). Real-time coordinates must be recorded.
- The use of a pole (with or without bipod) for the rover unit is required. Tripod is not suitable for RTK rover, because it requires frequent relocation.
- The use of a 2-meter fixed height pole or tripod is recommended for the base station.
- If we use a radio as the communication link, one base station can support an unlimited number of rovers. If we use a different communication method, for example, a cellular phone, it may be limited in the number of rovers the base can support.
- The benefits of RTK include very fast and efficient data collection that provides results in real time and in local coordinate systems.

- Typical achieved accuracy (averages): 1–2 cm + 1 ppm for horizontal measurements and 1.5–2.5 times greater in vertical measurements.

Another method of RTK is real-time network RTK. This technique is the one implemented by the real-time networks with baselines of up to 70–100 km or even more. Real-time network RTK surveys are similar in principal to conventional RTK surveys. Both are dependent on observations from base stations. The RTK networks extend the use of RTK to a larger area containing a network of reference stations. Operational reliability and accuracy will depend on the density and capabilities of the reference-station network. A continuously operating reference station network is a network of RTK base stations that broadcast corrections, usually over an internet connection. Accuracy is increased, in this case, because more than one station helps to ensure correct positioning and guards against a false initialisation of a single base station. RTK network control is an 'active' type of system. Ideally, these real-time network master stations would be part of the national and/or cooperative reference stations. If so, then they are monitored continuously and their integrity/accuracies are checked. The data is made readily available anytime. Also, any datum changes or readjustments of these master stations are handled seamlessly. The network administrators who then broadcast correctors to the rover(s) have the optimum values and performance of these master stations. There is no need to install a base station, thereby freeing up an additional receiver. However, this type of facility is not available in all countries. The method used to determine the position of the rover(s) depends on the configuration of the real-time network system. The derived solution may be a single baseline vector from one base station or the multiple baseline solutions resulting from a combined network solution broadcasted from a central computer.

Regarding the base-rover local RTK systems, one has to realise that different receiver manufacturers offer different combinations of facilities. They even use their own naming conventions rather than the names we have used in this book. For example, it was said that stop-and-go and continuous kinematic techniques are actually post-processed (PPK); however, Trimble offers stop-and-go RTK technique. We need to understand that the technique of survey is the same as it was explained in Section 11.6.4; however, instead of post-processing, the correction is performed in real-time via data link. Trimble also offers real-time continuous kinematic option (Trimble calls it *continuous topo*). Some other industry jargon refers to *RTK & logging, RTK & infill, RT differential & infill, RT differential & logging*. The RTK & logging survey type is similar to RTK except that raw GNSS data is recorded for the entire survey. This method is useful if raw data is required for quality assurance purposes. This raw data can be post-processed and compared with the RTK result. RTK & infill surveys are effectively a combination of RTK and PPK surveys. While the communication link between base and rover is available, the survey is performed in RTK. If the data link is temporarily lost, the survey becomes PPK. For the period of communication, interruption data is logged for later processing. When the communication link becomes available, the survey reverts back to RTK. When we collect points manually it is possible to monitor whether the initialisation has been lost or not. However, it is not possible to monitor when we are collecting the points

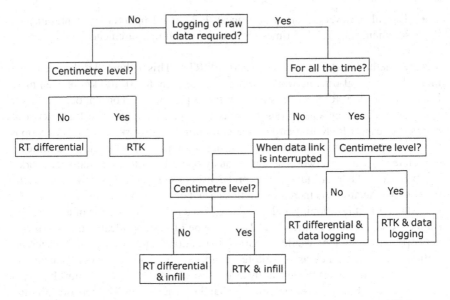

FIGURE 11.13 Choosing the appropriate real-time survey technique (Trimble 2006).

automatically on the move (vehicle mounted receiver to capture a linear feature). RTK & infill would be very useful in such cases (however, we recommend PPK in such cases). RT differential & logging and RT differential & infill are similar to the concepts of RTK & logging and RTK & infill, respectively, except that they use a code-based differential solution instead of carrier. Figure 11.13 shows a flowchart for selecting the appropriate method from the aforementioned RTK and differential (dynamic) techniques.

11.7   INITIALISATION TECHNIQUES

Before a feature of interest can be coordinated to an accuracy suitable for surveying applications, the carrier-phase integer cycle ambiguities must be determined. In the static survey, this is called ambiguity resolution. In a kinematic survey, this ambiguity resolution process is termed *initialisation*. The primary purpose of the initialisation procedure is to identify the integer ambiguity values.

How does a receiver solve the integer ambiguity? First, the receiver computes the pseudoranges based on code-based solution and determines its position. This position is not precise; however, it can define a search volume around this position. The actual point (possible solution) is somewhere within this volume. Consider this volume to contain a three-dimensional matrix of points (one of which is the possible solution); each one is associated with a combination of integers. Before the integer ambiguities are resolved, the position solution type is known as a float solution. This implies that the ambiguity values have a fractional part (values are non-integer). Then the receiver initiates an integer search—a series of computations of the rover position using every single integer combination. The correct combination will be identified by itself based on some statistical test defining the quality of the solution. Any such test is only possible if the receiver has more than one solution, which requires at least five visible satellites. Following the integer search, the integer ambiguities are resolved, and the RTK system is said to be initialised. The solution becomes a fixed solution (the integers are fixed). The position improves to centimetre level. There are four techniques that can be used to perform kinematic survey initialisation—*on-the-fly, static survey, known baseline*, and *antenna swap*.

11.7.1 ON-THE-FLY

On-the-fly (OTF) resolution (also called *fast ambiguity resolution*), computes the integer ambiguities while the receiver is in motion. To perform this efficiently, a minimum of five satellites are required, however, six or seven satellites are preferred. In addition, OTF techniques should not be attempted with single frequency receivers as the process is extremely inefficient. The advantage of this technique for surveyors is that previously surveyed marks are not required to be coordinated as frequently for initialisation and re-initialisation purposes (it is required in the case of known baseline procedure). An example of the flexibility of the OTF initialisation procedure is where a surveyor collects points, then passes under a bridge. By moving under the bridge, satellite tracking is interrupted (cycle slip occurs) and the survey must be re-initialised. A static survey could be performed on the other side of the bridge (refer to Section 11.7.2); however, this may be time-consuming if the distance to the reference receiver exceeds several kilometres. With the OTF technique, the surveyor can continue moving to the next feature of interest. While the surveyor is moving, the satellites will be re-acquired and the OTF resolution scheme will automatically begin to resolve the integer ambiguities.

In general, the ambiguities are safely resolved in less than 5 min. In most instances, 2 min of tracking six or seven satellites is sufficient. Once initialised, the survey can proceed as normal. Users should note that if there are only four satellites being tracked, the OTF technique cannot operate. In addition, if five satellites are observed or the satellite geometry is poor, initialisation times may exceed even 10 or 15 min. Before a kinematic survey is performed using OTF techniques, users should consult their equipment documentation for specific details of the supported functionality. If the receiver supports multiple GNSS constellations, the number of satellites should not be an issue.

11.7.2 STATIC SURVEY OF NEW POINT

The static survey initialisation is identical to a static survey performed to coordinate (positioning) points of interest. The rover is used to perform a short static survey. In order to coordinate features using static techniques, the observation period is designed to facilitate the identification of the integer ambiguities. Therefore, performing a static survey results in both the occupation of the mark as well as the ambiguity resolution. To perform a static initialisation efficiently, the base-rover separation should be kept short to minimise the static survey observation period.

Once sufficient measurements are considered to have been observed to complete the static survey, kinematic occupations can proceed. Some manufacturer implementations have certain restrictions when initialising using the static method. Users should refer to their equipment documentation for specific details.

11.7.3   KNOWN BASELINE OR KNOWN POINT

The known baseline initialisation technique requires knowledge of a previously determined base-rover baseline vector. This position vector can be derived from a previous static or kinematic survey. In addition, any point occupied in the current survey can also be used to initialise the survey. The rover is placed on a previously surveyed mark and the location is entered into the rover. Since the rover knows the position of the base receiver (and also of its own) it can determine the baseline vector. The known baseline procedure is based upon constraining the position vector and only estimating the integer ambiguities. Theoretically, one measurement epoch is sufficient to perform this estimation, however, 1 min of observation is generally recommended. Users should be aware that the points occupied after such an initialisation procedure are dependent on the success of the initialisation. As a result, it is best to favour a conservative approach when reinitialising surveys. Occupation periods used to initialise surveys should reflect this conservatism.

If cycle slip (lock lost) occurs for a satellite while the receiver is in motion, even though, the number of satellites tracked remains four or greater, the receiver position can be computed using the available satellites. This position is then constrained to enable the unknown satellite ambiguities to be determined. This occurs transparently to the surveyor and does not require modification of the field procedure. In instances where cycle slip occurs due to a decrease in number of satellites tracked below four, the known baseline technique can only be used by occupying a previously observed mark. This is not too difficult; we go back to the last surveyed mark, enter the coordinate and achieve re-initialisation within a minute.

11.7.4   ANTENNA SWAP

The antenna swap procedure is a technique which is used to initialise the integer ambiguities at the beginning of a kinematic survey. The main limitation of the antenna swap procedure is that the base and rover must be kept within ten meters of each other. In most implementations, two swaps of antenna are actually performed and this approach is recommended. The procedure is performed by placing the base and rover over well-defined marks and simultaneously collecting measurements for approximately 1 min. The two antennas are then swapped, such that the base receiver is now located over the rover mark, and vice versa (Figure 11.14). A further minute of observations are then collected. These two steps are sufficient to resolve the integer ambiguities, however, a further swap is strongly recommended. To complete the antenna swap procedure in this manner, the reference and rover antennas return to their original locations. The rover can then proceed to points of interest. It is vital that continuous tracking of the satellites occur during the antenna swap

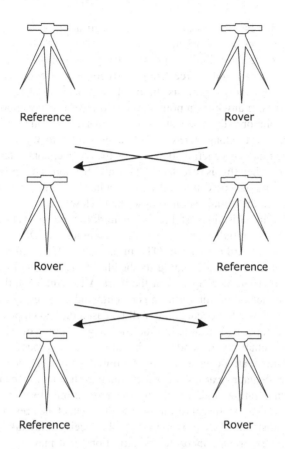

FIGURE 11.14 Antenna swap procedure for initialisation.

procedure. If a survey is to be performed by one person, the antenna swap procedure may be difficult to perform. The use of a third tripod can be used to accomplish the swap procedure.

As previously described, the biggest weakness of the antenna swap procedure is the need to remain adjacent to the reference receiver. This limits the applicability of the procedure in most surveying applications because, in the case of a cycle slip, one needs to come back at the base location for re-initialisation. In addition, the mechanism of the swap procedure is awkward, as the base receiver must be moved for every re-initialisation. As a result, this procedure has not been adopted by the modern receiver manufacturers.

11.7.5 Recommended RTK Initialisation Procedure

It was mentioned that OTF procedure is widely followed for RTK initialisation. But it is not the only procedure to follow. Selection of a procedure minimises the chance of surveying with a bad initialisation. A bad initialisation occurs when the integer ambiguities are not resolved correctly.

While performing the initialisation, the surveyor should choose a site that has a clear view of the sky and is free from obstructions that could cause multipath. Known baseline (or known point) is the quickest procedure of initialisation when this 'known' exists. If any known point is not available, we may choose either static survey of new point or OTF. The static survey takes a long time for initialisation if the baseline length is too long, whereas OTF takes just 5 min. Cycle slip or lock lost, which is very common, is generally detected by the receiver controller and automatically reinitialises when this is detected; but re-initialisation is not possible if we end the survey too soon. As a precaution, if we do not have any known mark, it is recommended to perform the initialisation as described below.

After the OTF initialisation, establishing a mark (say mark A) about 9 meters (30 feet) from where the initialisation occurred is recommended. Once this is done, the surveyor needs to discard the current OTF initialisation. If the surveyor is using an adjustable height range pole, changing the height of the antenna by approximately 20 cm is advised. Then reoccupation of the mark A to reinitialise the survey using the known point method of initialisation is recommended (entering the new antenna height is required). Further occupations will be based on known point initialisation procedure. Known point procedure has the advantage in terms of multipath.

Initialisation reliability depends on the initialisation procedure and also on whether or not multipath occurred during the initialisation. The occurrence of multipath at the GNSS antenna adversely affects the initialisation. If the initialisation is based on known point method, multipath can cause an initialisation attempt to fail. If it is based on OTF or static survey, it is difficult to detect the presence of multipath during the initialisation. If there is multipath, the receiver may take a long time to initialise, it may experience incorrect initialisation, or it may not initialise at all. Modern receivers are very reliable; if an incorrect initialisation occurs, it can be detected within 15–20 min and the receiver automatically discards the initialisation and issues a warning. Surveying with a bad initialisation will result in incorrect position recording. While performing an OTF initialisation, it is recommended to move the antenna around. This will reduce the effect of multipath.

11.8 PERSONNEL MANAGEMENT

The final issue to consider when planning a GNSS survey is personnel management. Let us consider a static survey, where four receivers are being used in a series of sessions as illustrated in Figure 11.15. Each session is highlighted with a different line-type in this figure. For this example, let us assume that each session is of 45 min in length, the baselines are less than ten kilometres, six satellites are available and dual frequency receivers are in use. The first session is planned to commence at 9.00 AM. The project requires four people, each with one receiver and sufficient batteries (if each receiver is to be monitored while it is running, to prevent theft). To operate efficiently, each person should have a vehicle. If this is the case, it can be assumed that the receiver can be stopped and packed away in the vehicle within

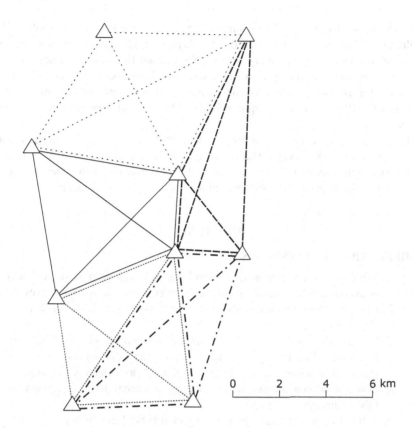

FIGURE 11.15 Static survey where four receivers are used in a series of sessions.

15 min. Allowing 45 min for travelling to the next point and a further 15 min to set up and start the survey, the second session will not commence until 11.00 AM. Using this procedure as a guide, four, or perhaps five, sessions can be performed in one day. The organisation of each of the receivers is vital to ensure that the survey proceeds smoothly. The importance of this aspect of GNSS surveying, particularly static surveying, cannot be underestimated. Additional considerations may apply if only two vehicles are used. This may limit the day to three sessions only. In such instances, the surveyor must weigh whether the rental of an additional vehicle justifies the additional productivity.

For kinematic surveys, the logistical problems of static surveys are not usually felt. Most often the base receiver is set up and one (or more) rover occupy points of interest. The area covered in a kinematic survey is usually much smaller than a static survey, therefore, timing and movement of personnel is not as critical. Consideration must be given to whether the reference receiver is to be monitored during the survey. In a kinematic survey, it is quite possible that the reference receiver will remain stationary for the entire day. Assigning a person to look after this receiver is not an efficient use of personnel. It may be more beneficial to establish reference receivers in secure locations in such circumstances (e.g., placing the base on the rooftop of a building and locking access to the roof). While planning for personnel management, we need to focus on two primary factors: observation time and accuracy requirement. The necessity of post-processing and involved time/personnel should also be considered. It is also advised to consider some additional time that may be required in case of malfunction of the equipment, blunders by the surveyors, local disturbances, etc.

In summary, there are a number of issues that need to be considered before embarking on GNSS surveys. Many of the issues require some thought to ensure appropriate application of GNSS technology. Users should take care when designing surveys in order to perform GNSS operations with a high rate of success.

EXERCISES

DESCRIPTIVE QUESTIONS

1. What do you understand by 'control survey', 'topographic survey', and 'stakeout'? Which survey techniques are good for these three surveys?
2. Compare the following GNSS survey techniques: DGNSS, RTK, PPK, static, and fast-static.
3. Discuss the relative advantages and disadvantages of the following: DGNSS, RTK, PPK, rapid static, static, and continuously operating static. Draw a flowchart for selecting an appropriate survey technique.
4. Discuss the considerations for selecting the appropriate equipment in GNSS survey.
5. What do you understand by planning of survey? Explain in brief.
6. Discuss the accuracy and obstruction factors on GNSS surveying.
7. Describe the guidelines for baseline length and occupation time.
8. What is data rate? Discuss the appropriate data rates for different surveying.
9. Why are redundant measurements necessary? Discuss different approaches for redundant measurements.
10. Briefly discuss the loop closure and baseline differences.
11. What do you understand by 'network adjustment'? Discuss the major elements of network adjustment.
12. How do we survey points and lines by using offset method? Explain with sketches.
13. What is float solution? How is this problem overcome?
14. Explain the classic static survey technique.
15. Explain rapid static survey technique.
16. Discuss stop-and-go survey technique.
17. Explain continuous kinematic survey technique. What are the differences between continuous kinematic and semikinematic?
18. Describe RTK surveying.

19. Discuss different initialisation techniques in kinematic survey.
20. What do you understand by 'personnel management' in survey? Explain in brief.

SHORT NOTES/DEFINITIONS

1. Controller
2. Optical plummet
3. Observation log
4. Obstruction
5. Pseudokinematic
6. Epoch rate
7. Loop closure
8. Error ellipse
9. Constrained adjustment
10. Line offset
11. Known baseline initialisation
12. On-the-fly
13. Antenna swap
14. Initialisation

12 Mapping with GNSS

12.1 INTRODUCTION

It was mentioned earlier that position information makes no sense without a common representation of the world. If we want to describe 'where we are', 'where we want to go', or 'where some point of interest is located', it must be defined with reference to a datum and a coordinate system. For most people, the map is the most widely used device for answering these questions, because it permits both qualitative (through the graphical representation) as well as quantitative (through coordinates) interpretation. Map also help us to understand spatial relationships among the neighbouring geographic features.

Maps have always played a special role in history. Early civilisations used maps for many of the same reasons we use them today: to depict the topographic form of territory, as an aid to navigation, to mark dangers, as an inventory of national assets, as evidence of ownership of land areas, to wage war, for the collection of taxes, to assist in development, to display spatial relationships, and many more. What has varied over time has been the accuracy of the maps, the methods of compilation, and the format of the map display.

The ingredients needed to make a map are: (1) a datum, (2) a projection, (3) data capture (survey) techniques, and (4) standards and specifications. The first and second have already been dealt with in Chapter 9. Spatial data capture techniques were explained in Chapter 11. In this chapter, we shall deal with the issues of standards and specifications relevant to map-making and other related matters.

12.2 INTEGRATION OF SURVEYING TOOLS

GNSS is a positioning technology that has revolutionised the surveying industry by providing the surveyor accurate and timely positioning data. For some geomatics practitioners, the advent of this technology causes wholehearted adoption for all surveying functions. It is important to realise, though, that GNSS is only one tool within the surveyor's toolbox. Therefore, a discussion of how to make sure it is being applied correctly to the task at hand is important. This section covers GNSS and other measurement technologies such as robotic total stations and digital levels. We shall also discuss how these combined technologies can be applied most effectively to meet our total surveying requirements.

When the surveyor's toolbox is integrated properly, it is possible to optimise the tool to the application easily. The process of selection itself becomes easier. The primary concept in tool integration in the field is to provide the ability to work independently of the instrument. By *instrument* we mean surveying equipment that is used to determine position, whether in one, two or three dimensions. The predominant instruments in use are optical total stations and real-time kinematic (RTK) GNSS, but other possibilities exist.

12.2.1 Achieving Instrument Independence

Instrument independence is achieved in two principal ways by having a data processing system that can work equally well, in real-time, within the operational parameters of any instruments: the first is by using the same data collection system in the field with each instrument; the second is by using digital media to transfer the data between sensors in the field and the computing system in the office. When thoroughly integrated with the office components of a surveying system, the toolbox enables the reduction, display, and analysis of data from different instruments. It also enables the seamless flow of data between these field systems and the office components.

When the integration is complete, the survey concept keys to each technology are handled in a consistent manner so that no corruption of the data occurs. With GNSS and optical total stations for example, integration of the data necessarily requires proper conversion of data points on an ellipsoid to a plane or vice versa.

12.2.2 GNSS Technology

GNSS, especially RTK GNSS is a widely accepted technology for achieving all aspects of the mapping function. It has been shown to improve productivity over optical total station of 50% to over 1000%. Capable of being used in all kinds of weather and at night, it accelerates the progression of the work, especially in projects where work continues in two or three shifts. Because GNSS is not limited by terrestrial line of sight and work overs a 10 km range from the reference station, accuracy as well as speed is enhanced. Since data collection is initiated at the antenna pole, which is located at the point of interest, quality of collected data is greatly improved and the occurrence of blunders reduced. Use of GNSS will be restricted under some circumstances (such as in urban canyons), however, so surveyors need to have an alternative system to reliably make a measurement when GNSS fails (cannot provide desired accuracy).

12.2.3 Optical Total Station Technology

Optical total station technology is a mature technology (Anderson and Mikhail 1998), that has been in reliable use for over three decades. It requires intervisibility between instrument station locations and between the instrument stations and the points being surveyed. The range of the electronic distance meter on the total station limits range. Usability is usually limited by weather and low light or night observations. Finally, because the data collection process occurs at the instrument (not at the point to be surveyed), the correlation between the surveyed points and the coded data is susceptible to errors.

Despite of all these, optical total stations are preferred by the surveyors in many instances. They are initialised relatively quickly, are easy to use (especially when the work involves just a few observations over a limited area), and are the 'standard' against all other types of sensors.

12.2.4 SERVO-DRIVEN AND ROBOTIC OPTICAL TOTAL STATIONS

The latest refinements to an existing technology (manual total stations) are servo-driven and robotic optical total stations (Anderson and Mikhail 1998). Because these tools use motors to aim and position the instrument, servo-driven instruments are particularly appealing where automatic pointing is desired and are gaining in popularity. Modern servo-driven total stations are *auto-tracking* capable, a further enhancement to the traditional servo-driven station, enabling the instrument to lock onto a target and follow it. Auto-tracking has provided several operational improvements: since the target is followed as it moves, the rod person seldom has to wait for the instrument person; aiming and focusing are eliminated from the manual operations required to take a reading; and errors in observations due to parallax error are eliminated as well.

Phenomenal increases in accuracy have been reported for this instrument as well. When active targets are used with the trackers in traversing and other control surveying operations, standard deviations in angle measurement of ±5 mm have been reported. Furthermore, in mapping or setting out operations, multiple rod persons can be used for higher productivity. A review of auto-tracking servo-driven instrument use compared to manually operated total stations showed an approximate 110% increase in the amount of data collected (Fosburgh and Paiva 2001).

Robotic optical total stations are another class of products for surveyors (Ghilani and Wolf 2008). When a communication link is added to facilitate the placement of a data collector/controller at the rod, so that one-person surveying is possible, it is referred to as a *robotic optical total station*. Robotic instruments have the advantages of being controlled from the rod, thus the coding and quality control is done at the point being measured, which greatly improves the usability of the data.

12.2.5 IMPACT ON SURVEYING OPERATIONS

Surveying operations for mapping thus have a minimum choice of three different kinds of technologies to use as instruments: RTK (GNSS), robotic and servo-driven total stations, and manual total stations. But none of these technologies offers an ultimate solution for the surveyor. Reliance on tried-and-true principles of surveying is still required. In fact, because of different basic principles underlying GNSS as compared with total stations, more understanding of the differences is required to properly assimilate and integrate the methodologies.

In the case of RTK, it is worth remembering that vectors are still being measured, similar to those measured with total stations; since line of sight is not required, it is easy to become careless and forget this fact. The concepts of traversing, strength of figure, and designed redundancies are still important with GNSS; and in many ways, because the distances involved are typically larger than with total stations, they are more important.

These new technologies have the added advantage of a much lower learning curve to be apparently productive. However, it is still the surveyor's job to ensure that the proper experience and judgment are applied together with basic surveying principles to ensure that the integration of these technologies is producing satisfactory results.

By combining the aforementioned options into an integrated system and selecting from them, the surveyor is able to be more productive and more accurate. Selection of the appropriate sensor enables most tasks to be done with more efficiency. The new integrated surveying technologies improve effectiveness in ways never before appreciated, including safety, improved ergonomics (to improve efficiency as well as health), and reduced fatigue. Fatigue may affect safety, and accuracy, particularly with respect to blunders. Combined with the advantages provided to surveyors, the new technologies improve the flexibility of the business to respond to needs by being able to manage productivity, accuracy, and quality control.

Finally, the increased flexibility can even add choices to the surveyor. All the sensors provide methods to do topographic surveys, short range hydrographic surveys, as-built surveys, construction layout, roadways and airports, and mining surveys. Depending on personnel, their training, the nature of the site and the surrounding, the density of points to be surveyed, the difficulty of access to the points, a variety of the technologies will have to be selected.

12.3   ACCURACY STANDARDS AND SPECIFICATIONS FOR SURVEY

One of the most important functions of geodesy is the determination of the precise position of points in relation to a well-defined reference system or datum. Hence geodesy is both the science of determining the 'figure of the earth' and the set of practical tools for capturing spatial data. However, it is customary to distinguish between those techniques which are 'geodetic', capable of the most precise position determination, from those which are not able to satisfy the same stringent accuracy requirements. There is, of course, a continuous spectrum of positioning accuracy requirements from high to low. This has long been recognised by the hierarchical system used in control surveying. Geodesy provides the most fundamental control or geodetic network (or framework of physical benchmarks), which is then progressively 'broken down' or 'densified' using less accurate techniques.

Network adjustment is actually a form of least squares solution in which the observed baseline vectors are treated as 'observations' in a secondary adjustment. It may be a *minimally constrained* network adjustment (refer to Chapter 11, Section 11.5.10) with only one station coordinate held fixed, or it may be constrained by more than one fixed (known) coordinate. In minimally constrained network adjustments, only one coordinate must be held fixed to its known value while all others are allowed to adjust. Typically, GNSS surveys measure more baselines than the minimum needed to coordinate all the points in the network. These extra 'observations' are redundant information that a minimally constrained network adjustment uses to derive optimum estimates of the coordinate parameters as well as valuable quality information in the form of parameter standard deviations and error ellipses (or ellipsoids) (Anderson and Mikhail 1998).

12.3.1 CLASS/ORDER OF SURVEY

Class of survey is a means of categorising the internal quality, or precision, of a survey. The number of categories, the notation applied, and the accuracy tolerances for the transition from one class to another are defined by individual nations. Typically, they are based on traditional geodetic surveying categories, supplemented by several extra categories of higher precision applicable to GNSS surveying and geodesy techniques and may be different for horizontal and vertical surveys. The attachment of a particular class 'label' (e.g., A, B, etc.) to a survey comprising a few or many points within a network, carried out using GNSS or any other technique, is performed as part of the process of network adjustment in which the relative error ellipses (in the horizontal case) or ellipsoids (in the 3D case) between coordinated stations are computed and compared with the accuracy standards that must be met for various categories of class.

In an analogous manner to 'class of survey', *order of survey* is a means of categorising the quality, or precision, of a static survey. However, it relates to the external quality, and is influenced by the quality of the 'external' network information. Typically, they mirror the categories of 'class of survey'; hence an 'A Class' survey generally corresponds to a 'first order' survey (not essentially). The labelling of a particular order (e.g., first, second, etc.) to survey points within a 'network' (whether it is carried out using GNSS or any other technique) is performed as part of the process of network adjustment in which the relative error ellipses (or ellipsoids) between coordinated stations are computed and compared with the accuracy standards that must be met for various categories of order. However, unlike the minimally constrained network adjustment that is a prerequisite for establishing the class of survey, the network adjustment must be constrained to the surrounding geodetic control. Hence a very high-quality GNSS network (therefore a high class survey) may be distorted to 'fit' the existing control which may have been determined using a lower class survey. The resulting order of the survey would have to match the lower of either the class of the GNSS survey or the class of the existing geodetic control. If the existing geodetic control is of a lower quality to what can be achieved using modern GNSS surveying techniques, then the geodetic control network must be upgraded or 'renovated' using more precise *GNSS geodesy* techniques (Bowditch 1995).

National or state geodetic authorities are responsible for the establishment, maintenance, and densification of the control framework. The greatest of care and the most precise positioning technology is used for this purpose. As we have seen, the control framework is the physical realisation of the geodetic datum, and hence, by making measurements to/from the control stations, we ensure that all our spatial information is in a consistent coordinate frame. Such a framework of points may be given a certain quality flag, for example 'first order' or 'second order', implying a certain relative accuracy (measured in terms of ppm) (Bowditch 1995).

12.3.2 POSITIONAL ACCURACY

Positional accuracy is one spatial data quality element with two sub elements: *absolute* or *external accuracy* and *relative* or *internal accuracy*. Absolute accuracy is the correctness of survey with respect to the exact location on the earth. While referring to the accuracies of geodetic control points, we mean absolute accuracy—how accurately they are providing the coordinates of the earth surface. Accuracy in relative terms means the size of the average or maximum error in the coordinate of one object relative to another, expressed as a ratio of error magnitude to point separation. Absolute accuracy concerns the accuracy of data elements with respect to a coordinate system, e.g., UTM. Relative accuracy is an evaluation of the amount of error in determining the location of one point or feature with respect to another. For example, the difference in elevation between two points on the earth's surface may be measured very accurately, but the stated elevations of both points with respect to the reference datum could contain a large error. In this case, the relative accuracy of the point elevations is high, but the absolute accuracy is low. Spatial data quality actually refers to the relative accuracy; and in surveying, it is measured as a ratio of error to the actual distance. For example, 1 cm error in the location of two objects 1 m apart is large, but if the two objects were 200 km apart and error was 1 cm, then the position fixing technique is a high accuracy one. Hence, a 'relative' measure of positioning accuracy would be to express the ratio in terms of *parts per million* (ppm). The ppm means an amount out of a million (just as per cent means out of a hundred). Thus 1 ppm error means one unit of error in one million units (1 cm in 1 million cm or 10 km); 5 ppm means five unit errors in one million units. Positional accuracy (absolute and relative) must be reported for both horizontal and vertical components of position.

In the case of mapping, we collect point features with a few epochs instead of one. One epoch means receiver has one position solution for a particular point feature. When we have overdetermined positions (more than the minimum number of required measurements—redundant data), they create an error ellipse (in 2D) or ellipsoid (in 3D). The error ellipse/ellipsoid is an expression of confidence around the most probable correct position. The ellipse is oriented along the direction of the greatest variance and the major and minor axes (*a* and *b*) are the variances along that direction and orthogonal to it (Fig. 12.1). The final solution (coordinate of the point feature) is recorded by determining the centroid of the ellipse/ellipsoid. The processing software can usually be directed to express the confidence as a percentage, roughly defined as the likelihood that the true position falls within the ellipse (95% or 90%, for example). Generally speaking, observing the error ellipses will help us to decide when further measurement would improve confidence or when a particular station or measurement may be erroneous. A rule of thumb is 'the smaller the ellipses, the more precise the measurements or confidence of the results'. Another general rule of thumb is 'the more circular the ellipses, the more balanced the measurements'.

In general, the highest orders are used to control the spatial errors across the distances usually represented by map sheets. For example, a map sheet 0.5 m across,

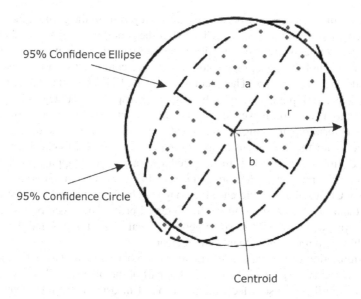

95% Confidence Ellipse

95% Confidence Circle

Centroid

FIGURE 12.1 Error ellipse.

at a scale of 1:100000, should have relative errors between two well-defined points on the map which are less than about 70 m with 90% probability/confidence if standard plotting quality criteria are used. In the case of a map at a scale of 1:10000, the relative accuracy requirement is ten times more stringent. However, the maximum separation of points may be of the order of 70 km in the former case (1:100000 scale) and 7 km in the latter case (1:10000 scale), but both imply a relative accuracy of only 1 part in 1000, well below of the level of geodetic accuracy. The geodetic network is, therefore, the very foundation of a map. However, often relative accuracy is of greater concern than the absolute accuracy. For example, a map can be accepted with the fact that their survey coordinates do not coincide exactly with the geodetic control points (absolute accuracy sacrificed); however, the error in sizes of parcels from a tax map can have immediate and costly consequences (relative accuracy sacrificed).

12.4 REMOTE SENSING AND PHOTOGRAMMETRIC CONTROL POINT

Air-borne (e.g., aeroplane or drone) and space-borne remote sensing images form the basis of many mapping applications with or without the involvement of stereo photogrammetry (see Bhatta 2020). However, remote sensing images are actually two-dimensional arrays of pixels stored in a rectangular coordinate system; they do not have any geographical coordinate. Therefore, these images are necessary to be assigned with geographical coordinates (latitude, longitude, and altitude). The process of assigning geographic coordinates to remote sensing images is called georeferencing. The georeferencing process involves identifying a few points on the image for which geographic coordinates are surveyed from the ground; these points are called *ground control points* (GCPs). After the georeferencing, we need to assess the accuracy of georeferencing. In this process we identify a few independent points on the image and their geographic coordinates are compared with the coordinates surveyed from the ground. These points are called checkpoints. The GCPs and independent check points relate to the absolute geospatial accuracy of a remotely-sensed dataset. Realise, however, that the checkpoint coordinates are also derived from some form of surveying measurement, and there is also some degree of error associated with them. It is customary to require that the GCPs/checkpoints be at least three times as accurate as the targeted accuracy of the mapping product being tested; for example, if an image product is specified to have 9 cm of horizontal error, then the GCPs used to georeference the image or checkpoints used to test the image should themselves contain no more than 3 cm of error. The check points or GCPs are surveyed with reference to some geodetic/local control points and they should themselves contain no more than 1 cm of error.

The resolution of remote sensing image is another deciding factor for the accuracy of survey; the error must be kept within half of the pixel size. The resolution of the GCP coordinates also affects the precision of measurements taken with it. For example, if we define the location of a GCP in UTM coordinates as 20,000 m east and 20,000 m north, it is actually an area of 1 m^2. A more precise specification would be 20,000.001 m and 20,000.001 m, which locates the position within an area of 1 mm^2.

NOTE

Photogrammetry is defined by the American Society of Photogrammetry (Thompson 1966) as the 'art, science, and technology of obtaining reliable information about physical objects and the environment through processes of recording, measuring, and interpreting photographic images and patterns of recorded radiant electromagnetic energy and other phenomena'. Photographs and digital images are the principal sources of information and are included within the domain of photogrammetry. It consists of making precise measurements from photographs/digital images to determine the relative location of points. This enables distances, angles, areas, elevations, sizes, and shapes of objects to be determined. The most common application of photogrammetry is the preparation of planimetric and three-dimensional topographic maps from aerial photographs or satellite images. Recent improvements in digital photogrammetry have made it so accurate, efficient, and advantageous that nowadays many mapping operations are performed in a photogrammetric environment.

12.5 INTELLIGENT MAP AND GNSS

A map is the two-dimensional representation of earth (or a part of it), a series of layers of spatial information overlaid one on the other (Fig. 12.2). An *intelligent map* may serve a certain purpose by including only those layers of spatial information

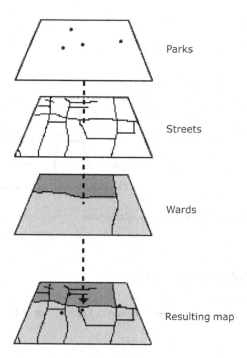

Parks

Streets

Wards

Resulting map

FIGURE 12.2   Spatial data layers overlain one on the other.

which are relevant to that application. Hence, we may have topographic maps, tourist maps, navigation maps, geological maps, census maps, political maps, and so on. Today, any kind of map can be electronically stored, and these 'digital' maps are highly customisable using anything from simple desktop mapping programs to sophisticated computer mapping software.

In Geographic Information System (GIS), special 'intelligence' is provided by attribute data linked to the spatial data and sophisticated query language to create digital maps (Bhatta 2020). GIS maps are able to support decision-making because of the ability to match the spatial characteristics of the data, such as position and topology of points, lines and polygons (i.e., how spatial features are connected or related to each other) with other descriptive data about the spatial objects. This auxiliary descriptive data about points, lines, and polygons on a map is part of the attribute data (Fig. 12.3).

Each layer of a map may have a set of global attribute data specifying, for example, the origin of the data set, when it was last updated, the purported accuracy of the data, how it was captured, etc. This is called *metadata* (Bhatta 2020). However, attribute data for specific features or points in a map (or map layer) are generally stored in a database table. The design of the table is generally different for different features. For example, a tree may have attribute data linked to it concerning species, height, girth, etc., while the attribute data linked to a road segment will have different entries such as width, date of construction, surface type, surface quality, etc. The

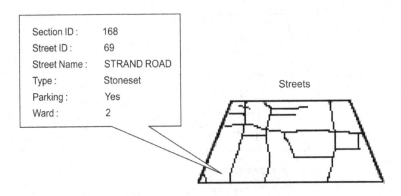

FIGURE 12.3   Spatial data linked with attribute data.

TABLE 12.1
Sample Attribute Table for a Tree

ID (as recorded by the receiver)	Species	Height	Girth	Age	Health Condition

---

ability to access this additional data simply by pointing to the map feature on the computer screen makes GIS a powerful query tool. Other queries could be 'display where such-and-such is located'. More complex questions such as 'where does a road A cross river B?' require the interrogation of more than one map layer. Therefore, survey of GIS map features is different than the traditional mapping survey. In the case of GIS map feature survey, we not only collect the spatial feature but also collect characteristics of the features. Low-cost receivers may not offer the ability to collect attribute data. In such cases, we use a hard copy observation log (refer to Chapter 11, Section 11.4) to collect attribute data; a process of writing down the ID of the feature (as recorded by the receiver) and the other attributes (as shown in Table 12.1. Later, this table can be encoded (entered) in the GIS. However, this inefficient approach can be replaced by designing a database in the office and transferring it to the receiver so that surveyor can enter attributes while collecting the data. Surveyors can take photos, shoot videos, record voices, and attach them to a surveyed feature. Advanced receivers offer this and other functions, like how many epochs must be collected to record a point, epoch rate, the time or distance interval for kinematic/dynamic linear feature survey, and so on. These functions help us to maintain consistency in survey if multiple devices and surveyors are involved.

The GIS map features are points, lines, and polygons. Points are used to represent features that are too small to be represented as areas; an example is a postbox. Lines are used to represent features that are linear in nature, for example, roads or pipelines. Areas are represented by a closed set of lines and are used to define features such as playgrounds, ponds, buildings, or administrative areas. For line and polygon features, we must use kinematic/dynamic survey; however, point features are also collected based on the same principle but different technique. We collect multiple epochs to record one point; 30–40 epochs at 1 sec rate for code-based solution and 10–15 epochs at 1 sec rate for carrier solution. This will create an error ellipse, the centroid of which is the solution. Line and polygon features may also be collected as a series of points (with multiple epochs); the receiver's software can connect these points to generate lines or polygons, as instructed.

GIS maps are an integral part of GNSS navigation. These can impact navigation applications in a number of ways. For example, given an 'intelligent' spatial dataset comprising of many layers (road network, topography, watercourses, property boundaries, etc.), with associated attribute data, it is possible to create the specialised maps which are central to the in-vehicle navigation system. Firstly, only those layers are selected which are relevant to road users, and those features of significance to road users may be highlighted in some way. These could be fuel stations, traffic lights, car parking, restaurants, etc. Such a customised map may then be converted into some convenient form for display within a vehicle. Such datasets are the electronic form of the street directory, and many of them have been developed to have almost the same look and feel of the paper equivalent.

A more sophisticated approach to compiling the vehicle navigation map is to go beyond just producing an electronic version of a paper map but to also make the attribute data available to the driver. Hence, in addition to compiling the spatial data in the manner described earlier, the attribute data is compressed into succinct descriptive information. For example, the name of the restaurant is included, the car park entrance is described, the brand of fuel/lubricants sold at the petrol station is included, etc. If the basic map data (spatial and attribute) have been collected carefully, and it is both current and comprehensive, then this process is not too difficult. However, general GIS map layers do not include some attribute data which are useful to drivers, such as which streets carry one-way traffic (and in which direction), the weight and speed restrictions on roads, where right turns (or left turns) are prohibited, which streets have median strips, and so on. The surveyor should be very careful regarding these features of transportation; a transportation engineer should be associated with the survey team in such cases. If these data are not available immediately in some cases, they will have to be obtained later in some way, and then regularly updated.

12.6   MAP-AIDED POSITIONING AND NAVIGATION

Digital maps are the basis of many navigation applications besides locating the vehicle in a map reference frame (Krakiwsky and Bullock 1994; Drane and Drane 1998). However, the accuracy of vehicle location depends on two primary things: (1) accuracy, correctness, and completeness of map (responsibility of the surveyor) and (2) accuracy of the navigation system's GNSS receiver. If both of the accuracies are sacrificed, it could become dangerous, even fatal for the drivers. There are several functions that digital road map data can perform to improve the reliability in the context of navigation and positioning applications:

Address-Matching. This function transforms receiver determined latitude and longitude into a street address, or vice versa. We know the address of our destination rather than its coordinates (lat/long). Hence, to navigate to an address requires that the positioning system be able to unambiguously convert the street address to a coordinate and vice versa. This would require that the combined accuracy (map and receiver) of the navigation system be better than 20 m. If we consider average accuracy of a navigation receiver is 10 m, the accuracy of the map must be kept within 10 m.

Map-Matching. This technique assumes the navigating vehicle is on a road. When a navigation system's receiver outputs coordinates that are not on a road segment as defined by the digital road map, the map-matching algorithm finds the nearest segment and 'snaps' the vehicle onto the road segment. Obviously map-matching requires maps with high positional accuracy to minimise incorrect road segment selections. The map-matching algorithm requires a minimum 10 m accuracy of the map.

Best-Route Calculation. This function supports the driver in planning the optimal travel route. A digital road map coupled with a best-route calculation algorithm provides an optimal route based on travel time, distance, or some other specified criteria. The result is a turn-by-turn description of a journey. Best-route calculation requires a high level of map information, for example, the digital road map must include traffic and turn restrictions so that the route selected is not illegal or dangerous. In this case, the accuracy of attribute is more important than spatial accuracy.

Route Guidance. This system supports drivers as they navigate along a route (selected by the driver, or by the best-route algorithm). Route guidance includes turn-by-turn instructions, street names, distances, intersections, and landmarks. This is particularly challenging in real-time because the algorithm must process position information, perform address and map-matching, and display the digital road map to the driver. A wrong road name in the attributes is never expected. If a road has an alternate name, it must be collected and entered; because the driver may know the alternate name only. If the driver misses a turn, the system must be able to compute a new best-route immediately and provide new guidance information.

Each of the aforementioned functions relies on specific features in the digital road map database; however, the most demanding are the last two: best-route calculation and route guidance. If a digital road map supports these functions it is said to be *navigable* (Krakiwsky and Bullock 1994). The digital road map database accuracy requirement would be similar for all the map-aided applications listed earlier; however, navigable digital road map databases are more complex because of the extensive additional information that must be stored. Hence, navigable digital road map databases are significantly more expensive to produce and maintain. The most sophisticated navigation map functions can therefore be supported only when accurate, complete, and seamless digital road map databases are available.

12.7   SCALE, DETAIL, ACCURACY, AND RESOLUTION OF MAP

Map scales may be confusing to those who are not trained in the basics of cartography or simply cannot 'read' a map. Earth features cannot be represented in their actual size. Reduction is necessary to display a representation of the earth's surface on a map. The scale of map refers to the ratio of distance on a map to the corresponding distance on the ground. The scale indicates exactly how much smaller than reality the map is. Map scale is the relation between the dimensions of a map and the dimensions of the earth. For example, 2 cm distance on the map corresponds to 1 km distance on the ground. Hence, the map scale can be calculated as follows:

$$\frac{\text{Distance on the map}}{\text{Distance on the ground}} = \frac{2\ \text{cm}}{1\ \text{km}} = \frac{2\ \text{cm}}{100000\ \text{cm}} = \frac{1}{50000} = 1:50000$$

From this example, on a 1:50000 scale map; 1 cm of map distance corresponds to 50,000 cm of ground distance. Since the scale ratio is a constant, it is true for whatever units in which the fraction is expressed. For example, maps designed at the ratio 1:10000 implies that 1 of any unit of measurement on the map corresponds to 10,000 of that same unit in reality.

Since map scale is a ratio, 1:10000 is larger than 1:50000 and the map with 1:10000 is a larger-scale map than a 1:50000 map. A large-scale map is one in which a given part of the earth is represented by a large area on the map. Large-scale maps generally show more detail than small-scale maps because, at a large scale, there is more space on the map to show features. Large-scale maps are typically used to show site plans, local areas, neighbourhoods, towns, and cities. A small-scale map is one in which a given part of the earth is represented by a small area on the map. Small-scale maps generally show less detail than large-scale maps, but cover larger parts of the earth. Maps with regional, national, and international extents typically have smaller scales. When we zoom-in on interactive map display software, the scale of the view becomes larger, and when we zoom-out the scale becomes smaller.

Level of detail on a map refers to the quantity of geographic information shown on the map. Large-scale maps typically show more details than small-scale maps. There is no standard rule for how many features and how much detail a map of a given scale should show. This is a cartographic decision depending on the purpose of the map and how many symbols can be drawn in the available space without visual clutter. On smaller-scale maps, there is simply not enough room to show all the available details, so features such as streams and roads often have to be represented as single lines, and area features like cities have to be shown as points. This is called generalisation. A surveyor must consider the map scale and level of detail before planning a survey.

Accuracy is defined as the closeness of measurements or estimates by computation to true values. Accuracy is generally represented by standard deviation of errors, which is the difference between measurements and the true value. The accuracy of a map generally depends on the accuracy of the original data used to compile the map, how accurately this source data has been transferred into the map, and the resolution at which the map is printed or displayed. However, it should be borne in mind that, as scales become smaller, one unit of map distance represents a very large distance on the ground. So, if one of the features shown on a very small-scale map is out of line by just a tiny part of a millimetre, it would still represent a substantial inaccuracy in reality.

The accuracy of the maps we create with GIS software depends primarily on the quality of the coordinate data in our spatial database. To create the spatial data, existing printed maps may have been digitised or scanned, and other original data, such as survey coordinates, aerial photographs and images, and data from third parties may have also been used. Our final map will therefore reflect the accuracy of these original sources. However, if we integrate the data from various sources they may not match properly because different data generation methods were not designed to be consistent with the errors. If the survey is being conducted to collect part of the features of a map, the surveyor must conduct a sample survey to check how they match with the other features of the map; otherwise, the entire survey data may need to be rejected due to low relative accuracy.

There are several aspects to map accuracy: (a) the accuracy with which the coordinates of a feature can be determined, and (b) the accuracy with which the coordinates of a feature must be known if it is to be correctly displayed on a map at a certain scale. Examples of varying accuracies with which map features can be surveyed are as follows (Drane and Drane 1998):

Control mark	1 cm
Corner of building	10 cm
Street intersection	1 m
River boundary	2–5 m
Forest boundary	10 m
Soil type demarcation boundary	50 m

The *resolution* of a map determines how accurately the features on the map can be depicted for a given scale. Resolution depends on the physical characteristics of the map, how it was made, what kinds of symbols have been used, and how it has been printed or displayed on-screen. For example, let us imagine we are making a map showing property boundaries at a scale of 1:50000 using a line symbol that prints out to be 0.5 mm wide. The width of this line symbol represents a corridor on the earth, which is almost 25 m ($0.5 \times 50,000$ mm = 25,000 mm) wide. Now, let us consider the problem—a property boundary of 25 m wide; no one would accept that. The resolution of that line symbol on a map that scales clearly has some impact on the map's level of detail and accuracy. However, from this discussion, it is clear that it would not be possible to recognise the positional inaccuracy of the property boundary if it is drawn with an error of up to 25 m (±12.5 m) if the map is plotted in 1:50000 scale.

Hence, depending on the map scale used to depict spatial data, accuracies may be either higher or lower than the plotting accuracy (plotting resolution). Map plotting accuracy is usually specified by requirements such as that 90% of 'well-defined' points should be within 0.5 mm of their correct position at map scale. This translates to the following accuracies (of the smallest plottable feature) (Drane and Drane 1998):

Map Scale	Accuracy Needed
1:1000	0.5 m (±0.25 m)
1:10000	5.0 m (±2.5 m)
1:25000	12.5 m (±6.25 m)
1:50000	25.0 m (±12.5 m)
1:100000	50.0 m (±25.0 m)

Similar to the plotting resolution, the resolution of the computer (or receiver's display) screen in use has an impact on the detail and accuracy of maps that we display on that screen. The screen uses pixels to draw the map and cannot draw features or parts of features that are less than a pixel wide. A map that is based on highly accurate spatial data will lose accuracy if printed or viewed on a low-resolution device. It should be remembered that the resolution of printers and plotters are generally higher than that of computer screens.

However, in a digital GIS database, there is no concept of scale but rather resolution expressed as pixel size (interval or pixels per inch), grid cell size or grid interval, ground (spatial) resolution for satellite images, and so on. Electronic vector maps rely on databases of object coordinates, and must be drawn on the computer screen 'on the fly'. The map scale may vary considerably through the process of 'zooming', and, when an area originally mapped at a relatively small scale is zoomed and displayed at a larger scale, the accuracy of the feature's position (based on the quality of the original survey) may no longer satisfy the plotting accuracy requirements. A surveyor must be aware of these issues and plan the survey accordingly.

EXERCISES

DESCRIPTIVE QUESTIONS

1. What do you understand by 'integration of surveying tool' and 'instrument independence'?
2. How do we integrate RTK, robotic & servo-driven total stations, and manual total stations in mapping applications?
3. Explain accuracy standards and specifications for mapping survey.
4. Explain class and order of survey.
5. What do you understand by 'positional accuracy' of a map? How is the mapping survey influenced by this?

6. Discuss 'remote sensing' and 'photogrammetric control point' survey.
7. Discuss the considerations of survey for an intelligent map.
8. What do you understand by 'map-aided positioning and navigation'? Explain briefly in the context of mapping survey.
9. Discuss the considerations of survey in the light of scale, level of detail, accuracy, and resolution of map.
10. What are the considerations to be followed for the survey of attribute data if the map is being prepared for navigation? What should the accuracy of a navigation map be?

SHORT NOTES/DEFINITIONS

1. Total station
2. Class of survey
3. Absolute accuracy of map
4. Relative accuracy of map
5. Parts per million
6. Error ellipse
7. Photogrammetry
8. Georeferencing
9. GCP
10. Electronic map
11. Attribute data
12. Address-matching
13. Map-matching
14. Map scale

Glossary

Absolute positioning: Mode in which a position is determined, using a single receiver, with respect to a well-defined coordinate system, typically a Geocentric system (i.e., a system whose point of origin coincides with the centre of mass of the earth). Also referred to as point positioning, single receiver positioning, standalone positioning, and navigation solution.

Accuracy: How close a fix comes to the actual position; a measure of how close an estimate of a GNSS position is to the true location.

Acquisition: The ability to find and lock on to satellite signals for ranging.

Albedo: The proportion of radiation reflected by a body compared to the amount incident upon it.

Algorithm: A special method used to solve a certain type of mathematical problem.

Almanac (GNSS): *See almanac data.*

Almanac (nautical): Also called **air almanac** A publication describing the positions and movements of celestial bodies for the purpose of enabling navigators to use celestial navigation.

Almanac data: Information, transmitted by each satellite, on the orbits and state (health) of every satellite in a GNSS constellation. Almanac data allows the GNSS receiver to rapidly acquire satellites shortly after it is turned on.

Almanac: An annual publication containing tabular information in a particular field or fields, often arranged according to the calendar.

Altitude: Elevation above or below a reference datum; the z-value in a spatial address.

Ambiguity resolution: If the initial integer ambiguity value for each satellite–receiver pair could be determined, then the ambiguous integrated carrier phase measurement can be corrected to create an unambiguous, and very precise (millimetre level observation accuracy), receiver–satellite distance measurement. A solution using the corrected carrier phase observations is known as an 'ambiguity-fixed' or 'bias-fixed' solution. The mathematical process or algorithm for determining the value for the ambiguities is *ambiguity resolution*. Tremendous progress has been made in ambiguity resolution techniques, making today's carrier phase-based GNSS systems very efficient by reducing the length of observation data needed and even allowing this process to occur while the receiver is itself in motion.

Ambiguity: Carrier phase measurements can only be made in relation to a cycle or wavelength of the carrier waves because it is impossible to discriminate different carrier cycles (they are all 'sine waves' if one ignores the modulated messages and PRN codes). Integrated carrier phase measurements may be made by those receivers intended for carrier phase-based positioning. In this case, the change in receiver–satellite distance can be measured by counting the number of whole wavelengths since initial signal lock-on and adding the instantaneous fractional phase measurement. However, such a measurement is a biased range or distance measurement because the initial number of whole (integer) wavelengths in the receiver–satellite distance is unknown. This unknown value is referred to as the 'ambiguity'. It is different for the different satellites, and different for different frequencies. It is, however, a constant if signal tracking continues uninterrupted through an observation session. If there is signal blockage, then a 'cycle slip' occurs, causing the new ambiguity after the cycle slip to be different from the value before. Cycle slip repair therefore restores the continuity of carrier cycle counts and ensures that there is only one ambiguity for each satellite–receiver pair.

Amplitude Modulation (AM): A method of encoding a message on the carrier signal by altering the height or amplitude of the signal while keeping its frequency constant.

Amplitude: Height of a radio wave as measured from an imaginary centre line to the wave peak. The maximum displacement of a peak from zero.

Antenna: That part of the GNSS receiver hardware which receives (and sometimes amplifies) the incoming signal. Antennas come in all shapes and sizes, but most these days use so-called 'microstrip' or 'patch' antenna elements. The geodetic antennas, on the other hand, may use a 'choke-ring' to mitigate any multipath signals.

Anti-Spoofing (AS): Is a policy of the US Department of Defense by which the P code is encrypted (by the additional modulation of a so-called 'W code' to generate a new 'Y code'), to protect the militarily important P code signals from being 'spoofed' through the transmission of false GPS signals by an adversary during times of war. Hence, civilian GPS receivers are unable to make direct P code pseudorange measurements.

Anywhere fix: A receiver's ability to achieve lock-on without being given a somewhat correct beginning position and time.

Apogee: The point of a satellite's orbit where the satellite is farthest from the earth.

Arcdegree (degree): A unit of angle; 1/360 of a circle.

Arcminute (minute): A unit of angle; 1/60th of an arcdegree.

Arcsecond (second): A unit of angle; 1/60th of an arcminute, or 1/3600th of a degree.

Atomic clock: A clock regulated by the resonance frequency of atoms or molecules. In GPS satellites, the substances used to regulate atomic clocks are caesium, hydrogen, or rubidium. *See also Caesium clock and Hydrogen maser.*

Attribute: Information about features of interest. Attributes such as date, size, material, colour, and so forth are frequently recorded during data collection.

Augmentation: Additional changes by which something increases, e.g., positioning accuracy.

Autocorrelation: *See code correlation.*

Availability: Defines that a particular location has sufficient satellites (above the specified elevation angle, and perhaps less than some specified DOP value) to make a GNSS position determination possible. It is the period, expressed as a percentage, when positioning from a particular system is likely to be successful. *See also outage.*

Azimuth: A direction in terms of a 360° radial measurement, measured along a horizontal plane in clockwise direction.

Band: A wavelength (or frequency) interval in the electromagnetic spectrum.

Baseline: A Baseline consists of a pair of stations for which simultaneous GNSS data have been collected. Mathematically expressed as a vector of coordinate differences between the two stations, or an expression of the coordinates of one station with respect to the other.

Base station: Also called a *reference station.* In GNSS navigation, this is a receiver that is set up on a known location specifically to collect data for differentially correcting data files of another receiver (which may be referred to as the 'mobile' or 'rover' receiver). In the case of pseudorange-based differential GNSS, the base station calculates the error for each satellite and, through differential correction, improves the accuracy of GNSS positions collected at unknown locations by another (roving) receiver. For GNSS surveying techniques, the receiver data from the base station is combined with the data from the other receiver to form double-differenced observations, from which the baseline vector is determined.

Baud rate: The transmission rate at which data flows between transmitter to receiver.

Bearing: The compass direction from a position to a destination. The 'north' direction is 'zero bearing', and the angle is measured clockwise through 360°.

Beat frequency: Also known as *intermediate frequency.* When two signals of different frequencies are combined, two additional frequencies are created. One is the sum of the two original frequencies. One is the difference of the two original frequencies. Either of these new frequencies can be called a *beat frequency.*

Beat: The pulsation resulting from the combination of two waves with different frequencies. It can occur when any pair of oscillations with different frequencies are combined. In GNSS, a beat is created when a carrier generated in a GNSS receiver and a carrier received from a satellite are combined.

BeiDou navigation satellite system: Chinese independent satellite navigation system. It is a truly global system consisting of 35 satellites. It passed through several developmental phases with different names (BeiDou-I, BeiDou-II, and now BeiDou-III; BeiDou-II was also known as Compass).

Between-epoch difference: The difference in the phase of the signal on one frequency from one satellite as measured between two epochs observed by one receiver. Because a GNSS satellite and a GNSS receiver are always in motion relative to one another, the frequency of the signal broadcast by the satellite is not the same as the frequency received. Therefore, the fundamental Doppler observable in GNSS is the measurement of the change of phase between two epochs.

Between-receiver single difference: Also called *between-receiver difference.* The difference in the phase measurement between two receivers simultaneously observing the signal from one satellite on one frequency. For a pair of receivers simultaneously observing the same satellite, a between-receiver single difference pseudorange or carrier phase observable can virtually eliminate errors attributable to the satellite's clock. When baselines are short, the between-receiver single difference can also greatly reduce errors attributable to orbit and atmospheric discontinuities.

Between-satellite single difference: Also called *between-satellite difference.* The difference in the phase measurement between signals from two satellites on one frequency simultaneously observed by one receiver. For one receiver simultaneously observing two satellites, a between-satellite single difference pseudorange or carrier phase observable can virtually eliminate errors attributable to the receiver's clock.

Bias: A systematic error. Biases affect all GNSS measurements and hence the coordinates and baselines derived from them. Biases may have physical origins, such as satellite orbits, atmospheric conditions, clock errors, and so forth. They may also originate from less than perfect control coordinates, incorrect ephemeris information, and so on. Modelling is one method used to eliminate or at least limit the effect of biases.

Binary Phase Shift Keying (BPSK): Also called *bi-phase shift keying, binary bi-phase modulation,* or *binary shift-key modulation.* The phase modulation technique used to transmit a GNSS signal. The phase of a carrier generated by a satellite is shifted by 180 degrees when there is a code or message binary signal level transition, either from 0 to 1 (normal to mirror image) or from 1 to 0 (mirror image to normal). *See also quadrature phase-shift keying.*

Binary: Numerical system using the base 2.

Bit: Contraction of binary digits (0s and 1s) which, in digital computing, represents an exponent of the base 2.

Block I, II, IIR, IIR-M, IIF, and IIIA satellites: Classification of the GPS satellite generations. Block I satellites were officially experimental. There were 11 Block I satellites. PRN4, the first GPS satellite, was launched on 22 February 1978. The last Block I satellite was launched on 9 October 1985, and none remain in operation. Block IIR and IIR-M satellites are operational satellites. Block IIF and IIIA will be the follow-on series of satellites.

Bluetooth: A wireless protocol utilizing short-range communications technology facilitating data transmission over short distances from fixed and/or mobile devices.

Blunder: Blunders are fundamental errors in the surveying process, usually caused by human error. They are also mistakes made during taking readings of, transcribing, or recording survey data.

Broadcast ephemeris: *See ephemeris.*

Caesium clock: Also known as *Caesium Frequency Standard* or *caesium oscillator,* it is an atomic clock that is regulated by the element caesium. Caesium atoms, specifically atoms of the isotope Cs-133, are exposed to microwaves, and the atoms vibrate at one of their resonant frequencies. By counting the corresponding cycles, time is measured.

Caesium: A soft silver-white ductile metallic element (liquid at or near room temperature); used in highly accurate atomic clocks.

Canada-wide DGPS (CDGPS): A satellite-based WAAS that was developed for the Canadian positioning market. It provides accuracy and coverage for positioning applications throughout Canada. *See also satellite-based augmentation system.*

Carrier beat phase: The phase of a beat frequency created when the carrier frequency generated by a GNSS receiver combines with an incoming carrier from a GNSS satellite. The two carriers have slightly different frequencies as a consequence of the Doppler shift of the satellite carrier compared with the nominally constant receiver generated carrier. Because the two signals have different frequencies, two beat frequencies are created. One beat is the sum of the two frequencies. One is the difference of the two frequencies. *See also beat frequency; Doppler shift.*

Carrier frequency: The frequency of an unmodulated output of a radio transmitter. The unmodulated frequencies of GNSS are called carrier frequencies, as its role is mainly to carry the information that is going to be the modulating data.

Carrier phase: Also known as *reconstructed carrier phase.* (1) The term is usually used to mean a GNSS measurement based on the carrier signal itself, rather than measurements based on the codes modulated onto the carrier wave. (2) Carrier phase can also mean 'a part of a full carrier wavelength'. A part of the full wavelength may be expressed in a phase angle from 0° to 360°, a fraction of a wavelength, cycle, etc.

Carrier signal: A radio-wave signal that conveys or carries information by means of modulation.

Carrier tracking loop: A feedback loop that a GNSS receiver uses to generate and match the incoming carrier wave from a satellite. *See also code tracking loop.*

Carrier: Also called *carrier wave.* An electromagnetic wave, usually sinusoidal, that can be modulated to carry information. Common methods of modulation are frequency modulation and amplitude modulation; however, in GNSS, the phase of the carrier is modulated to carry information.

Celestial navigation: Finding one's position in unknown territories for navigational purposes, where the sun, moon, and stars are used as points of reference. Celestial navigation is the process whereby angles between objects in the sky and the horizon are used to locate one's position on the globe.

Celestial object: A naturally occurring object in space such as a star, planet, etc.

Channel: A channel of a receiver consists of the circuitry necessary to receive the signal from a single satellite on one of the two carrier frequencies. A channel includes the digital hardware and software.

CHAYKA: A radio navigation system developed by the former Soviet Union; it is similar to American LORAN.

Checkpoint: A reference point from an independent source of higher accuracy used in the estimation of the positional accuracy of a data set.

Chip: *See code chip.*

Chipping rate: Also called *code frequency.* In a GNSS the rate at which chips, binary 1s and 0s, are produced. For example, 10.23 MHz for GPS P code.

Circular Error Probable (CEP): A statistical measure of the horizontal precision. The CEP value is defined as a circle of a specified radii that encloses 50% of the data points. Thus, half the data points are within a 2D CEP circle and half are outside the circle.

Class of survey: A means of categorising the internal quality, or precision, of a survey.

Clock bias: The difference between the receiver or satellite clock's indicated time and a well-defined time scale reference such as UTC or TAI.

Clock dithering: *See dithering.*

Clock drift: Refers to several related phenomena where a clock does not run at the exact right speed compared to another clock. That is, after some time the clock 'drifts apart' from the other clock.

Coarse Acquisition (C/A) code: Also known as *civilian access code.* The GPS and GLONASS signal for civilian use (also used by the military to get an initial fix).

Code chip: Each zero or one of the binary code. The term 'chip' is used instead of 'bit' to show that it does not carry any pieces of information. A PRN code consists of a sequence of chips.

Code correlation: The foundation of pseudoranges is the correlation of code carried on a modulated carrier wave received from a GNSS satellite with a replica of that same code generated in the receiver. This technique is called as code correlation.

Code Division Multiple Access (CDMA): A method whereby many radio signals use the same frequency, but each one has a unique code. Both GPS and Galileo use CDMA techniques with codes for their unique cross-correlation properties.

Code frequency: *See chipping rate.*

Code phase: Measurements based on the C/A code rather than measurements based on the carrier waves. Sometimes used to mean pseudorange measurements (by means of both C/A- and P code) expressed in units of cycles.

Code state: Transformed code chips from 0 and 1 to +1 and −1, respectively.

Code tracking loop: A feedback loop used by a GNSS receiver to generate and match the incoming codes, C/A or P codes, from a satellite. *See also carrier tracking loop.*

Code: Also called *coded information.* Used to define navigation codes and ranging codes.

Cold start: The power-on sequence where the GNSS receiver downloads almanac data before establishing a position fix.

Commercial Service (CS): The encrypted positioning service of Galileo; will be available at a cost and will offer an accuracy of better than 1 m. The CS can also be complemented by ground stations to bring the accuracy down to less than 10 cm.

Constellation: (1) The space segment or all satellites of a GNSS in orbit. (2) The particular group of satellites used to derive a position. (3) The satellites available to a receiver at a particular moment.

Continuous kinematic: Also known as *pure kinematic*. Type of GNSS positioning where at least one GNSS receiver continuously tracks GNSS satellites while moving with the objective of providing the position of the trajectory of the receiver.

Continuous-tracking receiver: *See multichannel receiver.*

Control network: Also known as *control*. A control network, or simply *control*, is a set of reference points of known geospatial coordinates. The higher-order (high precision, usually millimetre-to-decimetre on a scale of continents) control points are normally defined in both space and time using global or space techniques and are used for lower-order points to be tied into. The lower-order control points are normally used for engineering, construction, and navigation. The scientific discipline that deals with the establishing of coordinates on points in a high-order control network is called *geodesy*, and the technical discipline that does the same for points in a low-order control network is called *surveying*.

Control point: Also called a *control station*. A monumented point to which coordinates have been, or are being, assigned by the use of surveying observations.

Control segment: A worldwide network of GNSS monitoring and upload telemetry stations around the world. The tracking data is used by the master control station to calculate the satellites' positions (or 'broadcast ephemeris') and their clock biases. These are formatted into the navigation message which is uploaded periodically by the uploading stations.

Coordinate system: Also called *referential system*. A particular reference frame or system, such as plane rectangular coordinates or spherical coordinates, that uses linear or angular quantities to designate the position of points used to represent locations on the earth's surface relative to other locations or fixed references.

Coordinated Universal Time (UTC): International Atomic Time (TAI) with leap seconds added at irregular intervals to compensate for the earth's slowing rotation. Leap seconds are used to allow UTC to closely track the mean solar time at the Royal Observatory, Greenwich. *See also International Atomic Time.*

Coordinates: Linear or angular quantities that designate the position of a point in a given reference or grid system.

Compass: *See BeiDou Navigation Satellite System.*

Cospas-Sarsat: An international satellite-based search and rescue distress alert detection and information distribution system, established by Canada, France, the United States, and the former Soviet Union in 1979.

Course Deviation Indicator (CDI): A technique for displaying the amount and direction of cross-track error. *See also cross-track error.*

Course Made Good (CMG): The bearing from our starting point to our present position, commonly used in marine or air navigation.

Course Over Ground (COG): Our direction of movement relative to a ground position, used in navigation.

Course To Steer (CTS): The heading we need to maintain in order to reach a destination, used in navigation.

Course Up Orientation: Fixes the receiver's map display so the direction of navigation is always 'up'.

Course: The direction from the beginning landmark of a route to its destination (measured in degrees, radians, or mils); also, the direction from one route waypoint to the next waypoint in the route segment.

Crest: A point on a wave with the greatest positive value or upward displacement in a cycle.

Cross-Track Error (XTE/XTK): The distance we are off a desired course in either direction, commonly used in marine or air navigation.

Crystal oscillator: Also called *quartz crystal oscillator.* An electronic circuit that uses the mechanical resonance of a vibrating crystal of piezoelectric material to create an electrical signal with a very precise frequency. This frequency is commonly used to keep track of time (as in quartz wristwatches), to provide a stable clock signal for digital integrated circuits, and to stabilize frequencies for radio transmitters/receivers (such as in GNSS receivers).

Cutoff angle: Also referred to as *mask angle.* The elevation angle below which GNSS signals are not recorded due to an option set in the GNSS receiver or GNSS processing software.

Cycle ambiguity: Also known as *integer ambiguity.* When the number of whole wavelengths (the integer number of wavelengths) between a particular receiver and satellite is initially unknown in a carrier phase measurement, it is called the *cycle* (or *integer*) *ambiguity.*

Cycle slip: A discontinuity of an integer number of cycles in the carrier phase observable. A jump of a whole number of cycles in the carrier tracking loop of a GNSS receiver, usually the result of a temporarily blocked GNSS signal. A cycle slip causes the cycle ambiguity to change suddenly. The repair of a cycle slip includes the discovery of the number of missing cycles during an outage of signal.

Cycle: One cycle is a complete sequence of values, as from crest to crest in a wave (known as a *wave cycle*).

Data logger: Also known as *data recorder* and *data collector.* A data entry computer, usually small, lightweight, and often handheld. A data logger stores information supplemental to the measurements of a GNSS receiver.

Data message: Another term for *navigation message. See also navigation code.*

Data rate: *See epoch rate.*

Datum: A datum is a reference from which measurements are made. In surveying and geodesy, a datum is a reference point on the earth's surface against which position measurements and an associated model of the shape of the earth for computing positions are made. This is called geodetic datum. In geodesy we consider two types of datum, horizontal datum and vertical datum. *See also geodetic datum.*

Dead-reckoning: The technique of determining position by computing distance travelled on a given course. Distance travelled is determined by multiplying speed by elapsed time. The principle of a dead-reckoning system is the relative position fixing method, which requires knowledge of the location of a vehicle and its subsequent speed and direction (for example, the last position and velocity determination) in order to calculate its present position.

Demodulation: Separating coded data from the carrier signal.

Desired Track (DTK): The compass course between the 'from' and 'to' waypoints, used in navigation.

Differencing: Represents several types of simultaneous solutions of combined measurements.

Differential GNSS (DGNSS): (1) The term *DGNSS* is less commonly used to describe relative positioning. In this context it refers to more precise measurement to determine the relative positions of two receivers tracking the same GNSS signals in contrast to absolute or point positioning. *See also point positioning.* (2) DGNSS is more widely used in the practical field to describe an augmented extension of the GNSS system that uses land-based radio beacons to transmit position corrections to GNSS receivers based on pseudorange observable. One such implementation is DGPS, which reduces the effect of selective availability, propagation delay, etc. and can improve position accuracy to better than 10 m. *See also differential GPS.*

Differential GPS (DGPS): A method that improves GPS pseudorange accuracy. A GPS receiver at a base station, a known position, measures pseudoranges to the same satellites at the same time as other roving GPS receivers. The roving receivers occupy unknown positions in the same geographic area. Occupying a known position, the base station receiver finds corrective factors that can either be communicated in real-time to the roving receivers, or may be applied in post-processing.

Differential positioning: Precise measurement of the relative positions of two receivers tracking the same GNSS signals. May be considered synonymous with DGNSS. *See also relative positioning.*

Dilution Of Precision (DOP): An indicator of satellite geometry for a unique constellation of satellites used to determine a position. Positions tagged with a higher DOP value generally constitute poorer measurement results than those tagged with lower DOP. There are a variety of DOP indicators, such as GDOP (Geometric DOP), PDOP (Position DOP), HDOP (Horizontal DOP), VDOP (Vertical DOP), TDOP (Time DOP), etc. HDOP and VDOP are the uncertainty of a solution for positioning into its horizontal and vertical directions, respectively. When both horizontal and vertical components are combined, the uncertainty of three-dimensional positions is called PDOP. TDOP indicates the uncertainty of the clock. GDOP is combined component of PDOP and TDOP. RDOP includes the number of receivers, the number of satellites the receivers can handle, the length of the observing session, as well as the geometry of the satellites' configuration.

Dispersive medium: A medium in which different frequencies exhibit various behaviours. This is the case when the mathematical expression of the phenomenon observed is not a linear function of the parameter being considered.

Dispersive: That which becomes dispersed; that which causes dispersion. Different frequencies of electromagnetic waves behave (bend) differently in a dispersive medium.

Dithering: Also called *clock dithering*. Dithering is the introduction of digital noise into a system. Clock dithering is the process by which the US Department of Defense degrades the accuracy of the Standard Positioning Service (i.e., absolute positioning using C/A). It is the additional satellite clock bias induced in GPS by the 'selective availability' policy that cannot be corrected for by the broadcast navigation message clock correction parameters.

Doppler shift: Also called *Doppler Effect*. The apparent change in the frequency of a signal caused by the relative motion of the transmitter and receiver. In GNSS, a systematic change in the apparent frequency of the received signal caused by the motion of the satellite and receiver relative to one other.

Doppler-aiding: Describes a signal processing strategy that uses a measured Doppler shift to help the receiver smoothly track the GNSS satellite signal. This allows for more precise velocity and position determination, especially when the receiver is moving at high speed and/or in an erratic fashion.

Double difference: A GNSS observable formed by arithmetically differencing carrier phases that are simultaneously measured by a pair of receivers tracking the same pair of satellites. First, the phases obtained by each receiver from the first satellite are differenced. Second, the phases obtained by each receiver from the second satellite are differenced. And third, those differences are differenced. This procedure removes essentially all of the satellite and receiver clock errors. Although primarily used with carrier phases, the procedure can also be applied to pseudoranges.

Drift: A general tendency to change. In GNSS it is used to describe 'clock drift'.

drms (distance root mean square): Also known as *sigma value*. A measurement used to describe the accuracy of a fix. It is the square root of the sum of the squares of all radial errors surrounding a true point divided by the total number of measurements. If we double it, then it is called *2drms*. In GNSS positioning, 2drms is more commonly used. In practical terms, a particular 2drms value is the radius of a circle that is expected to contain from 95% to 98% of the positions a receiver collects in one occupation, depending on the nature of the particular error ellipse involved.

Dual-frequency: Refers to instrumentation that can make measurements on both L1 and L2 frequencies. Dual-frequency implies that advantage is taken of pseudorange and/or carrier phase on both L-band frequencies. Dual-frequency allows modelling of ionospheric bias and attendant improvement, particularly in long baseline measurements.

Dynamic positioning: A code-based positioning technique similar to kinematic positioning. *See kinematic positioning.*

Eccentricity: The measure of the degree of elongation of an ellipse.

Elapsed time: *See propagation delay.*

Electromagnetic spectrum: Continuous sequence of electromagnetic energy arranged according to wavelength or frequency.

Elevation angle: The angle from the GNSS receiver's antenna between the horizontal and the line of sight to the satellite

Elevation: The distance measured along the direction of gravity above a surface of constant potential. Usually the reference surface is the geoid.

eLORAN: Enhanced LORAN; comprises advancements in receiver design and transmission characteristics that increase the accuracy and usefulness of traditional LORAN. *See also LOng-range RAdio Navigation.*

Encryption: This is the process of translating data into a secret code for protection purposes. In the case of NAVSTAR GPS, the P code is encrypted by the additional modulation of a so-called 'W code' to generate a new 'Y code'.

Ephemeris (plural:: Ephemerides) A description of the path of a celestial body indexed by time (from the Latin *ephemeris*, meaning diary). The navigation message from each GNSS satellite includes a predicted ephemeris for the orbit of that satellite that is valid for the next few hours.

Ephemeris error: The difference between the satellite positions as calculated using its ephemerides and its 'true' position in space.

Epoch rate: The recording rate at which satellite measurements are stored. For example, a 15 sec epoch rate means one point will be stored every 15 sec.

Epoch: A particular instant of time or a date for which values of data are given, or a given period of time during which a series of events take place. In GNSS, the period or instant of each observation.

Equatorial orbital plane: Also called *equatorial plane.* The orbital plane containing the equator.

Error budget: A summary of the magnitudes and sources of statistical errors that can help in approximating the actual errors that will accrue when observations are made.

Error: *See Bias.*

Estimated Time Enroute (ETE): Used in navigation, the time it will take to reach our destination (in hours/minutes or minutes/seconds) based upon our present position, speed, and course.

Estimated Time of Arrival (ETA): Used in navigation, the estimated time we shall arrive at a destination.

European Geostationary Navigation Overlay Service (EGNOS): A satellite-based augmentation system developed by the European Space Agency (ESA), the European Commission (EC), and Eurocontrol (the European Organisation for the Safety of Air Navigation). *See also satellite-based augmentation system.*

Exosphere: The outermost layer of the earth's atmosphere.

Fast ambiguity resolution: Rapid static (fast-static) GNSS surveying technique utilizing multiple observables (dual-frequency carrier phase, C/A, and P codes) to resolve integer ambiguities with shortened observation periods.

The method may also be used for observations with the receiver in motion known as on-the-fly ambiguity resolution.

Fast-switching channel: *See multiplexing channel.*

Fidelity (of receiver): The completeness with which the essential characteristics of the original signal are reproduced.

Fix: A single position with latitude, longitude (or grid position), altitude (or height), time, and date.

Flattening: *See eccentricity.*

Frequency differential: The difference of signal propagation time for two different frequencies (e.g., L1 and L2) from a single satellite and measured by a single dual-frequency receiver, mainly used to remove ionosphere-related delays.

Frequency Division Multiple Access (FDMA): A method whereby many radios use the same code, but each one has a unique frequency. GLONASS uses FDMA techniques with frequencies as their unique identifier.

Frequency Modulation (FM): A method of encoding information about a carrier signal by altering the frequency while amplitude remains constant.

Frequency: The number of oscillations per unit time or number of wavelengths that pass a point per time.

Fundamental clock rate: The standard rate of the oscillators to which rate of all of the components of GNSS signals are expressed.

Galileo Galilei: Galileo Galilei (15 February 1564–8 January 1642) was a Tuscan (Italian) physicist, mathematician, astronomer, and philosopher who played a major role in the scientific revolution. His achievements include improvements to the telescope, subsequent astronomical observations, and support for Copernicanism.

Galileo In-Orbit Validation Element (GIOVE): The name for each satellite in a series being built for the European Space Agency to test technology in orbit for the Galileo positioning system. Giove is the Italian word for 'Jupiter'. The name was chosen as a tribute to Galileo Galilei, who discovered the first four natural satellites of Jupiter and later discovered that they could be used as a universal clock to obtain the longitude of a point on the earth's surface.

Galileo system time: *See GNSS time.*

Galileo Terrestrial Reference Frame (GTRF): A global geodetic datum defined and maintained for the Galileo satellite navigation system. As the control segment coordinates and the broadcast ephemerides are expressed in this datum, the Galileo positioning results are said to be in the GTRF datum.

Galileo: Galileo is a satellite-based navigation and positioning system currently being built by the European Union and European Space Agency. This is an alternative and complementary to GPS and GLONASS.

GEE: Gee means 'grid', i.e., the electronic grid of latitude and longitude. A British radio navigation system used in World War II, the first system based on the measurement of the difference of signal arrival times from two or more reference locations.

General theory of relativity: A physical theory published by Albert Einstein in 1916. In this theory, space and time are no longer viewed as separate but rather form a four-dimensional continuum called space-time that is curved in the neighbourhood of massive objects. The theory of general relativity replaces the concept of absolute motion with that of relative motion between two systems or frames of reference and defines the changes that occur in length, mass, and time when a moving object or light passes through a gravitational field. General relativity predicts that as gravity weakens the rate of clocks increases (they tick faster). *See also special theory of relativity.*

Geocentric datum: A datum which has its origin at the earth's centre of mass. The advantage of the geocentric datum is that it is directly compatible with satellite-based navigation systems.

Geodesy: The study of the size and shape of the earth's surface, the measurement of the position and motion of points on the surface, and the configuration and area of large portions of its surface.

Geodetic control: A network of sites for which precise positions and/or heights are known and for which the shape and size of the earth are taken into account.

Geodetic datum: A model defined by an ellipsoid and the relationship between the ellipsoid and the surface of the earth, including a Cartesian coordinate system. In modern usage, eight constants are used to form the coordinate system used for geodetic control. Two constants are required to define the dimensions of the reference ellipsoid. Three constants are needed to specify the location of the origin of the coordinate system, and three more constants are needed to specify the orientation of the coordinate system. In the past, a geodetic datum was defined by five quantities: the latitude and longitude of an initial point, the azimuth of a line from this point, and the two constants to define the reference ellipsoid.

Geodetic surveying: Surveying which takes into account the shape and size of the earth. The result of a geodetic survey is a continuous series of accurately marked points on the ground, to which topographic, land, and engineering surveys can be related to provide additional coordinated points for mapping and other purposes.

Geographic Information System (GIS): A computer-based information system that is capable of collecting, managing, and analysing geospatial data. This capability includes storing and utilising maps, displaying the results of data queries, and conducting spatial analysis.

Geoid: The equipotential surface of the earth's gravity field which approximates mean sea level more closely. The geoid surface is everywhere perpendicular to gravity. Several sources have contributed to models of the Geoid, including ocean gravity anomalies derived from satellite altimetry, satellite-derived potential models, and land surface gravity observations.

Geoidal height: The distance from the ellipsoid of reference to the geoid measured along a perpendicular to the ellipsoid of reference.

Geolocation: Geographic location, generally expressed in terms of geographic coordinates—latitude, longitude, and altitude.

Geomagnetic storm: *See magnetic storm.*

Geometric Dilution Of Precision (GDOP): *See dilution of precision (DOP).*

Geospatial positioning: A positioning system for determining one's position in geographic space.

GLObal NAvigation Satellite System (GLONASS): GLObal'naya NAvigatsionnaya Sputnikovaya Sistema—the Russian (formerly the Soviet Union) counterpart of GPS. It consists of a constellation of 24 satellites transmitting on a variety of frequencies in the ranges from 1597–1617MHz and 1240–1260MHz for navigation and positioning purpose.

Global Navigation Satellite System (GNSS): This is an umbrella term used to describe a generic satellite-based navigation/positioning system. It was coined by international agencies, such as the International Civil Aviation Organization, and refers to GPS, GLONASS, BeiDou, and Galileo, as well as any augmentations to these systems, regional satellite-based navigation systems, and to any future civilian developed satellite navigation system.

Global Orbiting Navigation Satellite System (GLONASS): *See GLObal NAvigation Satellite System (GLONASS).*

Global Positioning System (GPS): The NAVSTAR GPS satellite system developed and maintained by United States and in operation since the 1990s. A radio navigation system for providing the location of receivers with great accuracy.

GLObal'naya NAvigatsionnaya Sputnikovaya Sistema (GLONASS): *See GLObal NAvigation Satellite System (GLONASS).*

GLONASS time: *See GNSS time.*

GLONASS-K satellite: *See Uragan.*

GLONASS-M satellite: *See Uragan.*

GNSS augmentation: Augmentation of GNSS is a method of improving the navigation and positioning system's attributes, such as accuracy, reliability, and availability, through the integration of external information into the calculation process.

GNSS time: Every GNSS maintains a different time standard. Not only they are different from each other, they are also different than the Coordinated Universal Time (UTC) or International Atomic Time (TAI). GPS time, the scale for internal use in the GPS system, is similar to TAI (19 s behind TAI, a constant amount). The GPS system uses the UTC time reference maintained by United States Naval Observatory. GLONASS uses the UTC time maintained by Russia. Unlike GPS, GLONASS time is directly affected by the introduction of a leap second. Galileo also established a reference time scale, Galileo system time, to support system operations. The Galileo system plans to avoid the leap second by steering Galileo time towards TAI. Compass also maintains a different time scale, called BeiDou System Time, that is based on the UTC maintained by Chinese National Time Service Centre (NTSC).

GPS (system) time: *See GNSS time.*

GPS-Aided Geo Augmented Navigation (GAGAN): A implementation of satellite-based augmentation system by the Indian government. In Hindi language *gagan* means the sky. *See also satellite-based augmentation system.*

GPS III: GPS system with Block III satellites is commonly called *GPS III.*

GPS time: *See GNSS time.*

Graticule: A network of lines on a map or chart representing the parallels of latitude and meridians of longitude of the earth.

Gravitation: The force of attraction between all masses in the universe, especially the attraction of the earth's mass for bodies near its surface. The more remote the body the less the gravity; the gravitation between two bodies is proportional to the product of their masses and inversely proportional to the square of the distance between them.

Gross errors: Errors which result from some equipment malfunction or observer's mistake.

Ground control station: *See control segment.*

Ground station: *See control segment.*

Ground-Based Augmentation System (GBAS): A system that supports augmentation through the use of terrestrial radio messages. GBASs are commonly composed of one or more accurately surveyed ground stations, which take measurements concerning the GNSS and one or more ground based radio transmitters, which then transmit the information directly to the end user. GBASs can have local and regional coverage. Regional systems are commonly called Ground-based Regional Augmentation System.

Ground-based Regional Augmentation System (GRAS): *See ground-based augmentation system.*

Groundwave: A surface wave that propagates close to the surface of the earth.

Group delay: *See ionospheric delay.*

Heading: The direction in which a vehicle is moving. For air and sea operations, this may differ from actual Course Over Ground (COG) due to winds, currents, etc.

Hertz (Hz): One cycle per second. 1 kilohertz (kHz) = 1000 Hz; 1 megahertz (MHz) = 1000 kHz;1 gigahertz (GHz) = 1000 MHz.

Heuristic approach: Trial and error method.

High Precision Service (HPS): GLONASS positioning for the military at a higher level of absolute positioning accuracy; similar to GPS Precise Positioning Service (PPS).

Horizon: The apparent line that separates earth from sky—the meeting line of earth and sky that one sees.

Horizontal Dilution Of Precision (HDOP): *See dilution of precision (DOP).*

Hydrogen maser: An atomic clock. A device that uses microwave amplification by stimulated emission of radiation is called a *maser*. The microwave designation is not entirely accurate, since masers have been developed to operate at many wavelengths. In any case, a maser is an oscillator whose frequency is derived from atomic resonance. One of the most useful types of maser is based on transitions in hydrogen, which occur at 1421 MHz. The hydrogen maser provides a very sharp, constant oscillating signal and thus serves as a time standard for an atomic clock. The active hydrogen maser provides the best-known frequency stability that is commercially available. At a 1 hour averaging time the active maser exceeds the stability of the best-known caesium oscillators by a factor of at least 100, and the hydrogen maser is also extremely environmentally rugged.

Indian Regional Navigational Satellite System (IRNSS): An autonomous regional satellite navigation system developed by the Indian Space Research Organisation (under total control of the Indian government). The operational name of this system is NavIC (Navigation with Indian Constellation). *See also regional navigation satellite system.*

Inertial Navigation System (INS): A navigation aid that uses a computer and motion sensors to continuously track the position, orientation, and velocity (direction and speed of movement) of a vehicle without the need for external references. Other terms used to refer to inertial navigation systems or closely related devices include *inertial guidance system, inertial reference platform,* among others.

Initialisation: Before a feature of interest can be coordinated to an accuracy suitable for surveying applications, the carrier phase integer cycle ambiguities must be determined. In the static survey this is called ambiguity resolution. In a kinematic survey, this ambiguity resolution process is termed *initialization.*

Integer ambiguity: *See cycle ambiguity.*

Integrity: The ability of a system to provide timely warnings when it should not be relied on for navigation because of some inadequacy (e.g., in the event of a loss of navigation solution, excessive noise, or other factors affecting measured position).

Interference: Unwanted signals or any distortion of the transmitted signal that impedes the reception of the signal by the receiver. The intentional production of such interference to obstruct communication is called *jamming.* Unintentional interference is called *noise.*

Intermediate frequency: *See beat frequency.*

International Atomic Time (TAI): A high-precision atomic time standard that tracks proper time on earth's geoid. It is the principal realisation of terrestrial time and the basis for Coordinated Universal Time (UTC) which is used for civil timekeeping all over the earth's surface. *See also Coordinated Universal Time.*

International Civil Aviation Organization (ICAO): The ICAO, an agency of the United Nations, codifies the principles and techniques of international air navigation and fosters the planning and development of international air transport to ensure safe and orderly growth. Its headquarters are located in the Quartier International of Montreal, Canada.

International GPS Service (IGS): An initiative of the International Association of Geodesy, as well as several other scientific organisations, that was established as a service at the beginning of 1994. The IGS comprises of many component civilian agencies working cooperatively to operate a permanent

global GPS tracking network, to analyse the recorded data and to disseminate the results to users via the Internet. The range of 'products' of the
IGS include precise post-mission GPS satellite ephemerides, tracking station coordinates, earth orientation parameters, satellite clock corrections,
and tropospheric and ionospheric models. Although these were originally
intended for the geodetic community as an aid to carrying out precise surveys for monitoring crustal motion, the range of users has since expanded
dramatically, and the utility of the IGS is such that it is vital to the definition
and maintenance of the International Terrestrial Reference System (and its
various frame realisations ITRF92, ITRF94, ITRF96, etc.).

International Telecommunications Union (ITU): An international organisation
established to standardise and regulate international radio and telecommunications. It was founded as the International Telegraph Union in Paris on
17 May 1865. Its main tasks include standardisation, allocation of the radio
spectrum, and organising interconnection arrangements between different
countries to allow international phone calls.

Interoperability: A concept of the possibility of computing a position using satellites from different constellations. For example, a receiver that has the
capability to track both GPS and GLONASS satellites.

Ion: An electrically charged particle formed by either adding or taking away electrons from neutral particles.

Ionisation: The process by which atoms form electrically charged particles called
ions.

Ionised: The state in which an atom is missing on or more of its electrons, and is
therefore positively charged. An ionised gas is one in which some or all of
the atoms are ionised, rather than electrically neutral, and the ionised electrons behave as free particles.

Ionosphere: One of the upper layers of the atmosphere, above the stratosphere, starting at a distance of about 50 km from the earth's surface to 1000 km above
the earth's surface. This is a dispersive medium, ionised through the action
of the sun's radiation.

Ionospheric delay: Also known as *ionospheric refraction*. The difference in the
propagation time for a signal passing through the ionosphere compared with
the propagation time for the same signal passing through a vacuum. The
magnitude of the ionospheric delay changes with the time of day, the latitude, the season, solar activity, and the observing direction. There are two
categories of the ionospheric delay, phase and group. Group delay affects the
codes, the modulations on the carriers; phase delay affects the carriers themselves. Group delay and phase delay have the same magnitude but opposite
signs. Having a negative sign phase delay is actually phase advance.

Issue Of Data Clock (IODC): When the navigation message provides a definition
of the reliability of the broadcast clock correction.

Jamming: *See interference.*

Kalman Filter: An optimum mathematical procedure for recursively estimating
dynamically changing parameters, such as the position and velocity of a

vessel, from noise-contaminated observations. In GNSS, a numerical data combiner is used to determine an instantaneous position estimate from multiple statistical measurements on a time-varying signal in the presence of noise.

Keplerian elements: A set of six parameters that describe the position and velocity of a satellite in a purely elliptical (Keplerian) orbit. These parameters are the semi-major axis and eccentricity of the ellipse, the inclination of the orbit plane to the celestial equator, the right ascension of the ascending node of the orbit, the argument of perigee, and the time the satellite passes through the perigee.

Kinematic positioning: Positioning a continuously moving platform by using GNSS carrier-phase data while operating in a differential mode.

Kinematic surveying: A precision differential/relative GNSS surveying technique in which the roving user does not need to stop to collect precision information. Metre to centimetre-level accuracy is available in differential/relative mode using dual-frequency carrier-phase measurement techniques.

Kinematic: Pertaining to motion or moving objects.

L band: Radar wavelength region from 15 to 30 cm (frequency 1.0–2.0 GHz).

L6 signal: Galileo positioning system signal between the frequency range 1544–1545 MHz. This is a search-and-rescue (SAR)-specific band and is reserved specifically for that purpose.

Latency: Also called *age* or *time lapse*. The time it takes for a system to compute corrections and transmit them to users in real-time differential GNSS.

Latitude: The angle of a location on the earth's surface from the equator expressed in degrees north or south.

Leap second: One-second adjustment that keeps broadcast standards for time of day close to mean solar time. Broadcast standards for civil time are based on Coordinated Universal Time (UTC), a time standard which is maintained using extremely precise atomic clocks. To keep the UTC broadcast standard close to mean solar time, UTC is occasionally corrected by an intercalary adjustment, or 'leap', of one second. The timing of leap seconds is determined by the International Earth Rotation and Reference Systems Service (IERS).

Leitstrahl: *See Lorenz.*

Line Of Sight (LOS) propagation: Of an electromagnetic wave, propagation in which the direct transmission path from the transmitter to the receiver is unobstructed. LOS propagation is critical to a GNSS signal.

Local Area Augmentation System (LAAS): An all-weather aircraft landing system based on real-time differential correction of the GPS signal. It is a ground-based augmentation system. *See also ground-based augmentation system.*

Lock: Once correlation of the two codes (the codes sent by the satellite and the replica generated by the receiver) is achieved, it is maintained by a correlation channel within the GNSS receiver, and the receiver is sometimes said to have achieved lock or to be locked on to the satellites.

Longitude: The angle of a location on the earth's surface usually expressed in degrees east or west of the Greenwich Meridian.

LOng-range RAdio Navigation (LORAN): A radio navigation system developed and maintained by Americans and still in operation with some modifications. It is a network of radio signal transmitters in many areas of the globe that allows accurate position plotting. Loran transmitting stations around the globe continually transmit 90–100 kHz radio signals. Special shipboard Loran receivers interpret these signals and provide readings that correspond to a grid overprinted on nautical charts. By comparing signals from two different stations, mariners determine the position of the vessel.

Loop closure: A procedure by which the internal consistency of a GNSS network is discovered. A series of baseline vector components from more than one GNSS session, forming a loop or closed figure, is added together. The closure error is the ratio of the length of the line representing the combined errors of all the vector's components to the length of the perimeter of the figure.

LORAN-C: A developed version of LORAN. *See also LOng-range RAdio Navigation.*

Lorenz: A German radio navigation system developed in 1930s used for the safe landing of aircraft. Lorenz was the name of the company that established this system.

Lost lock: Once the correlation of the two codes (the codes sent by the satellite and the replica generated by the receiver) is somehow interrupted, the receiver is said to have lost lock. *See also lock.*

Luminosity: The rate at which the energy is radiated from an object.

Magnetic storm: A temporary disturbance of the earth's magnetosphere caused by a disturbance in space weather. It is caused by a solar wind shock wave which typically strikes the earth's magnetic field.

Mask angle: *See cutoff angle.*

Master control station: Also known as *system control station* or simply *control station*. It is the central facility in a network of worldwide tracking and uploading stations that comprise the GNSS control segment. In the master control station, the tracking information sent by the monitor stations are incorporated into precise satellite orbit and clock correction coefficients and then forwarded to the upload stations.

Medium earth orbit: Geocentric orbits ranging in altitude from 2000 km (1240 miles) to just below geosynchronous orbit at 35,786 km (22240 miles). Also known as an intermediate circular orbit.

Mesosphere: A portion of ionosphere from 50 to 85 km above the earth's surface. *See also ionosphere.*

Microwave band: The region of the electromagnetic spectrum in the wavelength range from 1 mm to 1 m.

Microwave zone: *See Microwave band.*

Mid-latitudes: Areas between 30° and 60° north and south of the equator.

Minimally constrained: A form of least squares solution in which the observed baseline vectors are treated as 'observations' in a secondary network adjustment (*see network adjustment*), and only one coordinate must be held fixed to its known value while all others are allowed to adjust. Typically GNSS surveys measure more baselines than the minimum needed to coordinate all the points in the network. These extra 'observations' are redundant information that a minimally constrained network adjustment uses to derive optimum estimates of the coordinate parameters, as well as valuable quality information in the form of parameter standard deviations and error ellipses (or ellipsoids).

Modem (Modulator/Demodulator): A device that converts digital signals to analogue signals and analogue signals to digital signals. Computers sharing data usually require a modem at each end of a phone line to perform the conversion.

Modernised GPS: The GPS system with Block IIR-M satellites called *Modernised GPS*.

Modulation: A method of encoding a message signal on top of a carrier, which can be decoded at a later time.

Monitor stations: Stations used in the GNSS control segment to track satellite clock and orbital parameters. Data collected at monitor stations are linked to a master control station at which corrections are calculated and from which correction data is uploaded to the satellites as needed.

Multichannel receiver: Also known as *parallel receiver*. A receiver with many independent channels. Each channel can be dedicated to tracking one satellite continuously.

Multi-functional Satellite-based Augmentation System (MSAS): A Japanese satellite-based augmentation system and similar to WAAS and EGNOS. The Japan Civil Aviation Bureau implemented the MSAS to utilise GPS for aviation. *See also satellite-based augmentation system.*

Multipath: Interference caused by reflected GNSS signals arriving at the receiver, typically as a result of nearby structures or other reflective surfaces. May be mitigated to some extent through appropriate antenna design, antenna placement and special filtering algorithms within GNSS receivers.

Multipath error: Errors caused by the interference of a signal that has reached the receiver antenna by two or more different paths. This is usually caused by one path being bounced or reflected. The impact on a pseudorange measurement may be up to a few metres. In the case of carrier phase, this is of the order of a few centimetres.

Multiplexing channel: Also known as *fast-switching, fast-sequencing,* and *fast-multiplexing*. A channel of a GNSS receiver that tracks through a series of a satellite's signals, from one signal to the next in a rapid sequence.

Multiplexing receiver: A receiver that tracks a satellite's signals in a rapid sequence and differs from a Multichannel Receiver in which individual channels are dedicated to each satellite signal.

Nautical mile: A unit of length used in sea and air navigation; 1.1508 mi or 1852 m or one minute of angle along a meridian on the earth.

Navigation code: Also called *navigation message* or *data message*. The message transmitted by each GNSS satellite containing system time, clock correction parameters, ionospheric delay model parameters, and the satellite's ephemeris data and health. The information is used to process GNSS signals to give the user time, position, and velocity.

Navigation message: *See navigation code.*

NAVigation Satellite Timing And Ranging Global Positioning System (NAVSTAR GPS): *See Global Positioning System.*

Navigation satellite: Satellite that transmits coded information that is used for navigation and positioning of a receiver. *See NAVSAT.*

Navigation solution: *See absolute positioning.*

Navigation: (1) The act of determining the course or heading of movement. (2) The process of planning, reading, and controlling the movement of a craft, vehicle, person, or object from one place to another.

Navigation with Indian Constellation (NavIC): *See Indian Regional Navigational Satellite System.*

NAVSAT: *See Transit.*

Navy Navigation Satellite System (NNSS): *See Transit.*

Network adjustment: A form of least squares solution in which the observed baseline vectors are treated as 'observations' in a secondary adjustment (*see minimally constrained*). It may be a minimally constrained network adjustment with only one station coordinate held fixed, or it may be constrained by more than one fixed (known) coordinates. The latter is typical of a GNSS survey carried out to densify or connect some newly coordinated points to a previously established control or geodetic framework (*see datum*).

Network: An interconnected system of things.

Noise: *See interference.*

Obliquity: In general, the angle between the equatorial and orbital planes of a body or, equivalently, between the rotational and orbital poles.

Observable: In GNSS, the signal whose measurement yields the range or distance between the satellite and the receiver. There are two types of observables in GNSS: the pseudorange and the carrier phase. *See also pseudorange and carrier phase.*

Observation log: Hard copy data sheets used for writing the GNSS readings and additional information. This serves as the backup for the data and a much easier way to write down notes.

Observations: Measurements carried out using GNSS.

Observing session: Continuous and simultaneous collection of GNSS data by two or more receivers.

Occupation time: Also called *observation time* (period). The time required to solve integer ambiguities.

Octant: An instrument used to measure angles of celestial bodies in celestial navigation.

OMEGA: A radio navigation system for aircraft, operated by the US in cooperation with six partner nations. OMEGA employed hyperbolic radio navigation techniques and the chain operated in the very low frequency portion of the electromagnetic spectrum between 10 and 14 kHz.

On-the-fly (OTF): A method of resolving the carrier phase ambiguity very quickly. The method requires a dual-frequency receiver capable of making both carrier phase and precise pseudorange measurements. The receiver is not required to remain stationary, making the technique useful for initializing in carrier phase kinematic GNSS.

Open Service (OS): (1) The free positioning service of Galileo for anyone to access. The OS signals will be broadcast in two bands, at 1164–1214 MHz and at 1559–1591 MHz. Receivers will achieve an accuracy of less than 4 m horizontally and less than 8 m vertically if they use both OS bands. Receivers that use only a single band will still achieve an accuracy of less than 15 m horizontally and less than 35 m vertically. (2) Freely available EGNOS augmentation service, which allows high accuracy with GPS, that can reach 1–3 m horizontally and 2–4 m vertically.

Orbit: Path of a satellite around a body such as the earth, under the influence of gravity.

Orbital eccentricity: The degree to which an elliptical orbit is elongated.

Orbital elements: Also called *orbital parameters*. A set of parameters defining the orbit of a satellite.

Orbital error: *See ephemeris error.*

Orbital inclination: The angle between the satellite's orbital plane and the earth's equatorial plane.

Orbital period: The time taken by a satellite to make one revolution around the earth.

Orbital perturbation: A phenomenon whereby the orbit of an object or celestial body is altered by one or more external influences. In the case of satellites, orbital perturbation may be a consequence of atmospheric drag (if the satellite is within the earth's atmosphere) or solar radiation pressure, whilst in the case of celestial bodies, orbital perturbation may be caused by comets, asteroids, solar flares, and other celestial phenomena.

Orbital plane: The plane in which a satellite moves.

Order of survey: A means of categorising the quality, or precision, of a static survey. This is analogous to 'class of survey'; however, it relates to the external quality, and is influenced by the quality of the 'external' network information.

Oscillation: Repetitive variation, typically in time, of some measure about a central value (often a point of equilibrium) or between two or more different states. Familiar examples include a swinging pendulum and AC power. The term vibration is sometimes used more narrowly to mean a mechanical oscillation.

Outage: A loss of availability, due to either there not being enough satellites visible to calculate a position, or the value of the DOP indicator is greater than some specified value (implying that the accuracy of the position is unreliable). *See also availability.*

Parameters of the Earth 1990 (PE-90): A global geodetic datum defined and maintained by the former Soviet Union and now Russia. As the control segment coordinates and the broadcast ephemerides are expressed in this datum, the GLONASS positioning results are said to be in the PE-90 datum.

Parts per million (ppm): One part in a million. In GNSS, it is an amount of error proportional to the distance measured. For example, an error of ±3 mm for a measurement of 30 m means that the measurement is good only to 100 ppm.

Perigee: The point of a satellite's orbit where the satellite is closest to the earth.

Perturbation: *See orbital perturbation.*

Phase angle: The time difference between the same point on two different waves, usually measured in fractions of a cycle (radians or degrees).

Phase centre: The apparent centre of signal reception at an antenna. The phase centre of an antenna is not constant but is dependent upon the observation angle and the signal frequency.

Phase delay: *See ionospheric delay.*

Phase differencing: The technique of using different GNSS receivers at different locations to measure the phase angles of the carrier signal from the same satellite. These angles are compared by a communications link between the two locations if real-time operations are required.

Phase lock: The adjustment of the phase of an oscillator signal to match the phase of a reference signal. First, the receiver compares the phases of the two signals. Next, using the resulting phase difference signal the reference oscillator frequency is adjusted. When the two signals are next compared the phase difference between them is eliminated.

Phase Modulation: Encoding information on a carrier signal by changing the phase so that some segments of the carrier are out of phase while others are in phase.

Phase shift: Phase shift represents a fractional part of the wavelength measurement. This term is used to define the difference in phase angles by comparing the phase angle of the satellite signal to that of a replica of the transmitted signal generated by the receiver.

Phase-lock loop: Another term for *carrier tracking loop.*

Point positioning: *See absolute positioning.*

Points of reference: *See reference point.*

Polar orbit: An orbit that passes close to the poles. An orbit having orbital inclination 0.

Position dilution of precision (PDOP): *See dilution of precision (DOP).*

Position: The 2D or 3D coordinates of a point, usually given in the form of latitude, longitude, and altitude (or x, y, and z in Cartesian form).

Positioning: A process used to determine the location of one's position.

Post-processed differential GNSS: In post-processed differential GNSS, the base and roving receivers have no real-time active data link between them.

Instead, each records the satellite observations with a time tag that will allow differential correction at a later time. Differential correction software is used to combine and process the data collected from these receivers.

Post-processed GNSS: A method of deriving high-accuracy positions from GNSS observations in which base and roving receivers have no real-time active data link between them. They do not communicate as they do in Real-Time Kinematic GNSS. Each receiver records the satellite observations independently. Their collections are combined later. The method can be applied to pseudoranges to be differential corrected or carrier phase measurements to be processed by double-differencing.

Post-processing: A technique of differential correction of the recorded positioning information by a receiver for better accuracy. In this technique real-time differential correction is not made; instead the roving receiver records all of its measured positions and the exact time it made each measurement. Later, this data can be merged with corrections recorded at a reference receiver for a final clean-up of the data. Therefore, we do not require the radio link as in real-time systems.

Precise Positioning Service (PPS): GPS positioning for the military at a higher level of absolute positioning accuracy than is available to C/A code receivers, which relies on SPS (Standard Positioning Service). PPS is based on the dual-frequency P code.

Precision: Agreement among measurements of the same quantity; widely scattered results are less precise than those that are closely grouped. The higher the precision, the smaller the random errors in a series of measurements. The precision of a GNSS survey depends on the network design, surveying methods, processing procedures, and equipment.

Projection: A method by which features on a curved earth are translated to be represented on a flat map sheet. This involves converting from longitude and latitude to rectangular x and y coordinates.

Propagation delay: Also called *time shift* or *elapsed time*. The time elapsed between the instant a GNSS signal leaves a satellite and the instant it arrives at a receiver; used to measure the range between the satellite and receiver.

Protected (P) code: Also known as *precise code*, a binary code of GPS and modulated on the L1 and the L2 carrier using binary biphase modulations. In the case of GPS, the P code is sometimes replaced with the more secure Y code in a process known as anti-spoofing.

Pseudolite: Also called *pseudo-satellite*. A ground-based differential station which simulates the signal of a GPS satellite with a typical maximum range of 50 km. Pseudolites can enhance the accuracy and extend the coverage of the GPS constellation. Originally intended as an augmentation for Local Area Augmentation Systems to aid aircraft landings. However, pseudolites may also be used where signal obstructions are such that insufficient GPS satellites can be tracked. In fact, pseudolites are feasible in circumstances where no satellite signals are observable, e.g., for indoor applications.

Pseudorandom Noise (PRN): code A sequence of digital 1s and 0s that appear to be randomly distributed, but can be reproduced exactly. Binary signals with noise-like properties are modulated on the GNSS carrier waves, used to determine pseudorange.

Pseudorange: In GNSS a time biased distance measurement. It is based on code transmitted by a GNSS satellite, collected by a GNSS receiver, and then correlated with a replica of the same code generated in the receiver. However, there is no account for errors in synchronisation between the satellite's clock and the receiver's clock in a pseudorange. The precision of the measurement is a function of the resolution of the code; therefore a C/A code pseudorange is less precise than a P code pseudorange.

Public Regulated Service (PRS): A commercially available encrypted positioning service of Galileo. Positioning accuracy is comparable to Open Service (OS). Main aim is robustness against jamming and the reliable detection of problems within 10 sec. The PRS provides positioning information to users authorised by governments.

Quadrature phase-shift keying (QPSK): One type of phase-shift keying, used to transfer information. There are four possible phases which the carrier can have at a given time. These four phase shifts may occur at 0°, 90°, 180°, or 270°. In QPSK, information in conveyed through quadrature phase variations. In each time period, the phase can change once. Since there are four possible phases, there are 2 bits of information conveyed within each time slot. The bit rate of QPSK is, thus, twice the bit rate of BPSK. *See also binary phase shift keying.*

Quartz crystal clock: *See crystal oscillator.*

Quasi-Zenith Satellite System (QZSS): A proposed Japanese navigation satellite system, which is basically a combination of augmented GNSS and regional navigation satellite system. *See also GNSS augmentation; regional navigation satellite system.*

Radiation: Energy that is radiated or transmitted in the form of rays, waves, or particles.

Radio band: The region of electromagnetic spectrum ranging from a wavelength range of 1 mm to 30 km (frequency range 10 kHz–300 GHz).

Radio Direction Finder (RDF): A radio receiver that features a directional antenna and a visual null indicator for use in determining lines of position from radio beacons at known positions.

Radio navigation: Also expressed as *radionavigation.* The determination of position, direction, and distance using the properties of transmitted radio waves.

Radio signal: Signal transmitted/received in the radio band (a wavelength range of 1 mm–30 km) of electromagnetic spectrum.

Radiosonde: A small radio transmitter which is attached to a balloon and released to measure pressure, temperature, and humidity in the troposphere.

Range rate: The rate at which the range between a GNSS receiver and satellite changes. Usually measured by tracking the variation in the Doppler shift.

Range: The distance between two points, particularly the distance between a GNSS receiver and satellite.

Ranging code: One type of complex coded information, used to measure the distance (range) from the satellite to the receiver. These codes are also called pseudorandom noise (PRN) code.

Ranging: A technique used to determine the distance between a receiver and a known reference point (satellite).

Rapid static: A form of static GNSS positioning which requires minutes instead of hours of observations due to special ambiguity resolution techniques which use extra information, such as P code measurements or redundant satellites.

Real-time DGNSS: A base station computes, formats, and transmits pseudorange corrections via some sort of data communication link (e.g., VHF or UHF radio, cellular telephone, or FM radio). The roving receiver requires some sort of data link receiving equipment to receive the transmitted DGNSS corrections so that they can be applied to its current observations.

Real-Time Kinematic (RTK) Positioning: A method of determining relative positions between known control and unknown positions using carrier phase measurements. A base station at the known position transmits corrections to the roving receiver(s). The procedure offers high accuracy immediately in real-time. The results need not be post-processed. In the earliest use of GNSS, kinematic and rapid static positioning were not frequently used because ambiguity resolution methods were still inefficient. Later when ambiguity resolution such as on-the-fly became available, real-time kinematic and similar surveying methods became more widely used.

Real-time processing: Positions computed as soon as data is collected.

Receiver Autonomous Integrity Monitoring (RAIM): A form of receiver self-checking in which redundant pseudorange observations are used to detect if there is a problem or 'failure' with any of the measurements. Only four measurements (from four satellites) are needed to derive 3D coordinates and the receiver clock error, hence any extra measurements can be used for checking. Once the failed measurements have been identified they may be eliminated from the navigation fix.

Receiver clock errors: Errors due to the inaccuracy of the receiver clock in measuring the signal reception time.

Receiver Independent Exchange format (RINEX): A set of standard definitions and formats to promote the free exchange of GNSS positioning data and facilities the use of data from any receiver with any post-processing software package. The format includes definitions for three fundamental observables: time, phase, and range.

Receiver noise: A quantification of how well a GNSS receiver can measure code or carrier observations.

Receiver position differential: *See between-receiver single difference.*

Receiver: An instrument used to collect radio waves sent by a transmitter.

Reference point: A location used to describe location or position of another one.

Reflectivity: Is the ability of a surface to reflect incident energy.

Regional Navigation Satellite System: A navigation satellite system having regional coverage—not global. *See also Global Navigation Satellite System (GNSS).*

Relative accuracy: The accuracy of a measurement between two points (i.e., the accuracy of one point measured relative to another).

Relative dilution of precision (RDOP): *See dilution of precision (DOP).*

Relative positioning: The determination of relative positions between two or more receivers which are simultaneously tracking the same GNSS signals. One receiver is generally referred to as the *reference* or *base station*, whose coordinates are known in the satellite datum. The second receiver may be stationary or moving. However, its coordinates are determined relative to the base station. In carrier phase-based positioning, this results from the determination of the baseline vector which, when added to the base stations coordinates, generates the user's coordinates. In pseudorange-based GNSS positioning, the coordinates are derived from the roving receiver's observations after they have had the differential corrections applied (either in the real-time or post-processed mode).

Relativistic time dilation: Systematic variation in time's rate on an orbiting GNSS satellite relative to time's rate on earth. The variation is predicted by the special theory of relativity and the general theory of relativity as explained by Albert Einstein. *See also special theory of relativity, general theory of relativity.*

Reliability: The ability to perform a specific function without failure under specified conditions for a given length of time.

Rover: Any mobile GNSS receiver collecting data during a field session. The receiver's position may be computed relative to another, stationary GNSS receiver at a base station. May also be referred to as the mobile receiver or roving receiver.

Safety of Life service (SoL): A commercially available encrypted positioning service of Galileo. Positioning accuracy is comparable to Open Service (OS). SoL will provide integrity in addition to the OS. It means that a user will be warned when the positioning fails to meet certain margins of accuracy.

Satellite-Based Augmentation System (SBAS): A system that supports wide-area or regional GNSS augmentation through the use of additional satellite-broadcast messages with the goal of improving its accuracy, integrity, reliability, and availability. *See also GNSS augmentation.*

Satellite health: Information about the health of the satellite the receiver is tracking when it receives the navigation message allows it to determine if the satellite is operating within normal parameters. Each satellite transmits health data for all of the satellites.

Satellite navigation system: *See global navigation satellite system; regional navigation satellite system.*

Satellite position differential: *See between-satellite single difference.*

Search and rescue (SAR): The search for and provision of aid to people who are in distress or imminent danger.

Selective Availability: (SA) Intentional degradation of the absolute positioning performance capabilities of the GPS satellite system for civilian use (the Standard Positioning Service) by the US military, accomplished by artificially 'dithering' the clock error in the satellites. Has generally been mitigated through the use of relative positioning techniques. SA was activated on 25 March 1990, and was removed on the 1 May 2000 (midnight Washington, DC time).

Selectivity (of receiver): The ability to confine reception to signals of the desired frequency and avoid others of nearly the same frequency.

Sensitivity (of receiver): The ability to amplify a weak signal to usable strength against a background of noise.

Session: *See observing session.*

Sextant: An instrument used to measure angles of celestial bodies in celestial navigation.

Sidereal day: The length of a planet's rotation period relative to the stars. The amount of time it takes the earth to rotate 360 degree about its axis. It is about 1436.07 minutes.

Sigma value: *See standard deviation.*

Signal-to-noise (S/N) ratio: An electrical engineering concept, also used in other fields, defined as the ratio of a signal power to the noise power corrupting the signal. In less technical terms, signal-to-noise ratio compares the level of a desired signal (such as music) to the level of background noise. The higher the ratio, the less obtrusive the background noise is.

Single Difference: A GNSS observable formed by arithmetically differencing carrier phases that are simultaneously measured by a pair of receivers tracking the same satellite (between-receiver single difference), or by a single receiver tracking a pair of satellites (between-satellite single difference). The between-receiver single difference procedure essentially removes all satellite clock errors. The between-satellite single difference procedure essentially removes all receiver clock errors. Although primarily used with carrier phases, the procedure can also be applied to pseudoranges.

Single receiver positioning: *See absolute positioning*

Skywave: The propagation of electromagnetic waves bent (refracted) back to the earth's surface by the ionosphere.

Solar radiation pressure: In astronomy, solar radiation pressure is the force exerted by solar radiation on objects within its reach. Solar radiation pressure is of interest in astrodynamics, as it is one source of the orbital perturbations.

Space segment: The space-based component of GNSS, i.e., the orbiting satellites and their signals.

Spatial data: Also known as *geospatial data*. Information that identifies the geographic location and characteristics of natural or constructed features. This information may be derived from, among other things, remote sensing, mapping, and surveying technologies.

Spatial positioning: A method or approach for determining one's position in space.

Spatial: Synonymous with *geospatial*. Pertaining to space or geographic space.

Special theory of relativity: A theory developed by Albert Einstein, predicting, among other things, the changes that occur in length, mass, and time at speeds approaching the speed of light. General relativity predicts that as gravity weakens the rate of clocks increases and they tick faster. On the other hand, special relativity predicts that moving clocks appear to tick more slowly than stationary clocks, since the rate of a moving clock seems to decrease as its velocity increases. Therefore, for GNSS satellites, general relativity predicts that the atomic clocks in orbit on GNSS satellites tick faster than the atomic clocks on earth. Special relativity predicts that the velocity of atomic clocks moving at GNSS orbital speeds tick slower than clocks on earth. The rates of the clocks in GNSS satellites are reset before launch to compensate for these predicted effects. *See also general theory of relativity.*

Spread spectrum signal: A signal spread over a frequency band wider than needed to carry its information. In GNSS a spread spectrum signal is used to prevent jamming, mitigate multipath, and allow unambiguous satellite tracking.

Stability (of receiver): The ability to resist drift from conditions or values to which set.

Standard deviation: Also known as *1 sigma* or *1σ* or *1drms*. An indication of the dispersion of random errors in a series of measurements of the same quantity. The more tightly grouped the measurements around their average (mean), the smaller the standard deviation.

Standard Positioning Service (SPS): The civilian Absolute Positioning accuracy obtained by using the pseudorange data obtained with the aid of a standard single-frequency C/A-Code GPS or GLONASS receiver.

Static positioning: Location determination when the receiver's antenna is presumed to be stationary on the earth. In the case of pseudorange-based techniques this allows the use of various averaging techniques to improve the accuracy. Static Positioning is usually associated with GNSS surveying techniques, where the two GNSS receivers are static for some observation period which may range from minutes to hours (and even in the case of GNSS geodesy, several days).

Stop-and-go positioning: Also known as *semi-kinematic positioning.* This is a GNSS surveying technique for 'high productivity', which is used to determine centimetre accuracy baselines to static points, using site observation times of the order of 1 min. Only carrier phase that has been converted into unambiguous 'carrier pseudorange' is used, necessitating that the ambiguities be resolved before the survey starts (and again at any time the lock is lost). It is known as the 'stop & go' technique because the coordinates of the receiver are only of interest when it is stationary (the 'stop' part), but the receiver continues to function while it is being moved (the 'go' part) from one stationary setup to the next. As the receiver must track the satellite signals at all times, hence the transport of the receiver from one static point to another must be done carefully.

Telemetry, tracking, and control (TT&C) station: Monitor and upload stations are collectively also called TT&C station.

Thermosphere: A portion of ionosphere from 85 to 1000 km above the earth's surface. *See also ionosphere.*

Timation: An experimental satellite navigation system. A predecessor of today's GNSS.

Time differential: *See between-epoch difference.*

Time dilation: *See relativistic time dilation.*

Time dilution of precision (TDOP): *See dilution of precision (DOP).*

Time shift: *See propagation delay.*

Total electron content (TEC): Integral value of the electron density along a path. TEC is an important descriptive quantity for the ionosphere of the earth. TEC is the total number of electrons present along a path between two points, with units of electrons per square meter. Ground-based GNSS receiver measures total number of the electron along a path of the radio wave between satellite and receiver.

Track: (1) An ordered group of coordinate locations automatically accumulated in GNSS receiver's memory as our position changes. (2) Track (TRK) is the direction of movement relative to a ground position. Commonly associated with navigation applications.

Tracklog: An ordered sequence of coordinate measurements stored by a GNSS receiver.

Track point: A coordinate that is a member of a track. During a travel with a GNSS receiver, the receiver by default starts recording the values of the points in between the waypoints at a regular interval. *See also track.*

Transceiver: A device that has both a transmitter and a receiver which are combined and share common circuitry or a single housing. The term originated in the early 1920s. Technically, transceivers must combine a significant amount of the transmitter and receiver handling circuitry. Similar devices include transponders, transverters, and repeaters.

Transit: An early satellite based positioning system funded by the US Navy. It measured successive Doppler (frequency) shifts of signals transmitted from satellites in polar orbits to determine position. Under the name NAVigation SATellite (NAVSAT) or Navy Navigation Satellite System (NNSS), it would help guide both weekend sailors and commercial shipping crews until the mid-1990s.

Transmitter: An instrument used to transmit (send) radio wave.

Triangulation: The process of finding coordinates and distance to a point by calculating the length of one side of a triangle, given measurements of angles and sides of the triangle formed by that point and two other known reference points, using the law of sines.

Trilateration: A method of determining the relative positions of objects using the geometry of triangles in a similar fashion as triangulation. Unlike triangulation, which uses angle measurements to calculate the subject's location, trilateration uses the known locations of two or more reference points, and the measured distance between the subject and each reference point. To accurately and uniquely determine the relative location of a point on a

2D plane using trilateration alone, generally at least 3 reference points are needed. And for 3D trilateration, generally, 4 reference points are necessary.

Triple difference: Also known as *receiver-satellite-time triple difference*. The arithmetic difference of two double differences in carrier-phase observations. The triple-difference observable is free of integer ambiguities. It is a useful observable for determining initial, approximate coordinates of a site in relative GNSS positioning and for detecting cycle slips.

Trivial baseline: Trivial baselines are those baselines formed when more than two GNSS receivers are used simultaneously in the field to perform static GNSS surveys. For example, when 3 receivers at points A, B, C are deployed only 2 baselines are independent (either A-B & A-C, A-B & B-C, or A-C & C-B), with the other one being trivial.

Troposphere: The lowest layer of the atmosphere, and is directly in contact with the earth's surface. Its height varies from 7 to 14 km, depending on the observation point. In fact, it extends from the surface to about 7 km over the poles and 14 km over the equator.

Tropospheric delay: Also known as *tropospheric error*. Retardation of GNSS signals caused by elements in the troposphere such as temperature, air pressure, and water vapour. The tropospheric delay on GNSS signals is of the non-dispersive variety because it is not frequency-dependent and hence impacts on both the L1 and L2 signals by the same amount (unlike that within the ionosphere). The wet and dry components of the troposphere cause the delay to the signals.

Trough: A point on a wave with the greatest negative value or downward displacement in a cycle.

True range: Also known as *geometric range*. The exact distance between two points, particularly the distance between a GNSS receiver and satellite.

Ultra-high frequency (UHF): A selective range of wavelength (or frequency) from radio band of electromagnetic spectrum. Range of frequency is 300–3000 MHz and wavelength range is 100–10 cm.

Ultrakurzwellen-Landefunkfeuer: *See Lorenz.*

Universal Serial Bus (USB): A serial bus standard to interface devices to a host computer. USB was designed to allow many peripherals to be connected using a single standardised interface socket and to improve the plug-and-play capabilities.

Upload stations: The upload stations transmit (upload) ephemeris data and clock drift information sent by the master control station to each satellite at least once in a day. The satellites then send these information to the receivers over radio signals.

Uragan: Russian GLONASS satellites are called Uragan. Modernization of the GLONASS satellites called Uragan-M. The latest development of GLONASS satellites is Uragan-K.

User Equivalent Range Error (UERE): Also known as *user range error*. The contributions of all biases to each individual positioning measurement. The total UERE is the square root of the sum of the squares of the individual errors.

User segment: That component of a GNSS system that includes the user, user equipment, applications, and operational procedures.

Velocity-made-good (VMG): A term used in navigation. The speed at which we are progressing toward the destination along a desired course.

Vernal equinox: The moment when the sun is positioned directly over the earth's equator.

Vertical dilution of precision (VDOP): *See dilution of precision (DOP).*

VHF Omnidirectional Range (VOR): A type of radio navigation system for aircraft.

Warm start: Also sometimes called *hot start.* The ability of a GPS receiver to begin navigating using almanac information stored in its memory from previous use.

Wave cycle: *See cycle.*

Wave front: The forward side of any wave.

Wavelength: Distance between two successive wave peaks (crests) or other equivalent points in a harmonic wave.

Waypoint: Waypoints are locations or landmarks worth recording and storing in a GNSS receiver. These are locations we may later want to return to. They may be check points on a route or significant ground features (e.g., camp, the truck, a fork in a trail, or a favourite fishing spot). Waypoints may be defined and stored in the unit manually by taking coordinates.

Wide Area Augmentation System (WAAS): An air navigation aid developed by the Federal Aviation Administration to augment the GPS, with the goal of improving its accuracy, integrity, and availability. Essentially, WAAS is intended to enable aircraft to rely on GPS for all phases of flight, including precision approaches to any airport within its coverage area. It is a satellite-based augmentation system.

World Geodetic System 1984 (WGS84): A global geodetic datum defined and maintained by the US National Imagery and Mapping Agency. As the control segment coordinates and the broadcast ephemerides are expressed in this datum, the GPS positioning results are said to be in the WGS84 datum.

Y code: The encrypted P code of GPS. When anti-spoofing is on, the P code is encrypted into the Y code.

Zenith: The point in the celestial sphere that is exactly overhead.

Zero baseline: A zero baseline test can be used to study the precision of receiver measurements (and hence its correct operation), as well as the data processing software. The experimental setup, as the name implies, involves connecting two receivers to the same antenna. When two receivers share the same antenna, biases such as those which are satellite (clock and ephemeris) and atmospheric path (troposphere and ionosphere) dependent, as well as errors such as multipath 'cancel' during data processing. The quality of the resulting 'zero baseline' is therefore a function of random observation error (or noise), and the propagation of any receiver biases that do not cancel in double-differencing.

Zulu time: *See Coordinated Universal Time.*

References

Abidi, A.A., P.R. Gray, and R.G. Meyer 1999, *Integrated Circuits for Wireless Communications*, IEEE Press, 686 pp.

Agrawal, N.K. 2004, Essentials of GPS, *Geospatial Today*, Hyderabad, 130 pp.

Aigong, X. 2006, GPS/GIS Based Electronic Road Pricing System Design, in *Proceedings of Map Asia 2006*. Available online at http://www.gisdevelopment.net/application/utility/transport/ma06_28.htm, accessed on 25 May 2007.

Aldridge, R.C. 1983, *First Strike!: The Pentagon's Strategy for Nuclear War*, South End Press, 325 pp.

Anderson, J.M., and E.M. Mikhail 1998, *Surveying: Theory and Practice*, McGraw-Hill, 1200 pp.

Andrade, A.A.L. 2001, *The Global Navigation Satellite System: Navigating into the New Millennium*, Ashgate, 221 pp.

Appleyard, S.F., R.S. Linford, P.J. Yarwood, and G.A.A. Grant 1998, *Marine Electronic Navigation*, Routledge, 605 pp.

Audoin, C., B. Guinot, and S. Lyle 2001, *The Measurement of Time: Time, Frequency and the Atomic Clock*, Cambridge University Press, 335 pp.

Aydan, O. 2006, The Possibility of Earthquake Prediction by Global Positioning System (GPS), *Journal of The School of Marine Science and Technology Tokai University*, 4(3): 77–89. Available online at http://www2.scc.u-tokai.ac.jp/www3/kiyou/pdf/2007vol4_3/omar.pdf, accessed on 17 May 2020.

Bauer, A.O. 2004, *Some Historical and Technical Aspects of Radio Navigation, in Germany, over the Period 1907 to 1945*, Foundation for German Communication and Related Technologies, 28 pp. Available online at http://www.xs4all.nl/~aobauer/Navigati.pdf, accessed on 13 April 2020.

Bedwell, D. 2007, Where Am I?, *American Heritage Magazine*, 22(4). Available online at http://www.americanheritage.com/articles/magazine/it/2007/4/2007_4_20.shtml, accessed on 20 April 2009.

Bellavista, P., and A. Corradi 2007, *The Handbook of Mobile Middleware*, CRC Press, 1377 pp.

Betz, J.W. 1999, The Offset Carrier Modulation for GPS Modernization, in *Proceedings of The Institute of Navigation's National Technical Meeting*, January 1999.

Bhatta, B. 2020, *Remote Sensing and GIS*, 3rd edition, Oxford University Press, 732 pp.

Black, H.D. 1978, An Easily Implemented Algorithm for the Tropospheric Range Correction, *Journal of Geophysical Research*, 83(B4): 1825–1828.

Borre, K. 2001, *Autocorrelation Functions in GPS Data Processing: Modeling Aspects*, Helsinki University of Technology, 39 pp. Available online at http://gps.aau.dk/downloads/hut_sl.pdf, accessed on 13 April 2007.

Borre, K., D.M. Akos, N. Bertelsen, P. Rinder, and S.H. Jensen 2007, *A Software-Defined GPS and Galileo Receiver: A Single-frequency Approach*, Birkhauser, 176 pp.

Bossler, J.D., J.R. Jensen, R.B. McMaster, and C. Rizos 2002, *Manual of Geospatial Science and Technology*, CRC Press, 623 pp.

Bowditch, N. 1995, *The American Practical Navigator*, Bethesda, MD: National Imagery and Mapping Agency, 880 pp. Available online at http://www.irbs.com/bowditch, accessed on 9 February 2006.

Broida, R. 2004, *How to Do Everything with Your GPS*, McGraw-Hill Professional, 304 pp.

Brunner, F.K., and W.M. Welsch 1993, Effect of the Troposphere on GPS Measurements. *GPS World*, **4**(1): 42–51.

Bugayevskiy, L.M., and J.P. Snyder 1995, *Map Projections: A Reference Manual*, CRC Press, 328 pp.

Burkholder, E.F. 2008, *The 3-D Global Spatial Data Model: Foundation of the Spatial Data Infrastructure*, CRC Press, 392 pp.

Cameron, L. 2004, *The Geocaching Handbook*, Falcon, 113 pp.

Cao, C., and G. Jing 2008, COMPASS Satellite Navigation System Development, Presentation in the *PNT Challenges and Opportunities Symposium*, November 5–6, 2008, Stanford University. Available online at http://scpnt.stanford.edu/pnt/PNT08/Presentations/8_Cao-Jing-Luo_PNT_2008.pdf, accessed on 23 January 2009.

CASA 2006, *Navigation Using Global Navigation Satellite Systems (GNSS)*, Civil Aviation Safety Authority, Civil Aviation Advisory Publication, 56 pp. Available online at http://www.casa.gov.au/download/caaps/ops/179a_1.pdf, accessed on 23 January 2009.

Chatfield, A.B. 2007, *Fundamentals of High Accuracy Inertial Navigation: Progress in Astronautics and Aeronautics*, American Institute of Aeronautics and Astronautics, 339 pp.

Clarke, A.C. 2001, *A Space Geodesy*, New American Library, 320 pp.

Clarke, A.R. 1880, *Geodesy*, Clarendon Press, 356 pp.

Clarke, B. 1996, *GPS Aviation Applications*, McGraw-Hill, 303 pp.

Clausing, D.J. 2006, *The Aviator's Guide to Navigation*, McGraw-Hill Professional, 271 pp.

Colombo, O.L., and A.W. Sutter 2004, Evaluation of Precise, Kinematic GPS Point Positioning, in *Proceedings of the ION GNSS 2004*, Long Beach, CA, September 21–24, 2004, CD-ROM.

COMDTPUB 1992, *Loran-C Users Handbook*, 219 pp. Available online at http://www.loran.org/ILAArchive/LoranHandbook1992/LoranHandbook1992.htm, accessed on 23 January 2009.

Corazza, G.E. 2007, *Digital Satellite Communications*, Springer, 535 pp.

Cotter, C.H., and H.K. Lahiry 1992, *The Elements of Navigation and Nautical Astronomy: A Text-book of Navigation and Nautical Astronomy*, Brown, Son & Ferguson, 463 pp.

Cruz, P.J.S., D.M. Frangopol, and L.C. Neves 2006, Bridge Maintenance, Safety, Management, Life-Cycle Performance and Cost, in *Proceedings of the Third International Conference on Bridge Maintenance, Safety and Management*, Porto, Portugal, 16–19 July 2006, Taylor & Francis, 1085 pp.

Czerniak, R.J., and J.P. Reilly 1998, *Applications of GPS for Surveying and Other Positioning Needs in Departments of Transportation*, Transportation Research Board, US, 46 pp.

Dixon, K. 1995, Global Reference Frames with Time, *Surveying World*, September 1995.

Dixon, R.C. 1984, *Spread Spectrum Techniques*, John Wiley & Sons, 408 pp.

DMA 1984, *Geodesy for the Layman*, Defense Mapping Agency. Available online at http://www.ngs.noaa.gov/PUBS_LIB/Geodesy4Layman/geo4lay.pdf, accessed on 16 May 2020.

Dong, S., X. Li, and H. Wu 2007, About Compass Time and Its Coordination with Other GNSSs, in *Proceedings of 39th Annual Precise Time and Time Interval (PTTI) Meeting*, November 26–29, 2007, Long Beach, CA. Available online at http://tycho.usno.navy.mil/ptti/ptti2007/paper3.pdf, accessed on 10 October 2008.

Dong, S., H. Wu, X. Li, S. Guo, and Q. Yang 2008, The Compass and Its Time Reference System, *Metrologia*, **45**: S47–S50.

Drane, C.R., and C.R. Drane 1998, *Positioning Systems in Intelligent Transportation Systems*, Artech House, 369 pp.

Dutton, B., and T.J. Cutler 2004, *Dutton's Nautical Navigation*, Naval Institute Press, 664 pp.

Edward, L. 2003, GPS on the Web: Applications of GPS Pseudolites, *GPS Solutions*, **6**(4): 268–270. Available online at http://www.springerlink.com/content/qhm9wmkfg9aulx gc, accessed on 12 October 2006.

Einstein, A., and R.W. Lawson 2001, *Relativity: The Special and the General Theory*, Routledge, 166 pp.

El-Rabbany, A. 2002, *Introduction to GPS: The Global Positioning System*, Artech House, 176 pp.

Enrico, D.R., and R. Marina (Eds.) 2008, *Satellite Communications and Navigation Systems*, Springer, 768 pp.

ESRI 2004, *Understanding Map Projections*, Environmental Systems Research Institute, 113 pp. Available online at http://webhelp.esri.com/arcgisdesktop/9.2, accessed on 23 December 2007.

Farrell, J., and M. Barth 1998; *The Global Positioning System and Inertial Navigation*, McGraw-Hill Professional, 340 pp.

Fosburgh, B., and J.V.R. Paiva 2001, *Surveying with GPS*, Trimble, 7 pp. Available online at http://www.acsm.net/sessions01/surveygps.pdf, accessed on 23 December 2007.

Freeman, F.L. 2005, *Fundamentals of Telecommunications*, Wiley-IEEE, 675 pp.

French, R.L. 1995, *Land Vehicle Navigation and Tracking*, American Institute of Aeronautics and Astronautics, pp. 275–301.

Fu, Z., A. Hornbostel, J. Hammesfahr, and A. Konovaltsev 2003, Suppression of Multipath and Jamming Signals by Digital Beamforming for GPS/Galileo Applications, *GPS Solutions*, **6**(4): 257–264. Available online at http://www.springerlink.com/content/ tg7h66rh5mbngpmt, accessed on 15 May 2006.

Gao, G.X., A. Chen, S. Lo, D. Lorenzo, and P. Enge 2007, GNSS Over China: The Compass MEO Satellite Codes, *InsideGNSS*, **2**(6): 36–43. Available online at http://www.inside-gnss.com/node/155, accessed on 30 July 2009.

Gao, G.X., A. Chen, S. Lo, D. Lorenzo, T. Walter, and P. Enge 2008, Compass-M1 Broadcast Codes and Their Application to Acquisition and Tracking, in *Proceedings of the 2008 National Technical Meeting of the Institute of Navigation*, January 28–30, 2008, San Diego, CA. Available online at http://www.ion.org/search/view:abstract.cfm?jp=p& idno=7671, accessed on 27 March 2009.

Gaposchkin, E.M., and B. Kołaczek 1981, Reference Coordinate Systems for Earth Dynamics, in *Proceedings of the 56th Colloquium of the International Astronomical Union*, September 8–12, 1980, Warsaw, Poland, International Astronomical Union, 396 pp.

Garrison, E.G. 2003, *Techniques in Archaeological Geology*, Springer, 304 pp.

Gelb, A. (ed.) 1974, *Applied Optimal Estimation*, MIT Press, 374 pp.

Georgiadou, Y., and K.D. Doucet 1990, The Issue of Selective Availability, *GPS World*, **1**(5): 53–56.

Ghilani, C.D., and P.R. Wolf 2008, *Elementary Surveying: An Introduction to Geomatics*, Prentice Hall, 931 pp.

Gill, T.P. 1965, *The Doppler Effect: An Introduction to the Theory of the Effect*, Logos Press, 149 pp.

GLONASS ICD 2002, *GLONASS Interface Control Document*, Version 5.0, Coordination Scientific Information Center, Moscow, 51 pp. Available online at http://www.glonass-i anc.rsa.ru/i/glonass/ICD02_e.pdf, accessed on 15 August 2005.

Gopi, S. 2005, *Global Positioning System: Principles and Applications*, Tata McGraw-Hill, 337 pp.

Grelier, T., J. Dantepal, A. Delatour, A. Ghion, and L. Ries 2007, Initial Observations and Analysis of Compass MEO Satellite Signals, *InsideGNSS*, **2**(4): 39–43. Available online at http://www.insidegnss.com/node/463, accessed on 28 February 2009.

Grewal, M.S., L.R. Weill, and A.P. Andrews 2001, *Global Positioning Systems, Inertial Navigation, and Integration*, Wiley-IEEE, 392 pp.

Gurtner, W., G. Mader, and D. MacArthur 1989, A Common Exchange Format for GPS Data, in *Proceedings of the Fifth International Geodetic Symposium on Satellite Systems*, Las Cruces, NM, and reprinted in CIGNET Bulletin 2(3), May–June, 1989.

Gutierrez, C. 2008, *Advanced Position Based Services to Improve Accessibility*, Ciudad University, 10 pp. Available online at http://www.inredis.es/render/binarios.aspx?id= 105, accessed on 30 October 2009.

Han, S.C., J.H. Kwon, and C. Jekeli 2001, Accurate Absolute GPS Positioning Through Satellite Clock Error Estimation. *Journal of Geodesy*, **75**(1): 33–43.

Hatch, R.R. 1982, The Synergism of GPS Code and Carrier Measurements, in *Proceedings of 3rd International Symposium on Satellite Doppler Positioning*, NM, 8–12 February, 1982, pp. 1213–1231.

Hecks, K. 1990, *Bombing 1939–1945: The Air Offensive against Land Targets in World War Two*, Robert Hale Ltd., 219 pp.

Hoffmann-Wellenhof, B., H. Lichtenegger, and J. Collins 1994, *Global Positioning System: Theory and Practice*, 3rd ed., Springer-Verlag.

Hofmann-Wellenhof, B., K. Legat, M. Wieser, and H. Lichtenegger 2003, *Navigation: Principles of Positioning and Guidance*, Springer, 427 pp.

Hofmann-Wellenhof, B., and H. Moritz 2005, *Physical Geodesy*, Springer, 403 pp.

Hopfield, H.S. 1969, Two-Quartic Tropospheric Refractivity Profile for Correcting Satellite Data, *Journal of Geophysical Research*, **74**(18): 4487–4499.

ICAO 2005, Draft Galileo SARPS – Part A, Working Paper, *International Civil Aviation Organization NSP/WG1: WP35*, 12 pp.

IFATCA 1999, *A Beginner's Guide to GNSS in Europe*, International Federation of Air Traffic Controllers' Associations, Montreal, QC, 14 pp. Available online at http://www.ifatca.org/docs/gnss.pdf, accessed on 7 June 2005.

ISRO 2017, *IRNSS Signal-in-Space ICD for SPS Version 1.1*, ISRO Satellite Centre, Indian Space Research Organization, Bangalore, 72 pp.

Issler, J.L., G.W. Hein, J. Godet, J.C. Martin, P. Erhard, R. Lucas-Rodriguez, and T. Pratt 2003, Galileo Frequency & Signal Design, *GPS World*, **14**(6): 30–37.

Javad, and Nedda A. 1998, *A GPS Tutorial*, Javad Navigation Systems, 47 pp. Available online at http://www.javad.com/jns/index.html?/jns/gpstutorial, accessed on 20 September 2016.

Jespersen, J., and J. Fitz-Randolph 1999, *From Sundials to Atomic Clocks: Understanding Time and Frequency*, 2nd ed., Dover.

Jiang, Z., G. Zhang, Y. Gao, and W. Wang 2007, Progress in Research of Earthquake Prediction in China, *Geophysical Research Abstracts*, **9**: 11637.

Jones, A.C. 1984, *An Investigation of the Accuracy and Repeatability of Satellite Doppler Relative Positioning Techniques*, School of Surveying, University of New South Wales, 222 pp.

Kaler, J.B. 2002, *The Ever-Changing Sky: A Guide to the Celestial Sphere*, Cambridge University Press, 495 pp.

Kalman, R.E. 1960, A New Approach to Linear Filtering and Prediction Problems, *Journal of Basic Engineering*, **82**: 35–45.

Kaplan, E. (ed.) 1996, *Understanding GPS: Principles & Applications*, Artech House Publishers, 554 pp.

Kato, T., G.S. El-Fiky, E.N. Oware, and S. Miyazaki 1998, Crustal Strains in the Japanese Islands as Deduced from Dense GPS Array, *Geophysical Research Letters*, **25**(18): 3445–3448.

Keohane, E. 2007, Is GPS Technology the Next Marketing Breakthrough?, *DMNews*, December 27, 2007. Available online at http://www.dmnews.com/Is-GPS-technology-the-next-marketing-breakthrough/article/100222/, accessed on 17 April 2018.

Kline, P.A. 1997, *Atomic Clock Augmentation for Receivers Using the Global Positioning System*, PhD thesis, Virginia Polytechnic Institute and State University, 223 pp. Available online at http://scholar.lib.vt.edu/theses/available/etd-112516142975720/un restricted, accessed on 2 April 2003.

Klobuchar, J.A. 1987, Ionospheric Time Delay Algorithm for Single Frequency GPS Users, *IEEE Transactions on Aerospace and Electronic Systems*, **23**(3): 325–331.

Klobuchar, J.A. 1991, Ionospheric Effects on GPS, *GPS World*, **2**(4): 48–51.

Klobuchar, J.A. 1996, Ionospheric Effects on GPS, in *Gobal Positioning System: Theory and Applications Volume 1*, eds. B.W. Parkinson, J.J. Spilker Jr., *Progress in Astronautics and Aeronautics Volume 164*, American Institute of Aeronautics and Astronautics, Inc., pp. 485–514.

Kouba, J. 2003, A Guide to Using International GPS Service (IGS), NASA. Available online at http://igscb.jpl.nasa.gov/igscb/resource/pubs/GuidetoUsingIGSProducts.pdf, accessed on 30 November 2006.

Krakiwsky, E.J. 1991, GPS and Vehicle Location and Navigation, *GPS World*, **2**(5): 50–53.

Krakiwsky, E.J., and J.B. Bullock 1994, Digital Road Data: Putting GPS on the Map, *GPS World*, **5**(5): 43–46.

Krakiwsky, E.J., and J.F. McLellan 1995, Making GPS Even Better with Auxiliary Devices, *GPS World*, **6**(3): 46–53.

Langley, R.B. 1990, Why is the GPS Signal so Complex?, *GPS World*, **1**(3): 56–59. Available online at http://gauss.gge.unb.ca/gpsworld/EarlyInnovationColumns/Innov.1990.05-06.pdf, accessed on 16 August 2005.

Langley, R.B. 1991a, Time, Clocks, and GPS, *GPS World*, **2**(10): 38–42.

Langley, R.B. 1991b, The Orbits of GPS Satellites, *GPS World*, **2**(3): 50–53.

Langley, R.B. 1991c, The Mathematics of GPS, *GPS World*, **2**(7): 45–50.

Langley, R.B. 1991d, The GPS Receiver – An Introduction, *GPS World*, **2**(1): 50–53.

Langley, R.B. 1993a, The GPS Observables, *GPS World*, **4**(4): 52–59.

Langley, R.B. 1993b, Communication Links for DGPS, *GPS World*, **4**(5): 47–51.

Langley, R.B. 1994, RTCM SC-104 DGPS Standards, *GPS World*, **5**(5): 48–53.

Langley, R.B. 1997, The GPS Receiver System Noise, *GPS World*, **8**(6): 40–45.

Langley, R.B. 1998a, A Primer on GPS Antennas, *GPS World*, **9**(7): 50–54.

Ledvina, B., P. Montgomery, and T. Humphreys 2009, A Multi-Antenna Defense: Receiver-Autonomous GPS Spoofing Detection, *InsideGNSS*, **4**(2): 40–46.

Lee, L.S. 2008, LBS in Weather and Geophysical Services, *GISdevelopment*, **12**(6): 48–50.

Leick, A. 2004, *GPS Satellite Surveying*, 3rd ed., John Wiley & Sons, 435 pp.

Levy, L.J. 1997, The Kalman Filter: Navigation's Integration Workhorse, *GPS World*, **8**(9): 65–71.

Liu, J-H., and T-Y. Shih 2007, A Performance Evaluation of the Internet Based Static GPS Computation Services, *Survey Review*, **39**(304): 166–175.

Lohan, E.S., A. Lakhzouri, and M. Renfors 2007, Binary-Offset-Carrier Modulation Techniques with Applications in Satellite Navigation Systems, *Wireless Communications and Mobile Computing*, **7**(6): 767–779. Available online at https://onlinelibrary.wiley.com/doi/abs/10.1002/wcm.407, accessed on 30 October 2019.

Lombardi, M.A. 2002, Fundamentals of Time and Frequency, in *The Mechatronics Handbook*, ed. Robert H. Bishop, CRC Press, 1272 pp.

Manning, J., and B. Harvey 1994, Status of the Australian Geocentric Datum, *Australian Surveyor*, **39**(1): 28–33.

Minkler, G., and J. Minkler 1993, *Theory and Application of Kalman Filtering*, Magellan Book Company.

Misra, P.N. 1996, The Role of the Clock in a GPS Receiver, *GPS World*, **7**(4): 60–66.

Mittal, A., and A. De 2007, Integrated Balanced BPSK Modulator for Millimeter Wave Systems, *Active and Passive Electronic Components*, 2007: 1–4. Available online at http://www.hindawi.com/getpdf.aspx?doi=10.1155/2007/69515, accessed on 15 January 2008.

NRC 1997, *Precision Agriculture in the 21st Century: Geospatial and Information Technologies in Crop Management*, National Research Council (U.S.), National Academies Press, 149 pp.

Owings, R. 2005, *GPS Mapping: Make Your Own Maps*, Ten Mile Press, 374 pp.

Parkinson, B.W. 1994, GPS Eyewitness: The Early Years, *GPS World*, **5**(9): 32–45.

Parkinson, B.W., and J.J. Spilker 1996, *Global Positioning System: Theory and Applications*, American Institute of Aeronautics and Astronautics, 643 pp.

PCC 2000, *The Guide to Wireless GPS Data Links*, Pacific Crest Corporation.

Peters, J.W. 2004, *The Complete Idiot's Guide to Geocaching*, Alpha Books, 316 pp.

Petovello, M. 2008, GNSS Solutions: Quantifying the Performance of Navigation Systems and Standards of Assisted-GNSS, *InsideGNSS*, **3**(6): 20–50.

Pratt, T., C. Bostian, and J. Allnutt, 2003, *Satellite Communications*, 2nd ed., Wiley, 536 pp.

Prolss, G.W. 2004, *Physics of the Earth's Space Environment: An Introduction*, Springer, 513 pp.

Raquet, J.F. 2002, Multiple GPS Receiver Multipath Mitigation Technique, *IEE Proceedings - Radar, Sonar and Navigation*, **149**(4): 195–201. Available online at https://digital-libr ary.theiet.org/content/journals/10.1049/ip-rsn_20020495, accessed on 2 January 2020.

Richard, P., and D. Mathis 1993, Integration of GPS with Dead Reckoning and Map Matching for Vehicular Navigation, in *Proceedings of National Technical Meeting*, January 20–22, 1993, San Francisco, 21–24.

Richey, M. 2007, *The Oxford Companion to Ships and the Sea*, eds. I.C.B. Dear and Peter Kemp, Oxford University Press.

RMITU 2006, *Surveying Using Global Navigation Satellite Systems*, RMIT University, Department of Geospatial Science, Australia, 126 pp.

Roddy, D. 2006, *Satellite Communications*, McGraw-Hill Professional, 636 pp.

Ryan, R.C., and C. Popescu 2006, *Construction Equipment Management for Engineers, Estimators, and Owners*, CRC Press, 552 pp.

Rycroft, M.J. 2003, *Satellite Navigation Systems: Policy, Commercial and Technical Interaction*, Springer, 266 pp.

Saastamoinen, J. 1973, Contributions to the Theory of Atmospheric Refraction, *Bull Gtodesique*, **107**: 13–34.

Samama, N. 2008, *Global Positioning – Technologies and Performances*, John Wiley and Sons, 419 pp.

Seeber, G. 1993, *Satellite Geodesy: Foundations, Methods & Applications*, Walter de Gruyter, 531 pp.

Shanwad, U.K., V.C. Patil, G.S. Dasog, C.P. Mansur and K.C. Shashidhar 2002, Global Positioning System (GPS) in Precision Agriculture, in *Proceedings of Asian GPS*, New Delhi, 24–25 October 2002. Available online at http://www.gisdevelopment.net/proce edings/asiangps/2002/agriculture/gagri001.htm, accessed on 26 January 2006.

Sickle, J.V. 2004, *Basic GIS Coordinates*, CRC Press, 173 pp.

Sickle, J.V. 2008, *GPS for Land Surveyors*, 3rd ed., CRC Press, 338 pp.

Simsky, A., D. Mertens, J. Sleewaegen, W. Wilde, M. Hollreiser, and M. Crisci 2008, MBOC vs. BOC(1,1): Multipath Comparison Based on GIOVE-B Data, *InsideGNSS*, **3**(6): 36–39. Available online at http://www.insidegnss.com/node/765, accessed on 30 June 2009.

Siouris, G.M. 2004, *Missile Guidance and Control Systems*, Springer, 666 pp.

Smith, J.R. 1997, *Introduction to Geodesy: The History and Concepts of Modern Geodesy*, Wiley-IEEE, 224 pp.

Spilker Jr., J.J. 1980, GPS Signal Structure and Performance Characteristics. *Journal of Global Positioning Systems*, papers published in *Navigation*, reprinted by the (U.S.) Inst. of Navigation, **1**: 29–54.

Spilker Jr., J.J., and B.W. Parkinson (eds.) 1995, *Global Positioning Systems: Theory & Applications*, American Institute of Aeronautics & Astronautics (AIAA), 1995, Vol.1 694 pp., Vol.2 601 pp.

Srinivasan, A. 2006, *Handbook of Precision Agriculture: Principles and Applications*, Haworth Press, 683 pp.

Stansell, T.A. 1978, *The Transit Navigation Satellite System: Status, Theory, Performance, Applications*, Magnavox, 83 pp.

Stehlin, X., Q. Wang, F. Jeanneret, and P. Rochat 2000, Galileo System Time Physical Generation, in *Proceedings of 38th Annual Precise Time and Time Interval (PTTI) Meeting*, Torino, Italy, pp. 395–405. Available online at http://tycho.usno.navy.mil/ptti/ptti2006/paper37.pdf, accessed on 7 April 2005.

Sweet, R.J. 2003, *GPS for Mariners*, McGraw-Hill Professional, 176 pp.

Swider, R.J. 2005, Can GNSS Become a Reality?, *GPS World*, **16**(12): 20–20.

Takac, F., and M. Petovello 2009, GLONASS Inter-Frequency Biases and Ambiguity Resolution, *InsideGNSS*, **4**(2): 24–28.

Tapley, B.D., B.E. Schutz, and G.H. Born 2004, *Statistical Orbit Determination*, Elsevier Academic Press, 547 pp.

Tekinay, S. 2000, *Next Generation Wireless Networks*, Springer, 266 pp.

Thompson, M.M. (ed.), 1966, *The Manual of Photogrammetry*, American Society of Photogrammetry, 3rd ed., 1 & 2.

Titterton, D.H., and J.L. Weston 2004, *Strapdown Inertial Navigation Technology*, Institution of Engineering and Technology, 558 pp.

Torge, G., 1993, *Geodesy*, Walter de Gruyter, 531 pp.

Trimble 2006, *Trimble GPS Tutorial*, Trimble Navigation Ltd., electronic document. Available online at https://www.trimble.com/gps_tutorial, last accessed on 1 February 2020.

Trimple 2001, *Trimble Survey Controller User Guide, Version 10, Revision A*, Trimble Navigation Ltd., 478 pp. Available online at https://www.ngs.noaa.gov/corbin/class_description/controllerv10UserGuide.pdf, accessed on 18 May 2020.

Tsakiri, M., 2008, GPS Processing Using Online Services, *Journal of Surveying Engineering*, **134**(4): 115–125.

USACE 2002a, *Engineering and Design - Structural Deformation Surveying*, U.S. Army Corps of Engineers. Available online at http://www.usace.army.mil/publications/eng-manuals/em1110-2-1009/toc.htm, accessed on 17 August 2004.

USACE 2002b, *Engineering and Design - Hydrographic Surveying*, U.S. Army Corps of Engineers. Available online at http://www.usace.army.mil/publications/eng-manuals/em1110-2-1003/toc.htm, accessed on 17 August 2004.

USACE 2002c, *Engineering and Design - Photogrammetric Mapping*, U.S. Army Corps of Engineers. Available online at http://www.usace.army.mil/publications/eng-manuals/em1110-1-1000/toc.htm, accessed on 17 August 2004.

USACE 2003, *Engineering and Design - NAVSTAR Global Positioning System Surveying*, U.S. Army Corps of Engineers. Available online at http://www.usace.army.mil/publications/eng-manuals/em1110-1-1003/toc.htm, accessed on 17 August 2004.

USACE 2007, *Engineering and Design - Control and Topographic Surveying*, U.S. Army Corps of Engineers. Available online at http://www.usace.army.mil/publications/eng-manuals/em1110-1-1005/toc.htm, accessed on 17 August 2004.

Van Dierendonck, A.J. 1995, Understanding GPS Receiver Terminology: A Tutorial, *GPS World*, **6**(1): 34–44.

Van Dierendonck, A.J., A. Fenton, and T. Ford 1992, Theory and Performance of Narrow Correlator Spacing in a GPS Receiver, *Navigation*, **39**(3): 265–283.

Van Loan, C. 1992, *Computational Frameworks for the Fast Fourier Transform*, Society for Industrial and Applied Mathematics, 273 pp.

Vanicek, 1995, *Global Positioning Systems: Theory and Application*, American Institute of Aeronautics & Astronautics (AIAA), 1995, Vol.1 694 pp., Vol.2 601 pp.

Wang, G., M. Turco, T. Soler, T. Kearns, and Welch, J. 2017, Comparisons of OPUS and PPP Solutions for Subsidence Monitoring in the Greater Houston Area, *Journal of Surveying Engineering*, **143**(4): 123–129. doi: 10.1061/(ASCE)SU.1943-5428.0000241.

Ward, P.W. 1994, GPS Receiver RF Interference Monitoring, Mitigation, and Analysis Techniques, *Navigation*, **41**(4): 367–391.

Weill, L.R. 1997, Conquering Multipath: The GPS Accuracy Battle, *GPS World*, **8**(4): 59–66.

Wells, D. (ed.) 1986, *Guide to GPS Positioning*, 2nd ed., Canadian GPS Associates, 600 pp.

Wilde, W., F. Boon, J. Sleewaegen, and F. Wilms 2007, More Compass Points: Tracking China's MEO Satellite on a Hardware Receiver, *InsideGNSS*, **2**(6): 44–48. Available online at http://www.insidegnss.com/node/157, accessed on 17 August 2009.

Williams, A.B., and F.J. Taylor 1998, *Electronic Filter Design Handbook: LC, Active, and Digital Filters*, McGraw-Hill, 816 pp.

Yang, Q.H., J.P. Snyder, and W.R. Tobler 2000, *Map Projection Transformation: Principles and Applications*, CRC Press, 367 pp.

Zaharia, D. 2009, *GALILEO – The European Global Navigation Satellite System*, Wiley, 320 pp.

Zilkoski, D., and L. Hothem 1989, GPS Satellite Surveys and Vertical Control, *Journal of Surveying Engineering*, **115**(2): 262–282.

Zilkoski, D.B., E.E. Carlson, and C.L. Smith 2005, *Guidelines for Establishing GPS-Derived Orthometric Heights*, National Geodetic Survey, 25 pp. Available online at http://www.nj.gov/transportation/eng/documents/survey/pdf/GPSOrthometricHeights.pdf, accessed on 29 November 2019.

Index

Printed in the United States
by Baker & Taylor Publisher Services

Printed in the United States
by Baker & Taylor Publisher Services